Toxic and Hazardous Materials

Bibliographies and Indexes in Science and Technology

Publishing Opportunities for Energy Research: A Descriptive Guide to Selective
Serials in the Social and Technical Sciences
Roberta A. Scull, compiler

Toxic and Hazardous Materials

A SOURCEBOOK AND GUIDE TO INFORMATION SOURCES

Edited by James K. Webster

Bibliographies and Indexes in Science and Technology, Number 2

Greenwood Press
New York • Westport, Connecticut • London

225905   016.3631
T 755

Library of Congress Cataloging-in-Publication Data

Toxic and hazardous materials.

(Bibliographies and indexes in science and
technology, ISSN 0888-7551 ; no. 2)
   Includes index.
   1. Hazardous substances—Bibliography.
2. Hazardous substances—Information services—
Directories. I. Webster, James K. II. Series.
Z7914.S17T69 1987   016.3631'7   86-25710
[T55.3.H3]
ISBN 0-313-24575-4 (lib. bdg. : alk. paper)

Library of Congress Catalog Card Number: 86-25710
ISBN: 0-313-24575-4
ISSN: 0888-7551

First published in 1987

Greenwood Press, Inc.
88 Post Road West, Westport, Connecticut 06881

Printed in the United States of America

The paper used in this book complies with the
Permanent Paper Standard issued by the National
Information Standards Organization (Z39.48-1984).

10 9 8 7 6 5 4 3 2 1

Contents

Preface

We live in a world surrounded by chemicals, plastics, electronics and nuclear energy. Many of the products or by-products of these substances, although intended to be beneficial, can have adverse effects on people, animals and plant life.

This guide to information sources is aimed at all of those who have to deal with the problem of toxic and hazardous materials in the environment—librarians, scientists, engineers, academics, environmental lawyers and consultants, regulatory officials, and concerned citizens. The book is truly interdisciplinary in its scope and its content. The chapters cover every aspect of the subject—from technical and engineering topics to legal matters to public policy concerns.

The book's contributors are highly qualified information specialists and librarians who were each given considerable latitude in preparing their chapters, resulting in somewhat differing approaches. The editor, besides contributing Chapter 11, <u>Transportation of Hazardous Materials</u>, concentrated on maintaining the consistency of entries and on drawing together frequently-mentioned items into a <u>General Sources of Information</u> chapter (Chapter 1). This section contains entries that pertain to a number of topics; in most instances, they were removed from the individual chapters and grouped in Chapter 1. It is important, therefore, that readers interested in one specific topic also consult the General chapter to get the complete range of information sources.

The array of material included here makes this book much more than a bibliography. It is a sourcebook leading not only to printed items but to organizations and other producers of further information. Over 1600 information sources have been included: books, periodicals, newsletters, looseleaf services, reports, conference proceedings, indexes, audiovisual materials, data bases, associations, government agencies, research organizations, libraries, and more. The emphasis has been on currency, so, with rare exceptions, only material published since 1980 has been

included. We did not include references to separate periodical articles unless they were deemed to have some extra-
ordinary significance.

Preparing an index for a volume of such varied sources
was a major undertaking in itself. We agonized over
numerous questions: How should monographic materials be
indexed, by author or editor? What about second and third
authors and editors? What about books, reports and proceedings which have an organization as main entry? It was
finally decided to have index entries only for titles, that
being the one thing that all of these formats have in
common. There are entries in the index, <u>by title</u>, for
everything listed in this book, except for the handful of
separate periodical articles. Many items have been listed
under more than one entry where it seemed to be useful.
There are also a generous number of subject entries to help
the reader locate related material in different chapters.

Two people in particular have made exceptional contributions in bringing this book to completion. Maureen Glenn
has done an outstanding job on word-processing, proof-
reading, catching inconsistencies, and generally keeping the
project (and the editor) moving. And my long-time colleague
Margaret Schenk was instrumental in assembling Chapter 1 and
helping to bring the contributions of so many other people
into some semblance of uniformity. I am very grateful to
both of them, and to the contributors who produced the
excellent chapters that comprise the body of this book.

Contributors

Patricia Ann Coty
Western New York Library
   Resources Council
Lafayette Square
Buffalo, NY 14203
(716) 852-3844

Deborah Husted
Central Technical Services
Lockwood Memorial
   Library Building
SUNY at Buffalo
Buffalo, NY 14260
(716) 636-2787

Sharon A. Keller
Information Services Dept.
Health Sciences Library
SUNY at Buffalo
Buffalo, NY 14214
(716) 831-3337

P. J. Koshy
Science & Engineering Library
SUNY at Buffalo
Buffalo, NY 14260
(716) 636-2946

Erich J. Mayer
West Valley Nuclear Services
   Co., Inc.
P.O. Box 191
West Valley, NY 14171-0191
(716) 942-3235

Mary Frances Miller
Charles B. Sears Law Library
SUNY at Buffalo
Buffalo, NY 14260
(716) 636-2301

Susan M. Neumeister
Central Technical Services
Lockwood Memorial
   Library Building
SUNY at Buffalo
Buffalo, NY 14260
(716) 636-2787

Donna Serafin
Central Technical Services
Lockwood Memorial
   Library Building
SUNY at Buffalo
Buffalo, NY 14260
(716) 636-2784

James K. Webster
Science & Engineering Library
SUNY at Buffalo
Buffalo, NY 14260
(716) 636-2946

Margaret R. Wells
Silverman Undergraduate
   Library
SUNY at Buffalo
Buffalo, NY 14260
(716) 636-2943

Theresa L. Wolfe
Ecology and Environment, Inc.
195 Sugg Road
P.O. Box D
Buffalo, NY 14225
(716) 632-4491

Abbreviations and Acronyms

AAOHNC	American Association of Occupational Health Nurses
AAS	atomic absorption spectroscopy
ACEIH	American Conference of Governmental Industrial Hygienists
ACS	American Chemical Society
AIF	Atomic Industrial Forum
ANS	American Nuclear Society
AOAC	Association of Official Analytical Chemists
AOMA	American Occupational Medicine Association
APCA	Air Pollution Control Association
API	American Petroleum Institute
APTIC	Air Pollution Technical Information Center
ASI	Advanced Study Institute (NATO)
ASTM	American Society for Testing and Materials
AV	audio-visual
AWWA	American Water Works Association
BOHS	British Occupational Hygiene Society
BRS	Bibliographic Research Service
CAS	Chemical Abstracts Service, a division of the American Chemical Society
CDC	Center for Disease Control
CEC	Commission of European Communities
CELA	Canadian Environmental Law Association
CERCLA	(Superfund) Comprehensive Environmental Response Compensation and Liability Act
CFR	Code of Federal Regulations
CHEMTREC	Chemical Transportation Emergency Center
CHRIS	Chemical Hazards Response Information System
CIS	Congressional Information Services, Inc.
CRC	Chemical Rubber Company Press
DDT	dichlorodiphenyltrichloroethane
DHHS	Department of Health and Human Services
DIN	Deutsches Institut fuer Normung (German Standards Institute)
DNA	dioxyribonucleic acid
DOP	di-2-ethylhexylphthalate
DOT	Department of Transportation
EDF	Environmental Defense Fund
EDTA	ethylenediaminetetraacetate
EEC	European Economic Community

EHMI   Environmental Hazards Management Institute
EIC/Intel...   Environmental Information Center/Intelligence
EIS   Environmental Impact Statement
ELI   Environmental Law Insitute
EMBO   European Molecular Biology Organization
EPA   Environmental Protection Agency (U.S.)
EPRI   Electric Power Research Institute
ETS   environmental tobacco smoke
EXIS   Expert Information Systems
FAA   Federal Aviation Administration
FAO   Food and Agriculture Organization (United
                    Nations)
GPO   Government Printing Office (U.S.)
HMCRI   Hazardous Materials Control Research
                    Institute
IAEA   International Atomic Energy Agency
IAHS   International Association of Hydrological
                    Science
ILO   International Labor Office
ILSI   International Life Sciences Institute
IUE   International Union of Electronic, Electrical,
                    Technical Salaried and Machine Workers
NADP   nicotinamide adenine dinucletide phosphate
NAL   National Agricultural Library (U.S.)
NAS   National Academy of Science
NBS   National Bureau of Standards
NEA   Nuclear Energy Agency
NEPA   National Environmental Policy Act
NFPA   National Fire Protection Association
NICEM   National Information Center for Educational
                    Media
NIH   National Institute of Health
NIOSH   National Institute for Occupational Safety
                    and Health
NLM   National Library of Medicine
NOHS   National Occupational Hazard Survey
NPDES   National Pollutant Discharge Elimination
                    System
NSC   National Science Council
NTIS   National Technical Information Service
NRC   National Research Council
NTSB   National Transportation Safety Board
NWWA   National Water Well Association
OECD   Organisation for Economic Co-Operation and
                    Development
OHM-TADS   Oil and Hazardous Materials Technical
                    Assistance Documentation System
OSHA   Occupational Safety and Health Act (or
                    Administration)
PAH   polycyclic aromatic hydrocarbons
PB   Original meaning was "Publications Board", an
                    agency set up to make information on Axis
                    technology available after World War II.
                    PB's are now NTIS accession numbers to
                    identify non-military research reports
                    without accession numbers assigned by
                    another agency.
PCB   polychlorinated biphenyls

PCDD	polychlorinated dibenzo-p-dioxins
PCDF	polychlorinate dibenaofurans
PS	Published Search (NTIS)
PWR	pressurized water reactor
RCRA	Resource Conservation and Recovery Act
RNA	ribonucleic acid
RTECS	Registry of Toxic Effects of Chemical Substances
SAPL	Seacoast Anti-pollution League
SDC	Systems Development Corporation
SPAID	Society for the Prevention of Asbestosis and Industrial Diseases
STP	Special Technical Publication (ASTM series)
SW	report designation of the Office of Solid Waste, EPA
TDB	Toxicology Data Bank
TLD	thermoluminescent dosimeters
TCDD	2, 3, 7, 8-tetrachlorodibenzo-p-dioxin
TLV	threshold limit values
TSCA	Toxic Substances Control Act
UMI	University Microfilms International
UNSCEAR	United Nations Scientific Committee on the Effects of Atomic Radiation
USCEA	U.S. Committee for Energy Awareness
USGS	U.S. Geological Survey
VDT	video display terminal
WES	Waterways Experiment Station (U.S. Army Corps of Engineers)
WHO	World Health Organization
WPCF	Water Pollution Control Federation
WRC	Water Research Centre (Great Britain)

Toxic and Hazardous Materials

1.

General Sources of Information

Introduction

"Remodeling the natural world to man's convenience has
however, brought with it potentially harmful results..."
This was the introduction to a report in 1966 expressing
concern for the handling of toxicological information. A
report by that title was issued by the President's Science
Advisory Committee that year, indicating the need for
coordinating and providing access to such information.
Among the earliest sources were Biological Abstracts,
Chemical Abstracts, Excerpta Medica, and Index Medicus,
still in the forefront today but now offering computerized
counterparts. Information sources have had to keep pace
with increasing legislation, regulation and research in
toxic and hazardous materials. Broad involvement of
researchers in varied disciplines has triggered a proliferation of information in the past 20 years. Access to it is
important to those who are conducting research on processes
and products to improve the environment and those who will
make decisions on policy. Because the literature is so
extensive, the representative titles used in these chapters
have been confined generally to those dated 1980 or later.

The fields covered by the term, "toxic and hazardous
materials", include monitoring, disposal, effects on humans,
air, land and water and a host of more specialized and
specific areas such as oil spills, acid rain or radiation.
These are the subjects of the chapters of this publication.
The information sources included in them range from books to
data bases, and each chapter covers a selection of those
available in its subject area.

To avoid repetition, sources that contain information common
to all areas covered by the chapters in this book are
included in this introductory chapter. Like the chapters
following, they are arranged by format, i.e., books, periodicals, etc. This chapter represents the collective efforts
of the authors of all the others.

Books

Books, or monographs, are useful to researchers for their
background material, information organized to facilitate its
use, and for additional references to the literature.
Access is provided by the catalog of the library which the
researcher uses, usually with author/title and subject
approaches.

There are countless headings that can be used to search the
catalogs of libraries. The few listed here can be supplemented by the use of Library of Congress Subject Headings.

    Chemicals - Toxicology
    Environmental engineering
    Hazardous substances
    Hazardous wastes
    Industrial toxicology
    Liability for hazardous substances pollution damage
    Pesticides
    Pollution - Control
    Pollution - Prevention

A search of library catalogs using these and other subject
headings will turn up such generally useful titles as these:

ENVIRONMENTAL PROTECTION
Ortolano, Leonard. Environmental Planning and Decision-
    Making. New York: Wiley, 1984.

HAZARDOUS WASTES
Cope, C. B. and W. H. Fuller. Scientific Management of
    Hazardous Wastes. New York: Cambridge University
    Press, 1983.

INDUSTRIAL TOXICOLOGY
Kirsch-Volders, Micheline, ed. Mutagenicity, Carcinogenicity
    and Teratogenicity. New York: Plenum, 1984.

Researchers will find certain types of "books" especially
useful: handbooks, manuals, encyclopedias, dictionaries, or
directories. Examples are:

Feulner, John A., comp. Hazardous Materials. (Who Knows?
    Selected Information Resources...SL 83-1) Washington,
    D.C.: National Referral Center, Library of Congress,
    rev. July 1983.

Gough, Beverly E., ed. <u>World Environmental Directory</u>. 4th
    ed. Silver Spring, MD: Business Publishers, Inc.,
    1980.

    Lists over 40,000 citations to pollution control manu-
    facturers and other firms with environmental interests;
    lobbyists; consulting, design and research services;
    governmental agencies; and law firms, universities,
    libraries, periodicals, and organizations with environ-
    mental interests. Listings are by country. This dir-
    ectory is reissued with updates every few years and is
    an inexpensive yet comprehensive reference source.

<u>Hazardous Material Control Directory 1985</u>. Silver Spring,
    MD: Hazardous Material Control Research Institute,
    1985.

LaDou, Joseph, ed. <u>Occupational Health Law: a Guide for
    Industry</u>. New York: Dekker, 1981.

    Information on OSHA and other regulatory agencies.

Pfafflin, James R. and Edward N. Ziegler, eds. <u>Encyclopedia
    of Environmental Science and Engineering</u>. 2nd rev. ed.
    New York: Gordon & Breach, 1983.

RBUPC Office, The British Office, comp. and ed. <u>Research in
    British Universities, Polytechnics and Colleges. Vol.
    2. Biological Sciences</u>. Wetherby, West Yorkshire:
    J. W. Arrowsmith Ltd., 1982.

    Subject index accesses a listing of researchers.
    Supplies title of research, institution, specific
    description of the topic, sponsoring bodies and
    starting and estimated completion dates of projects.

Sax, Irving. <u>Dangerous Properties of Industrial Materials</u>.
    6th ed. New York: Van Nostrand Reinhold, 1984.

    Information on 19,000 common industrial and laboratory
    materials including synonyms, description, formulas and
    physical constants, hazard analysis and countermeasures.

Sittig, Marshall. <u>Handbook of Toxic and Hazardous Chemicals
    and Carcinogens</u>. 2nd ed. Park Ridge, NJ: Noyes, 1985.

    Presents chemical health and safety information and
    measurement techniques for over 800 toxic and hazardous
    chemicals and carcinogens.

Standard Methods for the Examination of Water and Waste-
     water. 16th ed. Washington, D.C.: American Public
     Health Association, 1985.

     Presents important improvements in methodology and
     appropriate techniques for assessment and control of
     water quality and pollution. Covers laboratory and
     quality assurance. Definitive work on methods of water
     analysis.

Toxic and Hazardous Industrial Chemical Safety Manual for
     Handling and Disposal with Toxicity and Hazard Data.
     Tokyo: International Technical Information Institute,
     1981.

     Synonyms, uses, properties, disposal and treatment of
     700 industrial chemicals.

To determine what books are available beyond those in the
researcher's library, the researcher can use Scientific and
Technical Books and Serials in Print, a spinoff from Books
in Print. These tools list titles which are currently
available for purchase. Headings under which the books are
listed correspond to the Library of Congress subject headings noted above.

Book publishers' catalogs and brochures supply titles
available or to be published. A list of publishers offering
books on the environment with their addresses appears in the
section on Publishers and Vendors.

Periodicals

Since information on hazardous and toxic materials is constantly expanding and changing, sources of current information are vital to researchers in these areas. One of the
best is the periodical article. Periodicals are generally
defined as publications issued at regular intervals, whether
they be quarterly or weekly. Their form may vary from brief
newsletter to scholarly journals.

Articles in them can be located by abstract or index services. Periodicals containing articles on hazardous or
toxic materials are numerous and varied. A few representative titles are included here. Others containing articles
on more specific subjects appear in the appropriate chapters.

Ambio: a Journal of the Human Environment. Royal Swedish
     Academy of Sciences/Pergamon. Bimonthly, published
     since 1972.

Covers environmental management and technology and the natural sciences. Documents environmental research and supplies news of field.

Atmospheric Environment. Pergamon. Monthly, published since 1967.

An international journal on air pollution, atmospheric chemistry, aerosols etc., publishing research papers, review articles and preliminary communications. Covers all aspects of air pollution including the administrative, economic and political issues. Book reviews, conference news, etc.

Environmental Periodicals. Environmental Studies Institute, Issued 9 times each year, published since 1972.

Current awareness service for serials in environmental fields.

Environmental Science and Technology. American Chemical Society. Monthly, published since 1967.

Includes feature articles, critical reviews, current research papers and short book reviews. Pollution Control Directory is a special annual issue.

EPA Journal. Environmental Protection Agency, Office of Public Affairs. Monthly, published since 1975.

Reports on research and regulations formulated by the EPA in response to hazardous substances pollution. Cleanup projects, new research on toxic substances, and information on Superfund are covered.

EPRI Journal. Electric Power Research Institute. Monthly except combined issues in Jan/Feb and July/August, published since 1976.

Lists new technical reports from EPRI. Articles on various phases of power. A description of the Institute appears in the section on Research Centers and Industrial Laboratories in this chapter.

Journal of Environmental Sciences. Institute of Environmental Sciences. Bimonthly, published since 1958.

Official publication of the Institute of Environmental Sciences. Technical articles dealing with environmental management, simulation and testing. Reports current trends and developments.

Journal of Environmental Systems. Baywood Publishing Co.,
    Inc. Quarterly, published since 1971.

    A journal focusing on problems and solutions related to
    the environment. Includes articles on environmental
    analysis, design and management.

Journal of Hazardous Materials. Elsevier. Bimonthly, pub-
    lished since 1975.

    Features research, case histories, and safety informa-
    tion on a variety of hazardous materials emergencies.
    Book reviews. A periodic supplement, Hazbits, offers
    timely news between issues of the Journal.

Pollution Engineering. Pudvan Publishing Co. Monthly,
    published since 1969.

    Monthly trade magazine, publishes periodic directories
    listing manufacturers, services and equipment. A
    "Consultants Services Telephone Directory", yellow-page
    style, appears in the May issues, and an "Environmental
    Control Telephone Directory" is in the October issues.

Many publish special issues regularly or irregularly. Those
listed below are typical.

"1984 Directory of Governmental Air Pollution Control
    Agencies." Journal of the Air Pollution Control
    Association 34 (April, 1984): 419-444.

    Annual directory of agencies and personnel at federal,
    state and provincial levels in the United States and
    Canada arranged alphabetically by state and then by
    county or province. Gives full name, address, and
    telephone number of the agency and names and titles of
    personnel.

"Hazardous Waste Sites in the United States", edited by H.
Pishdadazar and A. Alan Moghissi, appeared in 1981 as a
special issue of the journal, Nuclear and Chemical Waste
Management, vol. 1 no. 3-4. This list, arranged alphabetically by state, cites hazardous waste disposal site name,
owner's name, site address, information on disposed materials, quantity of disposed material, and disposal technique
used. The informaton in this list was synthesized from
surveys produced by Congress and the EPA.

Many newsletters supply current awareness coverage in
various aspects of environmental control. They emanate from
commercial publishers, government agencies, research centers
in industry or universities and public interest groups. A
few examples are listed.

Reuse/Recycle Newsletter. Technomic. Monthly, published
    since 1971.

    Various types of wastes and waste environment are
    covered.

Tox-Tips: Toxicology Testing in Progress: a Monthly Current
    Awareness Bulletin. National Library of Medicine.
    Toxicology Information Program. Monthly, published
    since 1976.

    Industry, government and academic institutions report
    on new testing projects and epidemiology studies to
    determine toxicity of chemical substances and other
    agents. Project descriptions, supporting and perform-
    ing organizations and principal investigators are
    indexed. Indexes cumulated quarterly.

Toxic Materials News: Weekly Business Newsletter Published
    from the Nation's Capital. Business Publications, Inc.
    Weekly, published since 1974.

    This weekly newsletter provides information on the
    latest EPA policy developments. Contains articles,
    brief reports, grants and contracts awards and
    announcements of conferences and workshops.

Among numerous current awareness services are the following:

Environment Reporter. Bureau of National Affairs. Weekly
    supplementation and reports, published since 1970.

    A 17-volume looseleaf service covering the major
    environment issues. Contains full text of federal and
    state laws and regulations. A comprehensive reference
    source with a cumulative digest index of cases covered
    since 1970.

Inside E.P.A. Weekly Report. Inside Washington Publishers.
    Weekly, published since 1980.

Ulrich's International Periodical Directory, published
annually, lists periodicals currently being issued with
addresses of publishers, frequency of publication, price,
abstract/index services indexing their articles. Arrangement of the directory is by subject, and there is a title
index. The following sections will produce the titles of
periodicals which cover research in areas of interest to
those working in hazardous and toxic materials:

Environmental studies
Fish and fisheries
Hazard ...
Industrial health and safety
Medical science - Radiology and nuclear medicine
Medical science - Respiratory diseases
Pollution ...
Toxic ...
Water resources

Reports and Documents

The first form in which most research results appear is the technical report. Government and other contracts usually specify reports to be issued at regular, stated intervals, and those done in conjunction with government contracts are deposited with the National Technical Information Service (NTIS) in Springfield, Virginia. NTIS is located at 5285 Port Royal Road, and the phone number is (703) 487-4785.

NTIS disseminates an index service publication:

Government Reports Announcements & Index. National
    Technical Information Service. Biweekly, published
    since 1965.

    NTIS is a central source for U.S. Government-sponsored
    research, development and engineering reports as well
    as foreign technical reports. Documents cited are
    available in microform and paper. Arranged under 22
    broad subject categories and their subcategories, the
    entries are arranged alphanumerically. Report number,
    author, title and other bibliographic information
    followed by brief abstracts. Keyword index, personal
    author and corporate author index, report number and
    contact number index with cumulated annual index. The
    computerized counterpart of Government Reports
    Announcements & Index permits searching back to 1964.
    Available via BRS, DIALOG and SDC, the content of the
    data base consists of reports, etc., generated by
    government-sponsored research.

The National Technical Information Service, in addition to the index cited above, offers 28 weekly abstract newsletters covering specific subject areas, which give summaries of documents from the EPA and other sources. Appearing with the same frequency as the regular issues, they are subject-oriented, and the appropriate one for researchers in hazardous toxic substances is Environmental Pollution and Control: an Abstract Newsletter, which is described as covering air noise, solid waste, water, pesticides, radiation, environmental health and safety, environmental impact statements.

NTIS also publishes <u>Catalog of Directories of Computer Soft-</u><u>ware Applications</u> which includes a directory of environmental pollution and control.

<u>NTIS Published Searches</u> are bibliographies developed by information specialists in the subject area by searching several important data bases such as Energy Data Base, <u>Pollution Abstracts</u>, Engineering Information Inc., <u>Inter-</u><u>national Aerospace Abstracts</u>, and NTIS. These are available from NTIS, Springfield, Virginia 22161, each for $35, currently, plus handling.

Proceedings

Many new concepts, procedures or products are introduced to the scientific/technical community by way of papers presented at conferences, symposia and other meetings. In some cases these papers are disseminated only to attendees, but most become printed proceedings of the meetings. They may appear as preprints at the meeting or in volumes distributed as late as a year or two after the meeting. Determining if proceedings have been published is possible by using one of the tools listed below. These tools locate the proceedings whether they are published as a separate proceedings volume, a technical report or a journal article. Relevant material can be identified by the use of such terms as health hazards, occupational exposure, toxicity, toxicology, industrial chemicals, chemical exposure, environmental health or radiation.

<u>Conference Papers Index</u>. Bethesda, MD: Cambridge Scientific
   Abstracts. Monthly, published since 1973.

   Lists papers presented at conferences covering science,
   technology and medicine. Available online from Dialog
   Information Service from 1973 on.

<u>Directory of Published Proceedings: Series SEMT, Science/</u>
   <u>Engineering/Medicine/Technology</u>. Harrison, NY: Inter-
   Dok Corp. Issued 10 times each year, published since
   1965.

   Chronological arrangement. Keyword indexes for name of
   conference, sponsors and location.

<u>Index to Scientific and Technical Proceedings</u>. Published
   monthly and cumulated annually by the Institute for
   Scientific Information, Inc., since 1978.

   The main entry gives the complete bibliographic
   description of each proceedings with titles and authors

of individual papers. It also has an elaborate subject
index for easy location of papers. When papers are
available through the Institute's Original Article Text
Service, that is also indicated at the end of the
entry. Online via the publisher's search service.

Proceedings in Print. Arlington, MA: Proceedings
    in Print, Inc. Bimonthly, published since 1964.

    Covers a wide range of proceedings.

Yearbook of International Congress Proceedings. Brussels:
    Union of International Associations. Annually, published
    since 1969.

    Covers reports and proceedings of international
    conferences.

To locate information through these indexes as well as the
others listed, it is necessary to use many terms to identify
relevant material. A list of those terms which should be
used include: health hazards, occupational health, occupational hygiene, occupational exposure, carcinogens, mutagens, teratogens, toxicity, toxicology, reproductive hazards,
industrial chemicals, chemical exposure, environmental
health, as well as specific causes of health hazards
(radiation, lead, uranium, asbestos, dust, dioxin, wood).

Some associations which sponsor conferences and other meet-
ings frequently publish their proceedings as series. The
American Chemical Society and the American Society for
Testing and Material are typical, producing these series:

    ACS Symposia Series
    ASTM Special Technical Publications (Several STP's are
    listed in other chapters.)

    The Institute of Environmental Sciences has published
    the proceedings of its annual meetings under various
    "theme titles" since 1954.

Reviews

A quick and easy way to acquire a bibliography or list of
references is to locate a review. There are annual review
series which select, organize and evaluate the current
literature and cover the significant developments in the
fields they review. Several of the series are produced by
Annual Reviews, Inc. These series and others published in
areas relevant to toxic and hazardous materials can be
located in:

Irregular Serials & Annuals: an International Directory.
    New York: R. R. Bowker Co. Published every two years
    since 1970.

    A comprehensive list of review series, conference
    proceedings, etc., issued less frequently then twice a
    year.

A detailed review of toxicological information and its
information resources has been written by H. M. Kissman and
P. Wexler. Titled "Toxicological Information", it appears
in Annual Review of Information Science and Technology,
1983, vol. 18, pp. 185-230.

Other relevant review series:

Advances in Ecological Research. New York: Academic. Vol.
    1, 1962- .

    Papers reflect the broad scope of ecological research
    and the complexity of the systems associated with it.

Advances in Environmental Science and Technology. New York:
    Wiley. Vol. 1, 1969- .

    Review articles covering both general and technical
    topics in all aspects of the field.

Other types of reviews are those that appear in periodicals
or reports and state-of-the-art surveys. They can be
accessed by use of this reference tool:

Index to Scientific Reviews. Philadelphia: Institute for
    Scientific Information. Semiannually, published since
    1974.

    Some of the terms to be searched in the permuterm
    subject index: Air pollutants; Air pollution;
    Environments; Hazards; Pollutants; Diseases; Water.

Two periodicals devoted to reviews in relevant areas are:

CRC Critical Reviews in Environmental Control. Boca Raton,
    FL: CRC Press, Inc. Quarterly, published since 1970.

CRC Critical Reviews in Toxicology. Boca Raton, FL: CRC
    Press, Inc. Quarterly, published since 1971.

Indexes and Abstracts

Reading or scanning the issues of specific periodicals regularly only partially fulfills the researcher's commitment to
acquiring current information. To make contact with other
publications less relevant than his/her regular list, the
researcher can use the abstract/index services. Through the
subject indexes common to most of them, the user can ferret
out pertinent and peripheral articles. Indexes give complete bibliographic details for locating articles.
Abstracts have the added advantage of providing a brief
description of results, methodology, etc.

Several services encompass all aspects of toxic and hazardous
substances; others tend to be more specific. The latter are
covered in the various chapters. Samples of those with
broad coverage and their computerized counterparts appear
below.

Applied Science and Technology Index. H. W. Wilson Co.
    Monthly, except July, published since 1958.

        Reviews over 300 English language periodicals. Subject
        fields indexed include many aspects of environment.
        The index is arranged alphabetically by subject with
        cross references and quarterly and annual cumulations.

Bibliographic Index: a Cumulative Bibliography of Biblio-
        graphies. H. W. Wilson Co. Published April, August
        with a bound cumulative each December, since 1937.

        This index is arranged alphabetically by subject.
        Bibliographies with 50 or more citations which have
        been published separately or as parts of books, period-
        ical articles and pamphlets are selected for inclusion.
        Scans approximately 2600 periodicals.

Biological Abstracts and Biological Abstracts/RRM.
        Philadelphia, PA. Biweekly, published since 1926.

        9000 primary journals, monographs, symposia, reviews,
        reports and other sources are abstracted. The data
        base, Biosis Previews, is available from 1969 on,
        through BRS and DIALOG.

Chemical Abstracts. Chemical Abstracts Service, American
        Chemical Society. Weekly, published since 1907.

        Weekly issues include keyword, patent and author
        indexes. Annual and collective indexes include author,
        general subject, chemical substance, formula, ring
        systems and patent indexes. The Index Guide, with its

supplements, gives cross-reference, synonyms and other
information, for using the chemical substance and
general substance indexes. Chemical Abstracts'
computerized counterpart, CA Search, can be searched
back to 1967.

Current Contents: Agriculture, Biology, and Environmental
     Science. Institute for Scientific Information. Weekly,
     published since 1970.

     Each issue reproduces the table of contents pages of
     the latest issues of over 100 journals and includes
     current book contents. The Title Word Index is helpful
     in locating the desired article or book. The Current
     Contents Address Directory, Science and Technology
     provides addresses of authors currently publishing in
     these fields. A document delivery service called "The
     Genuine Article" can supply tear sheets or photocopies
     of most articles, etc. appearing in Current Contents...
     issues.

Environment Abstracts. EIC/Intelligence. Monthly
     (bimonthly May/June, Nov/Dec), published since 1971.

     Cumulated annually into Environment Abstracts Annual
     and its companion volume Environment Index. Covers
     published studies such as conference papers, journal
     articles and reports on all aspects of environmental
     issues. Each issue has a review section followed by
     subject index, industry index, source index, and author
     index. Also lists conferences, and new books in print.
     Documents available from EIC/Intelligence are identi-
     fied. Environment Index has a useful directory section
     listing governmental and nongovernmental organizations,
     conferences, films, film distributors and books.
     Enviroline is the name of the data base counterpart of
     Environment Abstracts. Available through DIALOG and
     SDC, it is searchable from 1971.

Pollution Abstracts. Cambridge Scientific Abstracts.
     Monthly, published since 1970.

     Abstracts and indexes about 2500 worldwide publications
     on environmental pollution. Abstracts section is
     arranged under ten major headings with full citations
     and abstracts followed by subject and author indexes.
     The index section is cumulated annually. The
     controlled vocabulary gives related and updated terms.
     Some examples: Air pollution; Hazardous materials;
     Pollutant; Pollution; Toxic materials; Toxic wastes;
     Toxicology. The service is available online through
     BRS and DIALOG from 1970 to the present.

Public Affairs Information Service Bulletin. Public Affairs
    Information Service. Semimonthly, with quarterly and
    annual cumulations, published since 1915.

    Aimed at legislators, administrators, business communi-
    ty, researchers and students, this index covers world-
    wide publications in the English language on contempor-
    ary public issues. Approximately 1400 periodicals, the
    latest books, federal, state, local and foreign
    documents.

Science Citation Index. Philadelphia: Institute for
    Scientific Information. Quarterly, published since
    1961.

    Citation Index, arranged by name of cited author, lists
    subsequent publications which cite author's articles,
    etc. Source Index must be used to find joint authors.
    Source Index lists titles and all other bibliographic
    information. Permuterm Subject Index is a keyword
    index. The computerized counterpart of SCI can be
    searched back to 1974 via DIALOG.

Data Bases

Conference Papers Index
Producer: Cambridge Scientific Abstracts
Time span: 1973 to present
Coverage: International
Updated: Monthly
Vendor: DIALOG

    Scientific and technical papers presented at over 1000
    meetings are added each year.

ENVIROLINE
Producer: Environmental Information Center, Inc.
Time span: 1971 to present
Updated: Monthly
Vendor: SCD, DIALOG
Print counterpart: Environment Abstracts

    Environmental health is one topic covered in this data
    base which includes journals, government reports and
    documents, monographs, conference papers, newspapers,
    the Federal Register, and some films.

Environmental Bibliography
Producer: Environmental Studies Institute
Time span: 1973 to present
Updated: Bimonthly
Vendor: DIALOG

    Human ecology and health are among the fields covered.

GPO Monthly Catalog
Producer: Government Printing Office
Time span: 1976 to present
Coverage: United States government publications
Updated: Monthly
Vendor: BRS, DIALOG

    Reports, studies, fact sheets, maps, handbooks and
    conference proceedings issued by U.S. federal govern-
    ment agencies are included.

NTIS
Producer: National Technical Information Service
Time span: 1964 to present
Coverage: United States government reports
Updated: Biweekly
Vendor: BRS, DIAlOG, SDC

    Government-sponsored research which produces unclassi-
    fied reports provides the content for this data base.

Further information on data bases is obtainable through the
following directories:

CIS Search Sample. Falls Church, VA: Computer Sciences
    Corp., 1983.

Data Base Directory Service. White Plains, NY: Knowledge
    Industry Publ., Inc. Semiannually, published since
    1985.

    Service includes monthly newsletter. Published in
    cooperation with the American Society for Information
    Science.

Directory of Online Data Bases. New York: Cuadra/Elsevier.
    Published quarterly since 1979.

    Supersedes all preceding. Master index. Names of
    bases and agencies. Some under Environment, Hazar-
    dous..., Toxic ...

Besides the bibliographic data bases, there are numerous
data bases that provide numeric data or text of articles.
Among them:

dictionary: Chemname, Chemsis, Chemzero (DIALOG)
               Online chemical dictionaries.

               TSCA Inventory (DIALOG, CIS)
               Non-bibliographic dictionary listing of
               chemical substances from the initial
               inventory of the Toxic Substances Control
               Act Substance Inventory.

full-text: ACS Journals Database (BRS)
               Complete text of articles, communications,
               notes, etc., produced by the American
               Chemical Society.

Audiovisual Materials

If a picture is worth a thousand words, audiovisuals (AV)
are an ideal form for presenting the problems generated by
hazardous and toxic materials to all levels of audiences.
There are several publications that direct educators, associations, etc. in these fields to suitable films, slides,
etc.

The NICEM (National Information Center for Educational
Media) indexes are available as printed indexes and
data bases. The Index to Environmental Studies - Multimedia
is in its second edition.

Science Books & Films has been published by the American
Association for the Advancement of Science since 1965. Five
issues per year contain "critical reviews of books, films
and filmstrips in mathematics and the social, physical and
life sciences ..."

AVLINE is a data base which locates audiovisual programs in
the health sciences (see Chapter 7).

Dissertations

Universities throughout the world are the sites of research
on all aspects of toxic and hazardous materials research.
Thousands of students produce theses and dissertations in
academic institutions. The main source for locating such
valuable work is the Dissertations Abstracts International.
Many universities have agreed to deposit their dissertations

with University Microfilms International which sells them to anyone interested in their subject matter.

DAI is a monthly compilation of doctoral dissertations from 450 institutions in the United States and Canada abstracted and arranged by broad subject categories. Beginning with Volume 30 a mechanized keyword title index is included to facilitate the identification of desired titles. Section B covers those in science and engineering. Published monthly since 1938 by University Microfilms International, the issues have a keyword title index and an author index. The Comprehensive Dissertations Index covers dissertations accepted for doctoral degrees between 1861 and 1971.

An online search service offered by UMI called DATRIX DIRECT can be reached by calling Dissertation Database Specialist, (800) 521-0600, for an up-to-date computer printout on selected subject categories for a fee of $20. (US and Canada; $30, all others). Dissertation Abstracts International can also be searched online through DIALOG and BRS.

The company frequently distributes catalogs of selected dissertations. The catalog with the most pertinent information for researchers in toxic and hazardous materials is Ecology and the Environment: a Catalog of Selected Doctoral Dissertation Research from Abstracts 1980-84.

Microform or paper copies of dissertations can be ordered from University Microfilms International, P.O. Box 1764, Ann Arbor, Michigan 48106. To inquire about rush orders, methods of payment, etc., customers can call (800) 521-3042.

Government Organizations

Governments, especially the United States, are in the fore-front of research and development in the areas of hazardous and toxic materials. Agencies such as the Environmental Protection Agency, Fish and Wildlife Service, Occupational Health and Safety Administration produce data, reports and other publications and make them available to the general public. Accessing them is easiest when a depository library is convenient to the researcher. Here the researcher will find a broad and diverse selection of publications from a wide range of agencies.

The proliferation of documents and reports generated by U.S. government agencies and independent laboratories involved in government-sponsored research includes analyses of transportation accidents, new response methods for all types of hazardous materials emergencies, and other research on emergency response. Regular monitoring of government indexes and data bases covering reports and documents is essential for maintaining a continuing awareness of these materials. The Monthly Catalog of United States Government Publications

and its online equivalent, the GPO Monthly Catalog Data
Base, cover publications available from the Government
Printing Office and other U.S. government agencies.

Monthly Catalog of United States Government Publications.
    Washington, D.C.: Superintendent of Documents. Govern-
    ment Printing Office. Published monthly since 1895.

    Reports, studies, fact sheets, maps, handbooks and
    conference proceedings issued by U.S. federal govern-
    ment agencies are included. Available online through
    BRS and DIALOG, the data base can be searched from 1976
    on.

    The keyword index includes such terms as the following:
    Air quality ...; Environment law; Environment protec-
    tion; Hazardous substance; Hazardous wastes ...; Pollu-
    tion - Environmental aspects; Toxicology.

The best source for information on government agencies and
their addresses, organization, function, etc. is:

United States Government Manual. Washington, D.C.: Office
    of the Federal Register, National Archives and Records
    Service, General Services Administration. Published
    annually since 1895.

    The Manual has subject and agency indexes. "Describes
    the purposes and programs of most governmental agencies
    and lists top personnel."

Other directories containing information on government agencies include the title noted below. Others appear in the
"Research Centers" section of this chapter.

Kruzas, Anthony T. and Kay Gill, eds. Government Research
    Centers Directory. 2nd ed. Detroit, MI: Gale Research
    Co., 1982.

    A guide to 1642 research facilities owned and operated
    or substantially supported by the federal government,
    with addresses, telephone numbers, type of research
    activities and principal fields. Name and keyword
    index, agency index, and geographic index.

The agency whose activities are most universal in interest
to researchers in toxic and hazardous materials is the
Environmental Protection Agency (EPA). The EPA has many
different subdivisions involved in different aspects of
environmental protection and management. Information about
the EPA and its offices can be found in:

EPA Guidebook. Rockville, MD: Government Institutes.
    Published annually. Title varies.

There are ten regional offices of the EPA (Boston, New York,
Philadelphia, Atlanta, Chicago, Dallas, Kansas City, Denver,
San Francisco, and Seattle), which provide assistance to the
general public and waste practitioners, through the mail or
by telephone. The EPA, in addition to its ten regional
offices, has about 30 subdivisions to direct research and
communications in specified areas (i.e. the Office of Solid
Waste, the Office of Toxic Substances, the Office of Policy
Analysis). An Office of Pesticides and Toxic Substances of
the EPA coordinates activities under the Toxic Substances
Control Act (TSCA) with other agencies for the assessment
and control of toxic substances.

A concise chart of EPA organizations is printed in the
EPA Journal, vol. 10 no. 1, (January/February 1984), pp. 30-
31. As noted previously, information on RCRA and the
Superfund can be obtained by dialing the toll-free hotline
telephone number (800) 424-9346. For guidance on toxic
materials or TSCA, call (800) 424-9065. The number for
pesticides information is (800) 858-7878.

The EPA Research Program Guide, published in 1983 by the
EPA (EPA-600/9-83/-011), lists all of the research laboratories of EPA, and identifies the areas of research in which
each of the laboratories is involved.

The following is a comprehensive bibliography of the
Agency's reports, etc.:

EPA Publications: a Quarterly Guide. Washington, D.C.:
    Office of Planning and Management, published since
    1977.

    1970-76 publications listed in EPA Cumulative
    Bibliography 1970-76 (NTIS PB 265 920).

The major instruments of promulgation of EPA rules and
regulations are the Federal Register and the Code of Federal
Regulations, and these publications should be readily available in most libraries. Both have periodic subject indexes
which guide the researcher to applicable sections. Some
companies reprint selected topical Federal Register
announcements, such as the reprint service for Federal
Register coverage of hazardous materials, chemicals and
substances published by the Hazardous Materials Publishing
Company. The online computer service CSI Federal Register
Data Base (FEDREG), available through DIALOG and SDC, provides access to regulatory activity as published in the
Federal Register.

Index to the Code of Federal Regulations, published annually
by Congressional Information Service, Inc., offers easy

access to government regulations, the latest edition covering 1977-1984.

An invaluable comprehensive source of toxicity information has been published by another U.S. agency, the National Institute for Occupational Safety and Health:

Tatken, Rodger L. and Richard J. Lewis, Sr., eds. RTECS. Registry of Toxic Effects of Chemical Substances 1981-82. 3 vols. DHHS (NIOSH) Publication No. 83-197. Cincinnati: Department of Health and Human Services, Public Health Service, Center for Disease Control, National Insitute for Occupational Safety and Health, 1983.

218,746 listings of chemical substances are indexed by CAS registry number and by chemical substance name. Each entry lists the name, synonyms, and foreign names of the substance; irritation, mutation, reproductive effects, tumorigenic and toxicity data, and references; aquatic toxicity rating; review articles on the substance; federal standards and regulations; criteria documents for standards; and NTP, NIOSH and EPA status information about the substance. This book is an invaluable comprehensive source of toxicity information. RTECS has been available as a data base since 1979.

Other agencies whose activities are of interest to researchers in toxic and hazardous materials are:

Department of Defense. Hazardous Materials Technical
    Center.

Department of Labor. OSHA Technical Data Center.

    Deals with industrial toxicology, hazardous materials,
    carcinogens, etc.

Oak Ridge National Laboratory. Chemical Effects Information
    Center.

    Deals with toxic substances: production, use, properties, effects of all kinds. Services available to private industry as well as the military.

Canadian agencies are also active in environmental concerns. One such is located in Northern Ontario:

Ontario Ministry of the Environment
199 Larch Street
Sudbury, Ontario   P3E 5P9
Canada   (705) 675-4501

Areas of interest: atmospheric chemistry, impact on
aquatic ecosystems, impact on terrestrial ecosystems,
transportation & deposition.

Publications: In professional journals, proceedings and
technical reports.

Research Centers and Industrial Laboratories

The "invisible college" enables researchers in various
fields to communicate with others doing similar work. To
locate research centers or industrial or academic laboratories in which they are working, investigators have at their
disposal several directories:

Jaques Cattell Press, ed. Directory of American Research
and Technology. 20th ed. New York: Bowker, 1986.

Annual. An alphabetical listing of 10,991 parent
organizations, their divisions and subsidiaries with
information on fields of R & D and professional staff.
Editions 1-19 were entitled: Industrial Research
Laboratories of the U.S.

Watkins, Mary Michelle and James A. Ruffner, eds. Research
Centers Directory. 10th ed. 1985-86. Detroit, MI:
Gale Research Co., 1985.

Approximately 3800 academic and nonprofit research
organizations in the United States and Canada are
listed with complete address and telephone numbers,
research activities and fields of research publications
and services of each. Arranged alphabetically under 14
major subject headings. Alphabetical, institutional,
special capabilities and subject indexes. See also the
"government organizations" section of this chapter.

Among the laboratories and centers performing research in a
broad spectrum of environmental control are these:

Battelle - Pacific Northwest Laboratories
P.O. Box 999
Richland, WA 99352   (509) 375-2121

Areas of interest: atmospheric chemistry, economic
impact, emission, impact on aquatic ecosystems, impact
on environment, impact on terrestrial ecosystems,
transportation and deposition.

California Institute of Technology
Environmental Quality Laboratory
Pasadena, CA 91125   (818) 356-6783

    Studies on air pollution, energy and environment
    interaction, water resources and quality, disposal of
    wastewater in the ocean and hazardous substances.

Electric Power Research Institute
P.O. Box 10412
Palo Alto, CA 94303   (415) 855-2000

    The Institute was founded by the nation's electric
    utilities to improve power production and utilization
    through economically and environmentally acceptable
    programs.

Franklin Institute. Franklin Research Center
Information Management Department
20th & Race Streets
Philadelphia, PA 19103   (215) 448-1000

    Air/water pollution, occupational hazards, nuclear
    safety, etc.

Princeton University
Center for Energy and Environment Studies
Engineering Quadrangle
Princeton, NJ 08544   (609) 452-5445

    Regional, national and international studies of energy
    and environmental problems including hazardous waste
    management.

Those of more specific interests are identified in the
appropriate chapters.

Libraries, Information Centers

Libraries and information centers in the area of hazardous
and toxic materials have evolved from those established for
research in chemistry or health sciences, for example, or
have been recently established from a recognized need of
those working in these areas. Several directories identifying the libraries or centers, their holdings, personnel
and research interests are available. One such directory,
is described below:

Darnay, Brigitte T., ed. <u>Directory of Special Libraries and Information Centers</u>. 9th ed. Detroit, MI: Gale Research Co., 1985.

> "A guide to special libraries, research libraries, information centers, archives and data centers ..." Subject index contains entries such as air pollution, pollution, radioactive waste disposal, toxicology, water.

There are a number of federal libraries that can be contacted for help in finding information about hazardous waste disposal. The United States International Environmental Referral Center (c/o EPA, 401 M Street, SW, Room 2902 WSM, Washington, D.C. 20460), offers free publications which list other libraries and sources of information on environmental topics. Letters identifying the specific topic of research should be addressed to the Center.

The Office of Library Systems and Services of the EPA at 401 M Street, SW, Room 2404, Washington, D.C. 20460, offers a free <u>Guide to Library Sources and Services</u> which delineates the services offered by the EPA Headquarters in Washington. Additionally, each of the ten regional offices of the EPA has a library with limited public access. The regional offices may be contacted directly for further information about their holdings and policies.

<u>Environmental Protection Agency</u>
Headquarters Library
401 M Street, SW
Washington, D.C. 20460   (202) 755-0308

> Holdings: 8,000 books
>   25,000 bound reports and documents
>   300 microfilms
>   150,000 microfiche
>
> Subscriptions: 1,000 journals and serials
>   5 newspapers
>
> Library is unrestricted

Typical of EPA's regional libraries:

<u>Environmental Protection Agency</u>
Region VII Library
324 East 11th Street
Kansas City, MO 64106   (816) 374-3497

> Holdings: 2,500 books
>   17,000 technical reports
>   100,000 technical reports on microfiche
>
> Subscriptions: 250 journals and serials

(Other EPA libraries are described in several of the
following chapters.)

Library of Congress
John Adams Building, Room 5228
National Referral Center
Washington, D.C. 20540   (202) 287-5670

The Center advises requestors as to how to obtain
information on specific topics in science and technology and directs them to organizations or individuals
with specialized knowledge. The Center deals with the
biological, social, physical and engineering sciences
and relative technical areas and calls on information
resources in government, industry and academic institutions. Center is open to the public.

National Oceanic & Atmospheric Administration
Environmental Science Information Center
Environmental Data & Information Service
11400 Rockville Pike, Code D8
Rockville, MD 20852   (301) 443-8137

NOAA explores global ocean and conserves its resources,
develops beneficial environmental modifications and
evaluates their consequences.

National Safety Council
Library/Safety Research Information Service (SRIS)
444 N. Michigan Avenue
Chicago, IL 60611   (312) 527-4800

Information on all aspects of safety is collected,
organized, and disseminated. Information service is
available to the public.

Associations & Societies

Professional societies and other organizations offer a
multitude of services to their members and, often, non-
members. These include publications such as books, directories, journals, proceedings of conferences and other
meetings, speakers' bureaus, et al.

There are hundreds of associations dealing with toxic and
hazardous materials from scientific and technical to
citizens' groups. Some have members representing diverse
interests like the American Society for Testing and Materials; others, specific, like those listed in the chapters
following.

A key to locating associations of persons working in toxic
and hazardous materials and related fields is Encyclopedia
of Associations. Volume 1 contains a keyword index to the
names of the associations; volume 2, a geographic and executive index; and volume 3, an interedition supplement to
volume 1, titled New Associations. Keywords such as
Hazardous, Pollution and Toxic will help to identify relevant associations. The Encyclopedia has been published by
Gale Research Co., Detroit, Michigan, since 1956; annually
since 1975.

It is often useful to know what publications have been
produced by various associations. Such information is provided by Associations' Publications in Print, published by
R. R. Bowker Co., New York, annually since 1981. Volume 1
is a subject index; volume 2 contains publisher, title,
association name and acronym indexes.

Typical associations:

American Society for Testing and Materials
1916 Race Street
Philadelphia, PA 19103   (215) 299-5400

> ASTM creates voluntary standards, test methods, speci-
> fications, practices and definitions. The Society
> issues the Annual Book of ASTM Standards, several jour-
> nals and Special Technical Publications (STP's) most of
> which are the proceedings of symposia sponsored by the
> Society's committees. The many volumes of the Book
> cover the standards created for various materials and
> tests. Volume 11.04 of the 1984 Book, for example:
> Pesticides; Resource Recovery; Hazardous Substances and
> Oil Spill Response; Waste Disposal; Biological
> Effects... (87 standards).

American Society of Safety Engineers
850 Gusse Highway
Park Ridge, IL 60068   (312) 692-4121

> Individuals concerned with industrial safety and
> accident prevention. Annual conference. Publishes
> Professional Safety.

Conservation Foundation
1717 Massachusetts Avenue, NW
Washington, D.C. 20036   (202) 797-4300

> Pollution and toxic substances control a primary
> concern. Conducts research and educates the public on
> improvement of the environment.

Hazardous Materials Control Research Institute
9300 Columbia Boulevard
Silver Spring, MD 20910   (301) 587-9390

Offers membership to individuals and corporations that
have a vested interest in hazardous material management
and control.   The Institute offers conferences, seminars, equipment exhibitions, and publications, including the HMCRI Forum newsletter. The HMCRI is a nonprofit organization that seeks to "promote the establishment and maintenance of a reasonable balance between
expanding industrial productivity and an acceptable
environment".

Hazardous Materials Information Center
62 Washington Street   (203) 344-0419
Middletown, CT 06457   (800) 433-1116

More of a publisher than a library-type information
center. Most recent publications include the
following:

OSHA's Right to Know Hazardous Chemicals:
  A Consolidated List

Right to Know: A Guide to Federal and State Programs

Hazard Classification Systems: A Comparative Guide to
  Definitions and Labels

Directory of State Hazardous Waste Officials

Federal Hazardous Waste Regulations: A User-Friendly
  Text

Using the Hazardous Waste Manifest: A Manual of Federal
  and State Requirements

One of the newest services of one of the oldest American
associations is CA Selects, product of Chemical Abstracts
Service. Catalog of CA Selects: current-awareness publication. (dated 3/84). Includes: Environmental pollution;
Liquid waste treatment; Pollution monitoring; Solid and
radioactive waste treatment; Water treatment; Carcinogens;
Mutagens and teratogens; Chemical hazards; Health and
safety; Food toxicity.

2.
Publications on the Testing, Analysis, Monitoring and Sampling of Pollutants

Theresa L. Wolfe

Introduction

The detection of hazardous materials in the environment
becomes more complex daily. Faced with hundreds of new
chemical substances and increasingly sophisticated instrumentation, the analytical chemist is faced with a vast array
of sources from which to choose his methods. In addition,
the regulated community is faced with the need to comply
with the regulations of their corresponding jurisdiction on
the federal and state level. Information on the analysis of
environmental pollutants is found in the literature of many
fields, from agriculture to zoology, as well as in the
growing literature of hazardous materials management. This
review presents a selection of some of the sources available
to chemists, engineers and others involved with monitoring
and testing.

The section on regulations lists the parts of the <u>Code of
Federal Regulations</u> which contain analytical methods
required by the U.S. Government. In addition, each state
has recommended methods for use in their respective jurisdictions. A list of all state methods is far beyond the
scope of this discussion. For a specific state's regulations, consult the appropriate state code for monitoring and
testing methods.

The section on books and monographs is divided into three
parts. The first includes those sources which are devoted
solely to air, and the second is dedicated to water. The
third part includes those sources which list methods for
both air and water, as well as soil and food. Here also are
found works on instrumentation and quality control as well
as those works which discuss one specific pollutant or group
of pollutants such as asbestos, or heavy metals.

Although the emphasis of this book is on recent publications
(1980 or later), many of the sampling and analytical methods
presently in use are described in 1970's literature. A

judicious selection of these earlier publications is
included here.

Many journals include papers on testing and monitoring
methods to be used in their particular subject area. Good
sources are <u>Analytical Chemistry</u>, <u>Atmospheric Environment</u>,
<u>Bulletin of Environmental Contamination and Toxicology</u>,
<u>Ground Water</u>, and the journals of the Air Pollution Control
Association and the Water Pollution Control Federation.

Similarly, many conferences include papers on the subject.
The American Society for Testing and Materials publishes the
proceedings of many of these conferences in their Special
Technical Publication series. The Aquatic Toxicology and
Hazard Assessment symposia have become an annual publication.

The United States Government, with its many regulatory
agencies, is a prime source of information on environmental
testing and sample collection. The Environmental Protection
Agency alone publishes many books, as well as individual
methods of analysis. The section on documents contains only
a partial listing, including those which are recommended in
the <u>Code of Federal Regulations, Title 40</u>. For a complete
list of documents published by EPA consult the quarterly
<u>EPA Publications Bibliography</u>. New methods, both final and
proposed, are announced daily in the <u>Federal Register</u>. EPA
and other government agency publications are listed in many
published abstract services and online data bases.

Regulations

<u>Code of Federal Regulations. Title 40. Protection of
Environment</u>. Washington, D.C.: Government Printing
Office, issued annually on July 1.

<u>Part 61</u>: National emission standards for hazardous air
pollutants. Appendix A contains national emission
standards for hazardous air pollutants, compliance
status information. Appendix B lists test methods, and
Appendix C has quality assurance procedures.

<u>Part 136</u>: Guidelines establishing test procedures for
the analysis of pollutants in water.

<u>Part 141</u>: National interim primary drinking water
regulations monitoring methods.

<u>Part 143</u>: National secondary drinking water regulations
monitoring methods.

<u>Part 261</u>: Identification and listing of hazardous
waste. Appendix I gives representative sampling
methods, Appendix II lists extraction procedure

toxicity test procedures, and Appendix III offers
chemical analysis test methods.

Part 796-798: Toxic Substances Control Act test guidelines for chemicals. Gives testing, environmental
protection, chemical fate, environmental effects,
effects, and chemicals guides. Includes the OECD
guidelines.

Books and Monographs on the Measurement
of Air Pollution

Air Pollution Control Association. **Measurement and Monitoring of Noncriteria (Toxic) Contaminants in Air**.
Pittsburgh: APCA, 1983.

Offers 46 original technical presentations, emphasizing
accuracy and reliability of data, field applications,
cost benefit issues, and population exposures. Includes
indoor contaminants.

American Conference of Governmental Industrial Hygienists.
**Air Sampling Instruments for Evaluation Atmospheric
Contaminants**. 6th ed.   Cincinnati: ACGIH, 1983.

Includes sampling methods, instrument operation and
performance, sampling systems and components. The new
chapters include measurement processes, precision,
accuracy and validation, sampling in exposure chambers,
organic sampling, and vapors and particles.

Brenchley, D. R., et al. **Industrial Source Sampling**. Lancaster, PA: Technomic, 1973. 4th printing 1982.

This detailed text on all aspects of air pollution
sampling and analysis from industrial sources proceeds
through every step from preliminary planning through
advanced methods of testing and monitoring.

Camagni, P. and S. Sandroni, eds. **Optical Remote Sensing of
Air Pollution**. Lectures of a Course Held at the Joint
Research Centre, Ispra, Italy, April 12-15, 1983. New
York: Elsevier, 1983.

Discusses the use of optical methods and remote sensing
of air pollutants.

Cheremisinoff, P. N., ed.   **Air/Particulate Instrumentation
and Analysis**. Stoneham, MA: Butterworth, 1981.

Examines techniques and instrumentation for measuring
and analyzing emissions from industry and utilities,
including stack sampling, particulates, bag sequential
sampling of ambient air, PAH, electrostatic precipita-
tors, thermal anemometry, fluidic flowmeters, and environmental signal processing.

Cheremisinoff, P. N. and A. C. Morresi.   Air Pollution Sampling and Analysis Deskbook.   Stoneham, MA: Butterworth, 1978.

Examines and evaluates the known methods for sampling
and measuring gases, particles and anions.   Includes
elemental, organic, and inorganic gas analysis, stack
sampling, particulates and fuels, and comparison for
source emission limits.

Halliday, D. A.   Air Monitoring Methods for Industrial Contaminants.   Davis, CA: Biomedical Publications, 1983.

Presents a compilation of the known commercial devices
for quantitatively determining each of 200 different
chemical substances in the workplace environment.   Includes initial cost of the system, measurement technique used, duration of sampling, whether it is portable or fixed, useful concentration span, additional
cost per test, and manufacturers.

Keith, L. W.   Identification and Analysis of Organic Pollutants in Air.   Stoneham, MA: Butterworth, 1984.

Explores sampling and analytical techniques including
mass spectrometry, gas and liquid chromatography, x-ray
fluorescence and Fourier transform infrared spectrometry.   Provides a state-of-the-art review of the analysis of low level organic compounds present in the
environment and in emissions from combustion sources.

Murphy, C. H.   Handbook of Particle Sampling and Analysis
Methods.   Deerfield Beach, FL: Verlag Chemie International, 1984.

Presents an overall discussion of airborne particles
from the atmosphere or in smokestacks of the workplace.
Begins with a discussion of particles, continues with a
description of sampling instruments, and analytical
methods.

Noll, K. E. and T. L. Miller.   Air Monitoring Survey Design.
Stoneham, MA: Ann Arbor Science, 1977.

This comprehensive guide to air monitoring is supplemented with numerous tables of basic reference data and

schematics illustrating procedures and the function of equipment. Includes definitions, selection of methods, calibration of instruments, price ranges and sources for hardware, procedures for conducting meteorological surveys, and methods for validation of mathematical simulation models.

Powals, R. J., et al. Handbook of Stack Sampling and Analysis. Lancaster, PA: Technomic, 1978.

This guide gives 53 collection systems, 201 collection media and 25 analytical methods for the 692 most common air pollutants from stationary sources.

Ruch, W. E., ed. Chemical Detection of Gaseous Pollutants. Stoneham, MA: Ann Arbor Science, 1982.

This guide to over 1,250 tests for the detection of 152 different industrial gases and air pollutants presents abstracts of articles arranged in alphabetical order by pollutant, from acetic acid to zinc oxide.

Stern, A. C. Air Pollution: A Comprehensive Treatise. Five volumes. Orlando, FL: Academic, 1977.

Volume 3, Measuring, Monitoring and Surveillance of Air Pollution, discusses sampling and analysis of particulates, gases, hydrocarbons, and sulfur and nitrogen compounds. Presents methods for surveillance of ambient air from stationary and mobile sources.

Verner, S., ed. Sampling and Analysis of Toxic Organics in the Atmosphere. Philadelphia: American Society for Testing and Materials, 1981.

Provides methods for sampling toxic organics in ambient and workplace atmospheres and analyzing PCBs, PAH and POM.

Books and Monographs on the Measurement of Water Pollution

American Public Health Association, American Water Works Association, and Water Pollution Control Federation. Standard Methods for the Examination of Water and Wastewater. 16th ed. Washington, D.C.: American Public Health Association, 1985.

Contains the EPA-recommended methods for the analysis of pollutants in water as listed in the Code of Federal Regulations.

American Society for Testing and Materials. Annual Book of ASTM Standards. Section 11 - Water and Environmental Technology. Philadelphia: ASTM, issued annually.

Called the "Bible of the Water Testing Business". Volume 11.01, Water (I) and Volume 11.02 Water (II) contain methods for the analysis of water. Volume 11.03 deals with atmospheric analysis and occupational health and safety. Volume 11.04 covers pesticides, resource recovery, hazardous substances and oil spill response, waste disposal and biological effects.

American Society for Testing and Materials. Power Plant Water Analysis Manual. 1st ed. Philadelphia: ASTM, 1984.

Provides 46 test methods and practices for rapid analysis of power plant water. Is also intended for use by any plant performing water chemical analysis for effluents.

Cairns, J., et al. Biological Monitoring in Water Pollution. New York: Pergamon, 1982.

Shows how to use biological monitoring to determine environmental changes in water caused by pollution.

Canter, L. W. River Water Quality Monitoring. Chelsea, MI: Lewis Publishers, 1985.

Discusses principles for planning and conducting studies, field work, selection of parameters to be monitored, location of stations, frequency of sampling, sample collection and analysis, and report writing.

Chau, A. S. and B. K. Afghan. Analysis of Pesticides in Water. Boca Raton, FL: CRC Press, 1982.

Volume 1 covers the environmental impact of pesticides, principles and practices of analysis, identification by chemical derivatization techniques, and the chemistry of cyclodiene insecticides. Volume 2 covers chlorine- and phosphorus-containing pesticides and phenoxyalkyl acid herbicides. Volume 3 covers nitrogen-containing pesticides, carbamates, and substituted urea and triazine herbicides.

Cooper, W. J., ed. Chemistry in Water Reuse. Stoneham, MA: Butterworth, 1981.

> Volume 1: Evaluation of the Published Literature in the Water Reuse/Recycle Area. Includes chapters on monitoring, quality assurance, the removal of toxic chemicals by membrane processes, disinfection, and swimming pool water. Volume 2: Analytical Methods and Characterization: Survey of Procedures for Determining Organics in Water. Includes chapters on the concentration of organics for toxicity testing, carbon adsorption, treatment techniques, and viruses and parasites.

Culp, G., ed. Trihalomethane Reduction in Drinking Water: Technologies, Costs, Effectiveness, Monitoring, Compliance. Park Ridge, NJ: Noyes, 1984.

> Part III covers monitoring and compliance to federal drinking water regulations, including laboratory certification criteria.

Food and Agriculture Organization. Manual of Methods in Aquatic Environment Research. Part 9 - Analyses of Metals and Organochlorine in Fish. FAO Fisheries Technical Paper 212. Ann Arbor, MI: Unipub, 1983.

> Details analytical methods for measuring contamination of fish, crustacea, and bivalves with heavy metals, organochlorines and methyl mercury. Describes sample collection and storage and preparation of samples for analysis.

Futoma, D. J., et al. Analysis of Polycyclic Aromatic Hydrocarbons in Water Systems. Boca Raton, FL: CRC Press, 1981.

> Discusses sampling and pre-concentration methods and methods of analysis currently used and those which are less widely used.

Grasshoff, K., et al. Methods of Seawater Analysis. 2nd ed. Deerfield Beach, FL: VCH Publishers, 1983.

> Presents analytical procedures for the determination of pollutants and trace metals in seawater.

Guseva, J. H., et al. Groundwater Contamination and Emergency Response Guide. Park Ridge, NJ: Noyes, 1984.

> Part I assesses methodology for investigating and evaluating possible sources of contamination. Part II includes state-of-the-art equipment and monitoring methods.

Keith, L. H. <u>Advances in the Identification and Analysis of
    Organic Pollutants in Water</u>. Stoneham, MA: Butter-
    worth, 1981.

    Volume 1 discusses protocols, high-resolution gas chro-
    matography, stable labeling, microextraction, resin
    adsorption, high-performance liquid chromatography and
    derivatization. Volume 2 covers computerized data,
    Grob closed-loop stripping, specialized purging tech-
    niques, and the analysis of surface and drinking water
    and industrial wastewater.

Lawrence Berkeley Laboratory. <u>Instrumentation for Environ-
    mental Monitoring. Volume 2. Water</u>. 2nd ed. Somerset,
    NJ: Wiley, 1985.

    Presents methods and instruments used to monitor
    pollutants in treated, raw, waste, or drinking water.
    Includes a listing of commercially available
    instruments and their specifications.

Minear, R. A. and L. H. Keith. <u>Water Analysis</u>. Orlando,
    FL: Academic, 1982.

    Volume 1: <u>Inorganic Species, Part 1</u>. Presents a
    thorough discussion of the chemical analysis of inor-
    ganic constituents in water and a survey of general
    water quality analysis techniques used to measure indi-
    vidual inorganic species. Volume 2: <u>Inorganic Species,
    Part 2</u>, 1984. Volume 3: <u>Organic Species</u>, 1984.

Morrison, R. D. <u>Ground-water Monitoring Technology</u>.
    Prairie du Sac, WI: Timco Mfg., Inc., 1984.

    First two parts provide information on monitoring in
    the vadose zone and saturated zones, and the third part
    gives sampling equipment.

National Water Well Association. <u>A Guide to the Selection
    of Materials for Monitoring Well Construction and
    Ground Water Sampling</u>. Worthington, OH: NWWA, 1983.

    Provides a description of ground water sampling devices
    and materials used in monitoring well construction.

Parsons, T. R., et al. <u>A Manual of Chemical and Biological
    Methods for Seawater Analysis</u>. New York: Pergamon,
    1984.

    Serves as an introduction to the quantitative analysis
    of seawater. Provides biological and chemical tech-
    niques which require a minimum of professional training
    and avoid the use of expensive equipment.

Pipes, W. O. <u>Bacterial Indicators of Pollution</u>. Boca Raton,
    FL: CRC Press, 1982.

    Presents the use of bacterial indicators to assess
    water quality and health risks. Costs of monitoring
    are also discussed.

Rand, G. M. and S. R. Petrocelli. <u>Fundamentals of Aquatic
    Toxicology: Methods and Applications</u>. New York:
    Hemisphere, 1984.

    Includes descriptions of the principles, methods, and
    procedures used in acute subchronic and chronic aquatic
    toxicity-testing studies, as well as methods of sub-
    lethal toxicity testing. Also discusses the biological
    effects of metals, PAH, pesticides and inorganics.

Rice, R. G., ed. <u>Safe Drinking Water: the Impact of Chem-
    icals on a Limited Resource</u>. Chelsea, MI: Lewis
    Publishers, 1984.

    Part IV includes papers on monitoring techniques for
    groundwater quality near sources of contamination and
    the NBS standard reference materials for use in vali-
    dating water analyses.

Scalf, M., et al. <u>Manual of Ground Water Sampling Proce-
    dures</u>. Worthington, OH: National Water Well Associa-
    tion, 1981.

    Describes procedures for sampling ground water for
    microbial and inorganic and organic chemical parame-
    ters.

Shuckrow, A. J. <u>Hazardous Waste Leachate Management Manual</u>.
    Park Ridge, NJ: Noyes, 1982.

    Describes management options for hazardous waste
    leachate from disposal sites. Chapter 7 discusses
    monitoring, sampling, analysis, and safety considera-
    tion.

Van Loon, J. C. <u>Chemical Analysis of Inorganic Constituents
    of Water</u>. Boca Raton, FL: CRC Press, 1982.

    Presents procedures, reagents, equipment and calcula-
    tions for the analysis of chemicals in water by the
    following methods: EDTA complexometric titration,
    atomic adsorption spectrophotometry, atomic emission
    spectrometry, ion-selective electrode methods, ion
    chromatography, and flame photometry.

Books and Monographs on the Measurement of
Pollution of the Environment

American Society for Testing and Materials. _Annual Book of
ASTM Standards. Section 14 - General Methods and
Instrumentation_. Philadelphia: ASTM, issued annually.

Volume 14.01 includes molecular and mass spectroscopy,
chromatography, resinography, temperature measurement,
microscopy, and computerized systems. Volume 14.02
covers general test methods, nonmetals, sections on
hazard potential of chemicals, and particle size
measurement.

American Society for Testing and Materials. _Compilation of
Methods for Emission Spectrochemical Analysis_. 7th ed.
Philadelphia: ASTM, 1982.

Contains all current standards developed by ASTM Committee on emission spectroscopy and found in the Annual
Book of ASTM Standards, and many suggested methods.

American Society for Testing and Materials. _Manual on Practices in Molecular Spectroscopy_. 4th ed. Philadelphia:
ASTM, 1980.

Contains current ASTM standards on molecular spectroscopy.

American Society for Testing and Materials. _Standards on
Chromatography_. Philadelphia: ASTM, 1981.

Contains 103 ASTM chromatographic methods and practices.

Association of Official Analytical Chemists. _Official
Methods of Analysis_. 14th ed. Arlington, VA: AOAC,
1984.

Contains over 1700 methods universally accepted by the
courts in the United States and Canada. National and
state laws often stipulate the use of these methods.
Gives a step-by-step format, reagents, apparatus, and
alternative methods. Covers a wide vaiety of substances, including foods, drugs, pesticides, hazardous
materials, and water. Supplemented by periodic updates
included in the purchase price of the book.

Baasel, W. D. _Economic Methods for Multipollutant Analysis
and Evaluation_. New York: Dekker, 1985.

Introduces a comprehensive, integrated, low cost method that combines sampling, chemical analysis and bioassays for the preliminary assessment of effluent streams and their effect on water, air and land. Chapters include The Environmental Assessment Process, Sampling, Chemical Analysis, Bioassays, Multimedia Environmental Goals, Models, Economic Assessments, and Risk Assessments.

Benson, R. C. <u>Geophysical Techniques for Sensing Buried Wastes and Waste Migration</u>. Worthington, OH: National Water Well Association, 1982.

Describes six techniques for sensing buried wastes and waste migration, including metal detection, magnetometry, ground penetrating radar, electromagnetics, resistivity, and seismic refraction.

Bergman, H. L., et al., eds. <u>Environmental Hazard Assessment of Effluents</u>. Elmsford, NY: Pergamon, 1985.

Presents methods to assess the hazards of complex pollutants that may enter aquatic ecosystems. Major sections cover the background of effluent monitoring, biological effects testing, exposure assessment, and hazard assessment.

Berman, E. <u>Toxic Metals and Their Analysis</u>. Heyden International Topics in Science. Somerset, NJ: Wiley, 1980.

Chapters deal individually with 31 metals and metalloids with toxic properties. The element's biochemical role, toxicology, distribution in the body, normal concentration in the diet, and methods for analysis are described.

Beyermann, K. <u>Organic Trace Analysis</u>. Somerset, NJ: Wiley, 1984.

Offers a complete description of the field of organic trace analysis, including sampling, storing and stabilizing the sample, analysis, ways to improve laboratory practices, and ways to avoid common errors.

Bowman, M. C. <u>Handbook of Carcinogens and Hazardous Substances: Chemical Trace and Analysis</u>. New York: Dekker, 1982.

Presents a state-of-the-art overview of the toxicological and chemical aspects of numerous pollutants including pesticides, polynuclear aromatic hydrocarbons, metals and metalloids. Gives specific procedures for a variety of substrates.

Boyle, T. P., ed. <u>Validation and Predictability of Laboratory Methods for Assessing the Fate and Effects of Contaminants in Aquatic Ecosystems</u>. Special Technical Publication 865. Philadelphia: American Society for Testing and Materials, 1985.

Presents fifteen technical papers comparing laboratory and modeling tests with studies under similar conditions in the environment.

Brodsky, A., ed. <u>CRC Handbook of Radiation Measurement and Protection</u>. Boca Raton, FL: CRC Press.

Volume I, <u>Physical Science and Engineering Data</u>, 1979, provides the basic scientific and engineering information used in radiation protection, including regulatory limits of exposure, data and methods for estimating exposures and natural radiation. Volume II, <u>Biological and Mathematical Data</u>, 1982, presents the basic anatomical, physiological and ecological data, fundamentals on radiation toxicology in humans and compilations of pharmacokinetic data. Includes mathematical and statistical tables and minicomputer programming for radiation measurement.

Brune, D., et al. <u>Nuclear Analytical Chemistry</u>. Deerfield Beach, FL: VCH Publishers, 1984.

Reviews the application of nuclear techniques to chemical analysis particularly to elements at low and trace concentrations.

Cairns, J. <u>Multispecies Toxicity Testing</u>. Elmsford, NY: Pergamon, 1985.

Discusses the scientific, technical, regulator, and industrial use of multispecies testing. Sections include quality assurance and the complexity of river ecosystems.

Cheremisinoff, P. N. and H. J. Perlis, eds. <u>Analytical Measurements and Instrumentation for Process and Pollution Control</u>. Lancaster, PA: Technomic, 1981.

This illustrated guide to the major analytical methods and instruments has industrial, envionmental and medical applications. Includes the analysis of environmental pollutants, drugs, products, toxic and hazardous materials, foods, gases, clinical specimens, etc. Discusses luminescence spectrometry, liquid chromatography, GC/MS trace organic analysis, x-ray diffraction, polarigraphic determination, instrumentation and control, gas pressure regulation, gas mixtures for instrumental analysis, a method for the determination

of heavy metals in marine sediments, and the instrumentation and control of ph.

Cheremisinoff, P. N. <u>Management of Hazardous Occupational
Environments</u>. Lancaster, PA: Technomic, 1984.

Reviews analytical methods for hazards in the workplace, including dust, gas, noise, biological and
infectious materials, flammable materials and chemicals.

Choudhory, G., et al. <u>Chlorinated Dioxins and Dibenzofurans
in the Total Environment</u>. Stoneham, MA: Butterworth,
1983.

Reviews new information on specific isomers of these
compounds including the synthesis, chromatographic
separation, and problems and solutions for sampling and
analysis. Includes information on exposure to humans.

Dermer, O. C., et al. <u>Biochemical Indicators of Subsurface
Pollution</u>. Stoneham, MA: Ann Arbor Science, 1980.

Provides a guide to the detection and measurement of
biochemical indicators of pollution such as enzymes,
ATP, and nucleic acids, for both soils and water.

D'Itri, F. M. and M. A. Kamrin. <u>PCBs: Human and Environmental Hazards</u>. Stoneham, MA: Butterworth, 1983.

Provides a broad overview of PCBs including chemical
analysis and monitoring, metabolism, biotransformation,
toxicology and persistence. Also includes data on
human health roles of federal and state agencies in
regulating PCBs.

Dzubay, T. G. <u>X-ray Fluorescence Analysis of Environmental
Samples</u>. Stoneham, MA: Ann Arbor Science, 1977.

Discusses trace element analysis of air and water pollution by x-ray fluorescence. Includes sample collection and preparation and calibration standards.

Erickson, M. D. <u>Analytical Chemistry of PCBs</u>. Stoneham,
MA: Butterworth, 1985.

A thorough review of the chemistry of PCBs. Major
chapters cover analytical procedures from ANSI, ASTM
and EPA: sample collection and storage; extraction;
determination; and quality assurance.

Everett, L. G., et al. Vadose Zone Monitoring for Hazardous Waste Sites. Park Ridge, NJ: Noyes, 1984.

Provides a compendium of monitoring techniques used in a vadose zone monitoring program. Appendix provides a monitoring methods index.

Hammons, A. S. Methods for Ecological Toxicology: a Critical Review of Laboratory Multispecies Tests. Lancaster, PA: Technomic, 1981.

Gives a review of laboratory tests and test systems for effects of chemical substances on ecosystems. Includes aquatic and terrestrial systems as well as mathematical models.

Hanai, T. CRC Handbook of Chromatography. Boca Raton, FL: CRC Press, 1982.

Contains tables and figures for paper, thin-layer, column liquid, and gas chromatography of phenolic compounds and organic acids.

Harris, J. C., et al. Combustion of Hazardous Wastes: Sampling and Analysis Methods. Park Ridge, NJ: Noyes, 1984.

Addresses the sampling and analysis methods to be used when measuring the levels of principal organic hazardous constituents in the streams of an incinerator facility for the purpose of calculating its destruction and removal efficiency. Includes inlet waste, stack gas, process water, fly ash and bottom ash.

Hassan, S. S. M. Organic Analysis Using Atomic Adsorption Spectrometry. New York: Wiley, 1984.

Presents methods for determining compounds of nitrogen, phosphorus, arsenic, oxygen and sulfur, vitamins, biological materials and food products.

Jungreis, E. Spot Test Analysis: Clinical, Environmental, Forensic, and Geochemical Applications. New York: Wiley, 1985.

Provides fast and inexpensive spot and screening tests for biochemical components, forensic substances, geochemical composition, air and water pollutants, soil chemistry, and food adulteration and composition. Includes proteins, urine components, drugs, explosives, toxic agents, bacteria, and others.

Kaplan, H. L., et al. Combustion Toxicology: Principles and Test Methods. Lancaster, PA: Technomic, 1983.

Provides a state-of-the-art review of combustion toxicology with emphasis on laboratory test methods, including DIN, FAA, NBS, USF, and radiant heat test methods.

Kateman, G. and F. W. Pijpers. Quality Control in Analytical Chemistry. New York: Wiley, 1981.

Presents techniques to improve the quality of analytical methods, sampling, and data handling.

Katz, S. A. and S. W. Jeniss. Regulatory Compliance Monitoring by Atomic Absorption Spectroscopy. Deerfield Beach, FL: VCH Publishers, 1983.

Serves as a manual for the measurement of pollutants by AAS. Includes methods for monitoring air, liquid, solids, sludges, sediments and soils, as well as methods for trace analysis of metals in animal tissues and body fluids.

Klement, A. W., ed. CRC Handbook of Environmental Radiation. Boca Raton, FL: CRC Press, 1982.

Reviews techniques and methods currently used to estimate and measure exposure to environmental radiation and radioactivity in situations such as transportation and waste disposal.

Kuwana, T., ed. Physical Methods in Modern Chemical Analysis. 3 volumes. Orlando, FL: Academic, 1984.

This three-volume set presents a step-by-step approach to selected physical/instrumental methods of chemical analysis, including gas chromatography, laser resonance spectroscopy, flame fluorescence and plasma spectroscopy, and atomic adsorption spectroscopy. Discusses the theory and principles of each method and the instrumentation and equipment required.

Lawrence, J. F., ed. Trace Analysis. 3 volumes. Orlando, FL: Academic, 1981.

These three volumes present detailed information on the applications of analytical chemistry in the detection, identification, and quantification of trace quantities of substances in many different sample types. Includes state-of-the-art discussions of selected topics including the use of high performance liquid chromatography for organic and inorganic trace analysis. Volume 3 is

divided into two sections: biological fluids and
tissues, and environmental analysis.

Lawrence Berkeley Laboratory Environmental Instrument Survey. Instrumentation for Environmental Monitoring.
Volume 1: Radiation. 2nd ed. New York: Wiley, 1983.

Reviews techniques and instruments for characterizing
and quantifying hazardous radioactive pollutants.
Gives programs for monitoring nuclear reactors, fuel
reprocessing plants, and nuclear waste and/or mining
sites. Includes methods for detecting and measuring
alpha, beta, gamma, and x-radiation, and biological and
ecological effects.

Liu, D. and B. J. Dutka, eds. Toxicity Screening Procedures
Using Bacterial Systems. Drug and Chemical Toxicology
Series, Volume 1. New York: Dekker, 1984.

Presents a review of conventional and novel microbial
techniques used in toxicity screening. Outlines numerous industrial and government agency experiences illustrating procedures and applications.

Long, F. A. and G. E. Schweitzer, eds. Risk Assessment at
Hazardous Waste Sites. Washington, D.C.: American
Chemical Society, 1982.

Addresses the problems of monitoring hazardous waste
sites to support risk assessments. Includes geophysical
methods currently in use and discusses incorporating
risk assessment into the RCRA regulatory process.

Lyman, W. J., et al., eds. Handbook of Chemical Property
Estimation Methods: Environmental Behavior of Organic
Compounds. New York: McGraw-Hill, 1982.

Contains selected estimation methods for physioco-
chemical properties of single-component organic
chemicals, including solubility in water and solvents;
adsorption coefficient for soils; bioconcentration
factor in aquatic organisms; rate of hydrolysis,
aqueous photolysis, and biodegradation; atmospheric
residence time; activity coefficient; boiling point;
volatilazation from water and soil; flash points;
interfacial tension with water; liquid viscosity; and
dipole moment.

MacDonald, J. C., ed. Inorganic Chromatographic Analysis.
New York: Wiley, 1985.

Explains how to monitor inorganic effluents using
chromatography. Includes the detection of binary metal

and metalloid compounds, organometallic compounds, metal chelates, crown ethers, and other compounds.

Martin, M. H. and P. J. Coughtrey. _Biological Monitoring of Heavy Metal Pollution: Land and Air_. New York: Elsevier, 1982.

Discusses the use of vegetation for monitoring airborne heavy metal deposition, plants and animals as monitors of soil contamination, and the use of imported biological materials for monitoring. Other chapters cover the biological indicators of natural ore bodies, and a historical review of monitoring. Includes both a species and subject index.

McCrone, W. C. _The Asbestos Particle Atlas_. Stoneham, MA: Ann Arbor Science, 1980.

Presents a thorough discussion of asbestos analysis from a description of asbestos building materials and the composition of asbestos and talc, through sampling and analytical methods. The recommended method, polarized light microscopy, is discussed in detail. Many color plates aid in the identification of asbestos, materials substituted for asbestos, other substances having similar composition, and substances often associated with asbestos as mined or as added commercially.

McKlveen, J. W. _Fast Neutron Activation Analysis: Elemental Data Base_. Lancaster, PA: Technomic, 1981.

Provides reaction and spectral information on the identification and assay of 92 elements. Includes fast neutron activation theory and equipment, predominant reactions and sensitivities, 14-MeV neutron activation data, reactions listed in order of increasing gamma ray energies, and interference reactions.

Mulik, J. D. and E. Sawicki. _Ion Chromatographic Analysis of Environmental Pollutants_. Volumes 1 and 2. Lancaster, PA: Technomic, 1979.

Presents 48 studies on the use of ion chromatography for the analysis of anions and cations.

Natale, A. and H. Levins. _Asbestos Removal and Control: An Insider's Guide to the Business_. Maple Shade, NJ: Asbestos Control Technology, Inc., 1984.

Includes descriptions of all procedures for sampling and analyzing suspected architectural asbestos materials.

National Fire Protection Association. Flammable and Combus-
    tible Liquids Code, NFPA 30. 2nd ed. Quincy, MA: NFPA,
    1984.

    Serves as a standard for legal regulations governing
    flammable and combustible liquids. Sections include
    the design, installation, and testing for tank and
    container storage, piping, valves and fittings, bulk
    and processing plants, service stations, refineries,
    chemical plants and distilleries.

National Research Council. Testing for Effects of Chemicals
    on Ecosystems. Washington, D.C.: National Academy
    Press, 1981.

    Reviews and evaluates test systems and recommends cri-
    teria for selecting appropriate test procedures. Dis-
    cusses the use of models.

National Research Council. Toxicity Testing: Strategies to
    Determine Needs and Priorities. Washington, D.C.:
    National Academy Press, 1984.

    Presents a state-of-the-art review of the testing of
    substances which have already been tested for toxicity,
    and proposes a system to determine priorities for those
    substances which should be tested.

National Research Council. Committee on Nonoccupational
    Health Risks of Asbestiform Fibers. Asbestiform Fibers:
    Nonoccupational Health Risks. Washington, D.C.:
    National Research Council, 1984.

    Comprehensive review of the relation of asbestiform
    fibers to disease, their major sources, and the extent
    of long-term risk to the public. Measurement techniques
    for asbestos dust in the workplace, as well as in the
    ambient environment, and the relationships among vari-
    ous measurement methods are included.

National Water Well Association. Geophysical Techniques for
    Sensing Buried Wastes and Waste Migration. Worthington,
    OH: NWWA, 1982.

    Describes the use of metal detection, magnetometry,
    ground penetrating radar, electromagnetics, resistivi-
    ty, and seismic refraction for sensing the presence and
    migration of buried wastes.

Organization for Economic Cooperation and Development. OECD
    Guidelines for Testing of Chemicals. Washington, D.C.:
    OECD, 1981.

Presents more than 50 tests for physical-chemical
properties of chemicals, their effects on biotic
systems, degradation and accumulation, and health
effects. Outlines test methods, procedures, and data
collection and reporting methods.

Rajhans, G. S. and J. Sullivan. Asbestos Sampling and
     Analysis. Stoneham, MA: Butterworth, 1981.

     Details methods for sampling, analyzing, identifying,
     and quantifying asbestos. Emphasis is placed on selec-
     tion of the most appropriate method.

Schweitzer, G. E. and J. A. Santolucito, eds. Environmental
     Sampling for Hazardous Wastes. Washington, D.C.:
     American Chemical Society, 1984.

     Discusses the problems of collecting, monitoring, and
     analyzing surface and subsurface samples. Reviews case
     studies of successful field programs for sampling diox-
     in, lead and cyanide.

Sittig, M. Handbook of Toxic and Hazardous Chemicals and
     Carcinogens. 2nd ed. Park Ridge, NJ: Noyes, 1985.

     Presents an alphabetical listing of almost 800 chemical
     compounds, including all EPA priority toxic pollutants,
     hazardous wastes, and hazardous substances, as well as
     most of the chmicals reviewed by NIOSH. Gives descrip-
     tion, code number, synonyms, exposure limits, incom-
     patibilities, harmful effects, symptoms, protective
     methods, and methods for determination in air and
     water. Includes hundreds of citations to secondary
     reference sources.

Slavin, W. Graphite Furnace AAS: A Source Book. Norwalk,
     CT: Perkin-Elmer, 1984.

     Presents a literature reivew of 988 references on the
     analysis of heavy metals.

Taylor, J. K. and T. W. Stanley. Quality Assurance for
     Environmental Measurements. Special Technical
     Publication 867. Philadelphia: American Society for
     Testing and Materials, 1985.

     Contains thirty-six papers on quality assurance in data
     quality assessment, the measurement of ambient air and
     water, and discharge monitoring. Special topics
     include meteorological measurements, the AMES test,
     atmospheric acid deposition, and personal exposure
     monitoring.

Van Loon, J. Analytical Atomic Absorption Spectroscopy.
     Orlando, FL: Academic, 1984.

     Gives useful, practical tips on the analysis of water,
     geological, organic, metal, air, petroleum, and indus-
     trial samples using atomic absorption. Written for the
     practicing analyst.

Van Loon, J. Selected Methods of Trace Metal Analysis:
     Biological and Environmental Samples. Somerset, NJ:
     Wiley, 1985.

     Includes only the most thoroughly evaluated and widely
     accepted laboratory methods of trace metal analysis
     using atomic absorption of ICP emission spectrometers.
     Details the preparation and analysis of plant and
     animal tissue, food, water, air, and soil.

Varma, A. CRC Handbook of Atomic Absorption Analysis. Boca
     Raton, FL: CRC Press, 1984.

     Presents a comprehensive review of over 10,000 referen-
     ces to the literature. Includes an introduction to the
     basic theory and instrumentation, preparation of stan-
     dard and sample solutions and discusses the causes of
     interferences. Volume I discusses AAS and general in-
     strumentation for Elements in Groups IA, IIA, IIIB, IVB,
     VB, VIB, VIIB, and VIII. Volume II covers elements in
     Groups IB, IIB, IIA, IVA, VA, VIA, and VIIA, as well as
     halogens, and zero group elements. Includes a list of
     hollow cathode lamps and manufacturers of spectrometers
     and accessories.

Walters, D. B. Safe Handling of Chemical Carcinogens, Muta-
     gens, Teratogens and Highly Toxic Substances. Stoneham,
     MA: Butterworth, 1980.

     Presents the logical sequence which should precede ex-
     periments with potentially hazardous materials. Covers
     laboratory design, methods for chemical monitoring,
     spill control, degradation, detoxification, deactiva-
     tion, and disposal.

Wang, J. Stripping Analysis. Deerfield Beach, FL: VCH
     Publishers, 1985.

     Discusses the stripping analysis process, how it works,
     and how it is applied in trace metals analysis.

Zweig, G. and J. Sharma, eds. Analytical Methods for Pesti-
     cides, Plant Growth Regulators and Food Additives.
     Orlando, FL: Academic, 1964- .

Series began in 1964. Volume 10, 1978, and Volume 11,
1980, give new and updated methods. Volume 12, 1982,
features high performance liquid chromatography. Volume
13, 1984, concentrates on synthetic pyrethroids and
other pesticides.

Proceedings

American Society for Testing and Materials.   Aquatic Toxi-
    cology and Hazard Assessment. Symposium. Philadelphia:
    ASTM, First, Special Technical Publication 634, 1977;
    Second, STP 667, 1979; Third, STP 707, 1980; Fourth,
    STP 737, 1981; Fifth, STP 766, 1982; Sixth, STP 802,
    1983; Seventh, STP 854, 1985; Eighth, STP 891, 1986.

    All feature papers on biological monitoring, chemical
    screening, environmental chemistry and fate modeling in
    aquatic hazard assessment.

American Society for Testing and Materials.   Toxic Materials
    in the Atmosphere: Sampling and Analysis. A Symposium
    Sponsored by ASTM Committee D-22 on Sampling and Analy-
    sis of Atmospheres, Boulder, Colorado, August 2-5,
    1981. Philadelphia: ASTM, 1982.

    Presents 16 papers on a variety of topics including the
    sampling of PCBs and diesel particulates, OSHA wipe
    sampling procedures, laser and calibration techniques.

Buikema, A. L. and J. Cairns.   Aquatic Invertebrate Bio-
    assays. Proceedings of an ASTM Conference, Blacksburg,
    Virginia, September 27-29, 1977. Special Technical
    Publication 715. Philadelphia: American Society for
    Testing and Materials, 1980.

    Focusses on toxicity testing, the development of test
    methods and interpretation of toxicity data.

Campbell, S. A. Sampling and Analysis of Rain. Special
    Technical Publication 823. Philadelphia: American
    Society for Testing and Materials, 1984.

    Presents nine papers describing techniques used to
    collect and analyze rain and quality assurance programs
    currently used by major acid precipitation measurement
    networks.

Conway, R. A. and B. C. Malloy, eds. Hazardous Solid Waste
    Testing: First Conference. Special Technical Publica-
    tion 760. Philadelphia: American Society for Testing
    and Materials, 1982.

    Provides methods for testing and identifying solid
    wastes in terms of disposal options. Sections include
    laboratory extractions and leaching procedures; large-
    scale leaching tests vs. laboratory tests; analysis of
    residues, extracts, soils, and groundwaters; evaluation
    of land disposal sites and materials; and risk assess-
    ment.

Conway, R. A. and W. P. Gulledge, eds. Hazardous and Indus-
    trial Solid Waste Testing: Second Symposium. Special
    Technical Publication 805. Philadelphia: American
    Society for Testing and Materials, 1983.

    Provides descriptions of new and improved means of
    evaluating hazardous and industrial solid wastes in
    terms of hazard degree and disposal options.

Gertz, S. M. and M. D. London, eds. Statistics in Environ-
    mental Sciences. Special Technical Publication 845.
    Philadelphia: American Society for Testing and
    Materials, 1984.

    Discusses the use of statistical and mathematical tech-
    niques to analyze and interpret environmental data for
    those who monitor to fulfill discharge permit require-
    ments and evaluate the toxicity of their products.

Himmelsbach, B., ed. Toxic Materials in the Atmosphere,
    Sampling and Analysis. Special Technical Publication
    786. Philadelphia: American Society for Testing and
    Materials, 1981.

    Based on the 1981 Boulder Conference. Contents include
    industrial hygiene sampling and analysis, continuous
    air monitors and portable instruments, passive monitor-
    ing devices, chemical characterization and analytical
    methods.

Jackson, L. P. and C. C. Wright. Analysis of Waters Associ-
    ated with Alternative Fuel Production. Special Techni-
    cal Publication 720. Philadelphia: American Society
    for Testing and Materials, 1981.

    Sixteen papers from a symposium discuss the analysis
    and monitoring of waters from coal gasification and
    geothermal and oil shale processes.

Kelly, J., ed. Effluent and Environmental Radiation
    Surveillance. Special Technical Publication 698.
    Philadelphia: American Society for Testing and Mater-
    ials, 1980.

    Addresses the techniques used in measuring radioactive
    effluent from facilities using nuclear materials, and
    the monitoring of the environment.

Leong, B. K. J., ed. Inhalation Toxicology and Technology.
    Lancaster, PA: Technomic, 1981.

    This proceedings of a symposium sponsored by Upjohn
    Company, October 1980, provides a broad review of the
    state of the art of laboratory analysis or airborne
    toxicants. Includes EPA regulatory guidelines for
    inhalation toxicity testing.

Levadie, B. Definitions for Asbestos and Other Health-
    related Silicates. Special Technical Publication 834.
    Philadelphia: American Society for Testing and Mater-
    ials, 1984.

    Presents eleven papers dealing with asbestos, silica
    and talc including their physiological impact, analyti-
    cal differences, and the need for regulation and con-
    trol mechanisms.

National Water Well Association. National Symposium on
    Aquifer Restoration and Ground-water Monitoring.
    Worthington, OH: NWWA.

    First in 1981, Second in 1982, Third in 1983. All
    present original research papers in the sampling, anal-
    ysis, and monitoring of groundwater, as well as model-
    ing and case studies.

National Water Well Association. Proceedings of the NWWA/API
    Conference on Petroleum Hydrocarbons and Organic Chemi-
    cals in Groundwater. Worthington, OH: NWWA, 1985.

    Describes the state-of-the-art of monitoring, contain-
    ing, and restoring groundwater polluted by spills and
    leaking underground storage tanks and transmission
    lines.

Shubert, L. E., ed. Algae as Ecological Indicators.
    Orlando, FL: Academic, 1984.

    Presents the proceedings of a symposium held by the
    American Institute of Biological Sciences at Oklahoma
    State University. Sections include freshwater, marine,
    and terrestrial ecosystems, toxic substances, heavy

metals and vitamins, modeling and industrial effluent
monitoring.

Steubing, L. and H. J. Jager, eds. <u>Monitoring of Air Pollu-</u>
<u>tants by Plants: Methods and Problems</u>. Hingham, MA:
Martinus Nijoff, Dr. W. Junk, 1982.

Proceedings of the International Workshop, Osnabruck,
F. R. G. September 24-25, 1981 on the use of bio-
indicators to quantify the impact of pollutants on the
ecosystem, and standardize monitoring techniques.

Van Hall, C., ed. <u>Measurement of Organic Pollutants in Water</u>
<u>and Wastewater</u>. Special Technical Publication 686.
Philadelphia: American Society for Testing and Materials, 1979.

Includes sample preservation, preparation, measurement,
confirmation, and validation.

Documents

Environmental Protection Agency. <u>Manual of Chemical Methods</u>
<u>for Pesticides and Devices</u>. 3rd update. Arlington,
VA: Association of Official Analytical Chemists, 1982.

Contains 317 analytical methods for 162 chemicals found
in commercial pesticide formulations. These methods
are tested by collaborative study, and upon achieving
official status, are removed from this manual and published in the <u>AOAC Official Methods of Analysis</u>. Used
for sampling air, water, soils, sediment, and dust.

Environmental Protection Agency. <u>Quality Assurance Handbook</u>
<u>for Air Pollution Measurement Systems</u>. Research
Triangle Park, NC: EPA, 1976, with periodic updates.

Volume I, <u>Principles</u>, EPA-600/7-76-005 contains general
guidance, discussions of quality assurance and simple
calculations. Volume II, <u>Ambient Air Specific Methods</u>,
EPA-600/4-72-027a, describes procedures and equipment
for sampling, monitoring, and analysis of sulfur, nitrogen dioxide, sulfur dioxide, carbon monoxide, ozone
and lead. Volume III, <u>Stationary Source Specific</u>
<u>Methods</u>, EPA-600/4-77-027b, patterned after Volume II,
contains specific measurement methods for stationary
sources.

Geological Survey. The Dangers and Handling of Hazardous
    Chemicals in the Geologic Laboratory. Circular 924.
    Washington, D.C.: USGS, 1984.

Goerlitz, D. F. and E. Brown. Methods for Analysis of
    Organic Substances in Water. Techniques of Water-
    Resources Investigations of the U.S. Geological Survey,
    Book 5, Chapter A3. Washington, D.C.: Government
    Printing Office, 1972.

Kopp, J. F. and G. D. McKee. Methods for Chemical Analyses
    of Water and Wastes. EPA-600/4-79-020. Washington,
    D.C.: Environmental Protection Agency, 1979.

    Lists the analytical methods recommended by the EPA for
    testing ground and surface waters and effluents. Gives
    procedures for measuring physical, inorganic and
    organic constituents.

Methods for Collection and Analysis of Aquatic Biological
    and Microbiological Samples. Techniques of Water-
    Resources Investigations of the U.S. Geological Survey,
    Book 5, Chapter A4. Washington, D.C.: Government
    Printing Office, 1977.

National Institute for Occupational Safety and Health.
    NIOSH Manual of Analytical Methods. Washington, D.C.:
    Government Printing Office, 1985. Subscription includes
    basic manual and periodic updates for an indeterminate
    period.

    Discusses over 500 methods and 737 analyses used by
    NIOSH, with a list of reagents and equipment needed,
    special safety precautions, and instructions for taking
    and handling samples.

National Institute for Occupational Safety and Health.
    NIOSH Manual of Sampling Data Sheets. Cincinnati:
    NIOSH, 1977. Supplement, 1978.

    Contains recommended sampling methodology for measuring
    occupational exposure to air contaminants. Includes
    standards, equipment, procedures, analytical methods
    and shipping instructions for samples.

Occupational Safety and Health Administration. Industrial
    Hygiene Technical Manual. Change 6, Volume VI. OSHA-
    3058. Washington, D.C.: OSHA, 1984.

    Presents current OSHA industrial hygiene practices and
    procedures. Includes chapters on standard methods for
    sampling air contaminants, analytical laboratory pro-

cedures, sampling for surface contamination. Appendices provide chemical information, tables, and a review of the use and calibration of instruments used.

Plumb, R. H. Procedures for Handling and Chemical Analysis of Sediment and Water Samples. Technical Report EPA/CE-81-1. Vicksburg, MI: Army Engineer Waterways Experiment Station, 1981.

Skougstad, M. W., et al., eds. Methods for Determination of Inorganic Substances in Water and Fluvial Sediments. Techniques of Water-Resources Investigations of the U.S. Geological Survey, Book 5, Chapter Al. Washington, D.C.: Government Printing Office, 1979.

Thatcher, L. L. and V. J. Janzer. Methods of Determination of Radioactive Substances in Water and Fluvial Sediments. Techniques of Water-Resources Investigations of the U.S. Geological Survey, Book 5, Chapter A5. Washington, D.C.: Government Printing Office, 1977.

Periodicals

There are no periodicals that deal exclusively with testing, monitoring, analysis and sampling of pollutants. However, many journals contain relevant articles on an occasional or regular basis.

Below is a listing of periodicals that have included information on these subjects. Many of these periodicals are described in detail in other chapters. See the index for specific titles.

American Industrial Hygiene Association Journal
American Laboratory
American Water Works Association Journal
Analyst
Analytica Chimica Acta
Analytical Chemistry
Analytische Chemie
Applied Spectroscopy
Archives of Environmental Contamination and Toxicology
Archives of Environmental Health
Association of Official Analytical Chemists Journal
ASTM Standardization News
ASTM Standards InfoBriefs
Atmospheric Environment
Atomic Spectroscopy
Bulletin of Environmental Contamination and Toxicology

Canadian Spectroscopy
Chemosphere
Chemtech
Chromatography Journal
Contributions of the Boyce Thompson Institute
Dangerous Properties of Industrial Materials
Ecotoxicology and Environmental Safety
Environmental Analyst's Bulletin
Environmental Geology
Environmental Pollution. Series B
Environmental Science and Technology
Environmental Technology Letters
Environmental Toxicology and Chemistry
Geotechnical Testing Journal
Ground Water
Groundwater Monitor
Ground Water Monitoring Review
Hazardous Waste and Hazardous Materials
Industrial and Engineering Chemistry, Analytical Edition
International Journal of Environmental Analytical Chemistry
Journal of Agricultural and Food Chemistry
Journal of Applied Ichthyology
Journal of Chromatographic Science
Journal of Chromatography
Journal of Great Lakes Research
Journal of Hazardous Materials
Journal of Industrial Hygiene and Toxicology
Journal of Research
Journal of Testing and Evaluation
Journal of the Air Pollution Control Association
Journal of the Environmental Engineering Division ASCE
Journal of the American Industrial Hygiene Association
Journal of the Association of Official Agricultural
   Chemistry
Journal of the Association of Official Analytical Chemists
Journal of the Water Pollution Control Federation
Marine Chemistry
Marine Environmental Research
National Bureau of Standards Technical Note
Oil and Gas Journal
Operations Forum
Pesticides Monitoring Journal
Pollution Engineering
Science
Science of the Total Environment
Spectroscopy Letters
Stack Sampling News
Toxic Substances Journal
Water and Sewage Works
Water Research
Water Technology

3.
The Effects of Hazardous Materials on the Aquatic Environment

Susan M. Neumeister

Introduction

A wide variety of hazardous substances enter the hydrosphere through leaching, runoff, precipitation, and dumping. In the past decade, much apprehension and concern about the impact of these toxic materials has been raised.

This chapter is concerned with the toxic and hazardous materials in water and their impact on animals, plants, and humans. All categories of water are included: oceans, rivers, lakes, etc., as well as groundwater, coastal water, and drinking water. The effects of thermal and oil pollution and pesticides as hazardous materials are also covered.

Research on the impacts on humans from drinking polluted water is cited. Sources which emphasize the chemical, biological, and physiological effects on fish and other marine animals, as well as aquatic plants are also included.

A number of important information sources are cited which should be given special attention. Allen Knight's Water Pollution: A Guide to Information Sources is an excellent guide to the pre-1980 literature on all aspects of water pollution. Also in the Book section, the six-volume set of Wastes in the Ocean is very useful.

Aquatic Toxicology in the Periodical section is a relatively new publication with valuable information on the subject of toxicology in the aquatic ecosystem.

The Proceedings sources include the Annual Symposium on Aquatic Toxicology and Hazard Assessment. Beginning in 1976, with the latest being the 8th annual in 1984, these published proceedings provide presentations that deal exclusively with the problems of toxicants in the water environment. Another source for proceedings is Water Science & Technology.

The Reports section provides citations to bibliographies
from selected online data bases, which are available from
NTIS (National Technical Information Service). Most are
updated annually.

An excellent source for reviews is <u>Journal of the Water
Pollution Control Federation</u>. Every June issue contains
abundant informative reviews on the worldwide water pollution literature from the previous year.

The references in this chapter are not exhaustive, but
hopefully cite the more important sources concerned with
toxic compounds and their effects in water.

Books

Ashworth, William. <u>The Late, Great Lakes: An Environmental
    History</u>. New York: Knopf, 1986.

Brooke, L. T., et al., eds. <u>Acute Toxicities of Organic
    Chemicals to Fathead Minnows (Pimephales Promelas)</u>.
    Superior, WI: Center for Lake Superior Environmental
    Studies, University of Wisconsin-Superior, 1984- .

Cairns, Victor W., et al., eds. <u>Contaminant Effects on
    Fisheries</u>. New York: Wiley, 1984.

    Deals comprehensively with the bioaccumulation, metabo-
    lism, and excretion of toxicants by fish. It addresses
    the tolerance of fish and fish populations to the
    stress of environmental alterations, and covers the
    changes in the biochemical and physiological systems
    associated with exposure to the common toxicants.

Duedall, Iver W., et al., eds. <u>Wastes in the Ocean</u>. New
    York: Wiley, 1983-1985.

    Considers the disposal of wastes in the ocean and
    evaluates the impact on human and marine life. V. 1 --
    <u>Industrial and Sewage Wastes in the Ocean</u>, v. 2 --
    <u>Dredged-Material Disposal in the Ocean</u>, v. 3 -- <u>Radio-
    active Wastes and the Ocean</u>, v. 4 -- <u>Energy Wastes in
    the Ocean</u>, v. 5 -- <u>Deep-sea Wastes Disposal</u>, and
    v. 6 -- <u>Near-shore Waste Disposal</u>.

Ecological Analysts, Inc. <u>The Sources, Chemistry, Fate, and
    Effects of Chromium in Aquatic Environments</u>.
    Washington, D.C.: American Petroleum Institute, 1982.

Forster, U. and G. T. W. Wittmann. <u>Metal Pollution in the</u> <u>Aquatic Environment</u>. 2nd rev. ed. Berlin: Springer-Verlag, 1981.

Discusses the concern about the capacities of toxic materials that water bodies can carry before the impact becomes fatal.

Gerlach, Sabastian A. <u>Marine Pollution: Diagnosis and</u> <u>Therapy</u>. Berlin: Springer-Verlag, 1981.

Examines oil pollution, discharges of wastes, radio-activity, contamination by heavy metals and industrial effluents. Includes an extensive bibliography.

Gilliom, Robert J., et al., eds. <u>Pesticides in the Nation's</u> <u>Rivers, 1975-1980, and Implications for Future</u> <u>Monitoring</u>. Washington, D.C.: G.P.O.; Alexandria, VA: For Sale by the Distribution Branch, U.S. Geological Survey, 1985.

Houston, Arthur H. <u>Thermal Effects Upon Fishes</u>. Ottawa: National Research Council of Canada, Associate Committee on Scientific Criteria for Environmental Quality, 1982.

IMCO/FAO/UNESCO/WMO/WHO/IAEA/UN Joint Group of Experts on the Scientific Aspects of Marine Pollution. <u>Thermal</u> <u>Discharges in the Marine Environment</u>. Geneva, Switzerland: United Nations Environment Programme, 1984.

Jackson, R. E., ed. <u>Aquifer Contamination and Protection:</u> <u>Project 8.3 of the International Hydrological Pro-</u> <u>gramme</u>. Paris: UNESCO, 1980.

Reviews and summarizes the theoretical and practical aspects of contaminant hydrogeology and groundwater quality management. Describes the physical and chemical processes controlling the transport of contaminants in groundwater systems.

Johnson, W. Waynon and Mack T. Finley. <u>Handbook of Acute</u> <u>Toxicity of Chemicals to Fish and Aquatic Inverte-</u> <u>brates: Summaries of Toxicity Tests Conducted at Colum-</u> <u>bia National Fisheries Research Laboratory, 1965-78</u>. Washington, D.C.: Department of the Interior, Fish and Wildlife Service, 1980.

Compiles the results of toxicity tests on fish and aquatic invertebrates. Studies include 1,587 acute toxicity tests on chemical effects in 28 species of

fish and 30 species of invertebrates. The handbook
helps in establishing water quality criteria and in
estimating potential environmental impacts of chemicals.

Knight, Allen W. and Mary Ann Simmons. Water Pollution:
A Guide to Information Sources. Detroit: Gale Research
Co., 1980.

Assists in locating water pollution literature. Includes indexes and abstracts, bibliographies, information services, government documents, dissertations,
conference proceedings, directories, translations, lex-
icons, encyclopedias and atlases, journals and newsletters, handbooks and audiovisuals.

Metelev, V. V., et al. Water Toxicology. New Delhi:
Amerind, 1983.

Moore, James W. and S. Ramamoorthy. Heavy Metals in Natural
Waters: Applied Monitoring and Impact Assessment. New
York: Springer-Verlag, 1984.

Reviews the principles and methods of monitoring and
assessing chemical pollutants in natural waters. Extensive information on the most common heavy metals is
included. Discusses the industrial uses, discharges,
chemical behavior, toxicity and impact of metals on
natural waters.

Moore, James W. and S. Ramamoorthy. Organic Chemicals in
Natural Waters: Applied Monitoring and Impact
Assessment. New York: Springer-Verlag, 1984.

Deals with organic compounds in natural waters outlined
in the priority pollutant list (EPA and Environmental
Contaminants Act [Canada]) which include aliphatic compounds, aromatic compounds, chlorinated pesticides,
petroleum hydrocarbons, phenols, polychlorinated
biphenyls, and polychlorinated dibenzo-p-dioxins. Most
of these chemicals are widespread in the environment
and toxic to fish and humans; many are mutagenic,
carcinogenic, and teratogenic. This volume, a companion to Heavy Metals in Natural Waters, provides a
unique review of the principles and methods of monitoring and assessing the pollution of natural waters. The
authors present extensive information for the most
common organic pollutants including production, uses,
discharges, physical and chemical behavior, toxicity,
residue dynamics and impacts.

Murty, A. S. Toxicity of Pesticides to Fish. Boca Raton,
FL: CRC Press, 1986.

National Research Council, Safe Drinking Water Committee.
    <u>Drinking Water and Health</u>. Washington, D.C.: National
    Academy Press, 1977- .

    Examines the health effects of biological and chemical
    contaminants in drinking water. It also evaluates
    health effects associated with products of water disin-
    fection such as chlorine, which may have carcinogenic
    by-products.

Nemerow, Nelson Leonard. <u>Stream, Lake, Estuary, and Ocean
    Pollution</u>. New York: Van Nostrand Reinhold, 1985.

    Emphasis is on biological, hydrological, and bio-
    chemical aspects of stream analysis. Landbased contam-
    inants and lake pollutants are discussed and analyzed.
    Large quantities of data are given in assessing the
    effect of comtaminants on watercourses.

Nriagu, Jerome O., ed. <u>Aquatic Toxicology</u>. New York:
    Wiley, 1983.

    Focuses on the toxicity of many inorganic and organic
    contaminants to the aquatic environment. It examines
    the response of organisms to particular toxicants and
    the biochemical and physiological changes induced by
    these toxicants.

Nriagu, Jerome O., ed. <u>Environmental Impacts of Smelters</u>.
    New York: Wiley, 1984.

    Focuses primarily on the effects of smelter emissions
    on the aquatic ecosystems. This 15th volume in <u>Advances
    in Environmental Science and Technology</u> deals with the
    devastation of aquatic ecosystems by mining, smelting,
    and ancillary lumbering activities. It provides an
    overview of the rates of influx of smelter emissions
    into the adjacent aquatic ecosystems and deals compre-
    hensively with the subsequent impacts on the biogeo-
    chemical cycling of nutrients and toxic metals, photo-
    synthetic activity, community stability and respira-
    tion, etc.

Nriagu, Jerome O. and Milagros S. Simmons, eds. <u>Toxic
    Contaminants in the Great Lakes</u>. New York: Wiley,
    1984.

    Examines the influx of industrial, municipal and agri-
    cultural wastes which have adversely affected the
    quality of the Great Lakes and their fish resources.
    Of primary concern is the problem of toxic contami-
    nants, their discharge into, movement through and their
    effects upon the Great Lakes. This 14th volume in
    <u>Advances in Environmental Sciences and Technology</u> pro-

vides comprehensive reviews of current information on
contaminants such as chlorinated hydrocarbons, toxa-
phene, mirex, and toxic metals.

Rand, Gary M. and Sam R. Petrocelli, eds. Fundamentals of
    Aquatic Toxicology. Washington, D.C.: Hemisphere,
    1985.

    Describes basic concepts and methodologies used in
    aquatic toxicity testing. Also describes sublethal
    effects testing and its utility in evaluating the less
    obvious effects of chemical exposure on aquatic organ-
    isms. Includes a summary of the available literature
    on the toxicity of generic types of chemicals to aqua-
    tic organisms. It also identifies the specific laws
    that provide the regulatory agencies with enforcement
    powers to control discharges into the aquatic environ-
    ment.

Thompson, J. A. J., et al. Organotin Compounds in the
    Aquatic Environment: Scientific Criteria for Assessing
    their Effects on Environmental Quality. Ottawa:
    National Research Council, 1985.

Yaron, B., et al., eds. Pollutants in Porous Media: the
    Unsaturated Zone Between Soil Surface and Groundwater.
    Berlin: Springer-Verlag, 1984.

    Focuses on groundwater quality control including sur-
    face interactions between pollutants and porous media,
    and new ways of modelling pollutant transport through
    porous media. Includes case studies dealing with pol-
    lution sources and management.

Periodicals

Aquatic Toxicology. Elsevier. Bimonthly, published since
    1981.

    Provides full papers, reviews and short communications
    on the assessment of toxicity in aquatic environments.
    Also includes papers dealing with adaptive responses of
    aquatic organisms due to excretion of toxicants in
    nearby waters.

Clean Water Report. Business Publishers, Inc. Biweekly,
    published since 1964.

    Provides information on water pollution control,
    drinking water supply and safety, and water resources

issues. Covers national policy, legislation, regulations, enforcement and litigation, state and local news
as well as research and technology news.

EWOOS: Environmental & Water Quality Operational Studies.
Army Engineer Waterways Experiment Station. Quarterly,
published since 1978.

Acts as an information exchange reporting research
results and information on new or improved technology
to solve selected environmental quality problems
associated with Civil Works activities of the corps in
a manner compatible with authorized project purposes.

Idaho Clean Water. Idaho Department of Health and Welfare,
Division of Environment. Quarterly, published since
1977.

Concerns water quality and pollution control in Idaho.
Also covers such topics as soil erosion, silviculture
and waste disposal. Informs readers of brochures,
pamphlets, and audiovisual materials pertaining to
water pollution.

International Water Report. Water Information Center, Inc.
Irregular, published since 1977.

Keeps the North American manager, scientist, tech-
nician, educator and legislator up-to-date on water
developments in Europe, Asia, Africa, South America and
Oceania. Reports on new products, pollution and waste
control, water supply, treatment and legislation.

Journal of the Water Pollution Control Federation. Water
Pollution Control Federation. Monthly, published since
1928.

Contains technical information on the design and management of water control facilities. Regular features
include "Meetings," "Washington Notebook," and "Recent
Books." See also "Reviews" section.

Marine Pollution Bulletin. Pergamon. Monthly, published
since 1970.

Includes news, comments, reviews, and research reports
on toxic substances and their effects on marine life.
Covers all aspects of the fight for life of the lakes,
estuaries, seas and oceans.

National Water Conditions. Water Resources Division, U.S.
   Geological Survey. Monthly, published since 1940.
   Formerly: Water Resources Review, July 1982.

   Describes the month's water conditions in the United
   States and Canada, compiling data on streamflow, ground
   water conditions, surface water, reservoirs, the flow
   of large rivers, water temperatures, and dissolved
   solids.

Newsletter, Gulf Breeze Laboratory. Environmental
   Protection Agency. Quarterly, published since 1973.

   Includes EPA research on effects of pesticides and
   other toxic substances on marine, estuarine and coastal
   ecosystems.

SAPL News. Seacoast Anti-Pollution League. Quarterly,
   published since 19__.

   Concerns the environmental protection of the New
   Hampshire and Massachusetts seacoast. Discusses
   nuclear power, waste disposal, land use, coastal zone
   management, and air and water quality. Particular
   attention is given to the Seabrook Nuclear Power Plant
   and its impact on the surrounding environment.

The Soil and Health Foundation--News. The Soil and Health
   Foundation. Quarterly, published since December 1972.

   Contains information relating to water and soil purity
   and research sponsored to study problems of pollution
   and health.

Washington Report. Washington Office, American Water Works
   Association. Nine times per year, published since
   1967.

   Contains news of political and legislative developments
   affecting the water supply industry. Covers such
   topics as water resources, pollution, programs for
   waste disposal, research in progress, and environmental
   issues.

Water Impacts. Institute of Water Research, Michigan State
   University. Monthly, published since December 1979.

   Reports on water quality, with articles on lake eutro-
   phication, groundwater contamination, stormwater runoff, acid rain and wastewater treatment.

Water News. Virginia Water Resources Research Center,
    Virginia Polytechnic Institute and State University.
    Monthly, published since April 1970.

    Focuses on water and land-related resource issues.
    Monitors national and state legislation affecting the
    environment, particularly the quality of water.
    Reviews the activities of the Center and other state
    and national environmental organizations. Emphasizes
    water resource issues in Virginia, but also discusses
    water issues in other states, as well as related
    environmental issues such as nuclear waste disposal,
    uranium mining, sewage treatment, and toxic substances.

Water Pollution Control: Journal of the Institute of Water
    Pollution Control. Institute of Water Pollution
    Control. Five times a year, published since 1901.

    Includes topics on marine pollution surveys, control of
    discharges to estuarial and coastal waters, tidal water
    pollution, and microbial pollution.

Water Research: the Journal of the Association on Water
    Pollution Research. Pergamon. Monthly, published
    since 1967.

    Includes articles on biodegradation in natural waters,
    nematodes, groundwater contamination, etc.

Water Resource Bulletin. American Water Resources Associa-
    tion. Bimonthly, published since 1965.

    Contains articles on all aspects of water resources
    including professional papers, technical notes, and
    discussions.

Water Science & Technology. Pergamon. Monthly, published
    since 1975.

    Contains proceedings of the meetings of the Inter-
    national Association of Water Pollution. See "Proceed-
    ings" section.

Waterline. International City Management Association.
    Bimonthly, published since May 1977.

    Covers wastewater construction grants, areawide water
    quality management and planning, and safe drinking
    water.

Abstracts and Indexes

Aquatic Sciences and Fisheries Abstracts. Cambridge
    Scientific Abstracts.   Monthly, published since 1971.

A monthly abstracting and indexing journal providing
coverage of world literature on pollution of aquatic
environments, physical environments of aquatic organisms, and all biological and ecological aspects of
marine freshwater and oceanography. Includes two sections: Part 1 - Biological Sciences & Living Resources;
Part 2 - Ocean Technology, Policy & Non-living Resources.

AWWA/AWWARF Merged Index. American Water Works Association.
    Semiannually, published since 1982.

A semiannual index of literature published by the
American Water Works Association and the AWWA Research
Foundation. The index provides coverage of selected
international English-language materials, primarily
those published in North America. The index covers all
aspects of water including drinking water industry,
water reuse and wastewater reuse, industrial water use,
water renovation and recycling, toxicological and medical related data, water pollution, health effects and
environmental problems.

Marine Pollution Research Titles. Marine Pollution Informa-
    tion Centre. Monthly, published since 1974.

A monthly current awareness bulletin providing references to the literature dealing with marine and estuarine
pollution. Covers materials published worldwide in all
languages and countries. Subjects include general
aspects of marine and estuarine pollution, detection,
analysis, oil, metals, pesticides, radioactivity, PCBs,
other chemicals, and domestic sewage.

S.A. Waterabstracts. South African Water Information
    Centre. Annually/biannually, published since 19__.

An abstracting and indexing bulletin covering South
Africa periodical and report literature on fresh water
and other related topics. Subject coverage includes
all aspects of water including wastewater, solid waste,
and pollution.

Selected Water Resources Abstracts. Department of the
    Interior, Office of Water Research & Technology.
    Bimonthly, published since 1968.

Bimonthly journal comprising abstracts and indexes of current and earlier literature pertinent to water resources. Subject coverage includes: nature of water, water cycle, water supply augmentation and conservation, water quality management, control and protection, and water resources planning and data.

Water Pollution Control Federation Conference - Abstracts of Technical Papers. Water Pollution Control Federation. Annually, published since 19__.

An annual publication which provides abstracts of papers on current research in pollution control technology and the effects of pollution presented at Water Pollution Control Federation conferences.

Water Quality Instructional Resources Information System. EPA Instructional Resources Center. Quarterly, published since 1979.

A quarterly loose-leaf service providing abstracts and indexes of English-language literature and non-print materials published in the United States dealing with wastewater treatment and water quality education and instruction. Subject areas are water quality education including, pesticides, hazardous wastes and public participation.

WRC Information. Water Research Centre. Weekly, published since 1927.

A weekly journal providing abstracts of current international scientific and technical literature on water, wastewater, and the aquatic environment. Subject coverage includes water quality, monitoring and analysis of water and wastes, sewage, industrial effluents and the effects of pollution.

Proceedings

Bahner, Rita Comotto and David J. Jansen, eds. Aquatic Toxicology and Hazard Assessment. 8th Annual Symposium, Fort Mitchell, Kentucky, April 15-17, 1984. Philadelphia: American Society for Testing and Materials, 1985.

Major papers presented: "Clean Water Act update: a new role for toxicology," "What industry is doing to protect aquatic life," and "An aquatic hazard assessment: Picloram."

Bates, J. M. and C. I. Weber, eds. Ecological Assessments
of Effluent Impacts on Communities of Indigenous Aquatic Organisms. Proceedings of a Symposium, Ft. Lauderdale, Florida, January 29-30, 1979. Philadelphia:
American Society for Testing and Materials, 1983.

Major papers presented: "Evaluation of the effects of
effluents on aquatic life in receiving waters--an overview," "Effects of copper on the Periphyton of a small
calcareous stream," and "Impact of trinitrotoluene
wastewaters on aquatic biota in Lake Chickamauga,
Tennessee."

Bermingham, N., et al, eds. Proceedings of the Seventh
Annual Aquatic Toxicology Workshop, November 5-7, 1980,
Montreal, Quebec. Ottawa: Ministere des peches et des
oceans, 1981.

Boyle, Terence P., ed. Validation and Predictability of
Laboratory Methods for Assessing the Fate and Effects
of Contaminants in Aquatic Ecosystems. Proceedings of
a Symposium, Grand Forks, North Dakota, August 8, 1983.
Philadelphia: American Society for Testing and Materials, 1985.

Major papers presented: "Effects of fluorene on survival, growth, reproduction, and behavior of aquatic
organisms in laboratory tests," and "Sediment toxicity,
contamination, and macrobenthic communities near a
large sewage outfall."

Carmichael, Wayne W., ed. The Water Environment: Algal
Toxins and Health. Proceedings, Wright University,
Dayton, Ohio, June 29-July 2, 1980.

In Environmental Science Research 20, (1981).

Major papers presented: "Water-associated human illness
in northeast Pennsylvania and its suspected association
with blue-green algae blooms," and "Potential for
groundwater contamination by algal endotoxins."

CEP Consultants Ltd. Heavy Metals in the Environment. 4th
International Conference, Heidelberg, Federal Republic
of Germany, September 1983. Edinburgh: DEP Consultants,
Ltd., 1983.

Major papers presented: "Metal pollutants in waters:
their effects, controlling factors and ultimate fate,"
"Mercury dynamics in estuarine sediments: a ten year
study," "Fate and effects of metals in the sea-surface
microlayer," and "Effects of heavy-metals on ETS-
activity in lake water."

Crusberg, T. C., et al., eds.   Water Quality and the Public Health.   Proceedings of the Conference, Worcester Massachusetts, May 1983.   Worcester, MA: Worcester Consortium Higher Education, 1984.

Major papers presented: "Acid precipitation and its effects on several streams feeding the Worcester (MA) reservoirs," and "The effects of Vermont's fertilizer industry on groundwater quality."

Ditri, F. M. and M. A. Kamrin, eds.   PCBs: Human and Environmental Hazards.   Proceedings of the International Symposium on PCBs in the Great Lakes, East Lansing, Michigan, March 17-19, 1982.   Ann Arbor, MI: Ann Arbor Science, 1983.

Major papers presented: "An overview of the scientific basis for concern with polychlorinated biphenyls in the Great Lakes," "The environmentalist's view of the PCBs problem in the Great Lakes," and "Effects of polychlorinated biphenyls on the expression of embryotoxicity caused by model teratogens."

Hagen, R. D. and H. G. Elderkin, eds.   Oceans 84: Conference Record, Vols. 1-2.   1984 Oceans Conference, Washington, D.C., September 10-12, 1984.   New York: Institute of Electrical and Electronics Engineers, 1984.

Major papers presented: "The future of metal determination in pollution studies," "Ocean disposal of New York City sewage sludge: a multi-media waste management assessment," and "Evaluation of the limiting permissible concentration concept in ocean dumping assessments."

International Association on Water Pollution Research and Control, Biennial Conference.   12th, Amsterdam, Netherlands, September 12-20, 1984.   Formerly Advances in Water Pollution Research: Proceedings of the International Conference Volumes 1-6, 1962-1972.

In Water Science and Technology 17 (1985).

Major papers presented: "Response of aquatic organisms to mixtures of toxicants," and "Environmental acidification and its consequences for the toxicity of weak acids and bases to aquatic organisms."

International Conference on Coal Fired Plants and the Aquatic Environment, Papers Presented.   Copenhagen, Denmark, August 16-18, 1982.

In Water Science & Technology 15 (1983).

Major papers presented: "Disposal of waste material: evaluation in the KHM project of short and long term effects," and "Aquatic effects of wet ash disposal and wet limestone scrubber systems."

International Symposium on Water Supply and Health, Papers Presented. Noordwijkerhout, Netherlands, August 27-29, 1980.

In Science of the Total Environment 18 (1981).

Major papers presented: "Lead in drinking water and health," "Asbestos and drinking water in Canada," "Health aspects of nitrate in drinking water," and "Toxicity assessment of organic compounds in drinking water in the Netherlands."

Ketchum, B. H., et al., eds. Ocean Dumping of Industrial Wastes. Proceedings of the 1st International Ocean Dumping Symposium, University of Rhode Island, October 10-13, 1978. New York: Plenum, 1981.

Major papers presented: "The effects of industrial wastes on marine phytoplankton," "The effects of pollutants on marine zooplankton at deep water dumpsite 106," and "Effects of ocean dumping on a temperate midshelf environment."

Kramer, C. J. M. and J. C. Duinker, eds. Complexation of Trace Metals in Natural Waters: Developments in Biogeochemistry, Vol. 1. 1st International Symposium on the Complexation of Trace Metals in Natural Waters, Den Burg, Netherlands, Netherlands Institute of Sea Research, May 2-6, 1983. The Hague: Martinus Nijhoff/ Dr. Junk Publ., 1984.

Major papers presented: "The relationships between metal speciation in the environment and bioaccumulation in aquatic organisms," and "The effect of natural complexing agents on heavy metal toxicity in aquatic plants."

Leppard, G. G., ed. Trace Element Speciation in Surface Waters and its Ecological Implications. Proceedings of the NATO Advanced Research Workshop, Genoa, Italy, November 2-4, 1981.

In Ecology 6 (1983).

Major papers presented: "Biological response to trace metals and their biochemical effects," and "Biological aspects of trace element speciation in the aquatic environment."

Mehrle, Paul, et al., eds. Toxic Substances in the Aquatic
     Environment: an International Aspect. Papers from an
     International Symposium held in Conjunction of the
     112th Annual Meeting of the American Fisheries Society,
     Hilton Head, South Carolina, September 24, 1982.
     Bethesda, MD: Water Quality Section, American Fisheries
     Society, 1985.

Ocean Pollution 1981, Papers Presented. Halifax NS, Canada,
     October 19-23, 1981.

     In Canadian Journal of Fisheries and Aquatic Sciences
     40 (1983).

     Major papers presented: "Particulate hydrocarbon and
     coprostanol concentrations in shelf waters adjacent to
     Chesapeake Bay," "Annual input of petroleum hydrocar-
     bons to the coastal environment via urban runoff," and
     "Heavy metals in tissues and organs of the Narwhal
     (Monodon-monoceros)."

Rasmussen, L., ed. Ecotoxicology. 3rd Oikes Conference,
     Copenhagen, Denmark, November 30-December 2, 1982.
     Stockholm: Publishing House of the Swedish Research,
     1984.

     Major papers presented: "Uptake of copper in the gills
     and liver of perch, perca-fluviatilis," "The effect of
     chromium on growth and photosynthesis of a submersed
     macrophyte, myriophyllum-spicatum," and "Uptake by
     aquatic organisms of Cl-36-labeled organic compounds
     from pulp mill effluents."

Sladecek, V., ed. International Association of Theoretical
     and Applied Limnology: Proceedings, V. 22, pt. 4. 1983
     Congress of the International Association of Theorecti-
     cal and Applied Limnology, France, 1983. Stuttgart: E.
     Schweizerbart Verlagbuch, 1985.

     Major papers presented: "Effects of acute parathion
     pollution on macroinvertebrates in a stream," and
     "Model for predicting acute toxicity of river water for
     fish."

Soule, D. F. and D. Walsh, eds. Waste Disposal in the
     Oceans. Proceedings of the Symposium on Ocean Disposal
     in the 1980s, sponsored by the Southern California
     Academy of Sciences, Los Angeles, California, April
     30-May 1, 1982. Boulder, CO: Westview Press, 1983.

     Major papers presented: "Physiological stress (scope
     for growth) of mussels in San Francisco Bay," "Asses-
     sing the effects of a coastal stream electric genera-
     ting station on fishes occupying its receiving water,"

"Effects of sewage disposal on the Polyschaetous Anne-
lids at San Clemente Island, California," and "Hazards
of onsite percolation and package treatment plants to
local coastal water quality: policy and planning solutions."

Vernberg, F. John, et al. Biological Monitoring of Marine
Pollutants. Proceedings of a Symposium on Pollution
and Physiology of Marine Organisms, Milford, Connecticut, November 7-9, 1980. Orlando, FL: Academic, 1981.

Major papers presented: "An overview of the acute and
chronic effects of first and second generation pesticides on an estuarine mysid" and "Factors affecting
trace metal uptake and toxicity to estuarine organisms."

Vernberg, W. B., et al., eds. Physiological Mechanisms of
Marine Pollutant Toxicity. Proceedings of the Sympo-
sium on Pollution and Marine Organisms, held at the
University of South Carolina, Columbia, South Carolina,
November 30-December 3, 1981. Orlando, FL: Academic,
1982.

Major papers presented: "Effects of sublethal concentrations of the drilling mud components, attapulgite
and Q-Broxin on the structure and function of the gill
of the scallop, placopecten magellanicus (Gmelin)," and
"Toxicity and accumulation of naphthalene in the mysid
neomysis americana (Smith) and effects of environmental
temperature."

Wong, C. S., et al., eds. Trace Metals in Sea Water.
Proceedings of a NATO Advanced Research Institute,
Erice, Italy, March 30-April 3, 1981. NATO Conference
Series, IV, Marine Sciences, Volume 9. New York:
Plenum, 1983.

Major papers presented: "The significance of the river
input of chemical elements to the ocean," and "Sensiti-
vity of natural bacterial communities to additions of
copper and to cupric ion activity: a bioassay of copper
complexation in seawater."

Reviews

Bonaventura, Joseph, et al. "Effects of Heavy Metals on the
Respiratory Proteins on Marine Organisms in Relation to
Environmental Pollution." Advances in Experimental
Medicine and Biology 148 (1982): 75-83.

Discusses the biochemical, physiological, and behavioral adaptations which have evolved to allow aquatic organisms to exist in toxic environments.

DeMayo, Adrian, et al. "Effects of Copper on Humans, Laboratory and Farm Animals, Terrestrial Plants, and Aquatic Life." CRC Critical Reviews in Environmental Control 12 (No. 3 1982): 183-255.

Constitutes the entire August issue of which almost one-half is devoted to the accumulation and toxicity of copper in aquatic environments. Adrian DeMayo has also co-written similiar reviews on the influence of lead and zinc on aquatic ecosystems.

DeMayo, Adrian, et al. "Toxic Effects of Lead and Lead Compounds on Human Health, Aquatic Life, Wildlife Plants, and Livestock." CRC Critical Reviews in Environmental Control 12 (No. 4 1982): 257-305.

Dillon, Peter J., et al. "Acidic Deposition: Effects on Aquatic Ecosystems." CRC Critical Reviews in Environmental Control 13 (No. 3 1984): 167-194.

Examines the literature regarding the influence of the atmosphere deposition of acidic substances on aquatic environments.

Hall, Lenwood W. Jr. and Dennis T. Burton. "Effects of Power Plant Coal Pile and Coal Waste Runoff and Leachate on Aquatic Biota: An Overview with Research Recommendation." CRC Critical Reviews in Toxicology 10 (No. 4 1982): 287-301.

Emphasis is to evaluate studies regarding the occurrence and toxicity of arsenic, cadmium, and selenium associated with coal effluent in the aquatic environment. Each metal, and its toxic effects, are reviewed separately.

Hall, Lenwood W. Jr., et al. "Power Plant Chlorination Effects on Estuarine and Marine Organisms." CRC Critical Reviews in Toxicology 10 (No. 1 1982): 27-47.

Analyzes chlorine toxicity data related to biofouling control in estuarine and marine thermoelectric power plants. It also identifies deficiencies in available information in order to recommend future research concerning the ecological effects of power plant cooling water chlorination.

Heckman, Charles W. "Pesticide Effects on Aquatic Habitats."
Environmental Sciences and Technology 16 (No. 1 1982):
A48-A57.

Emphasizes the effects of pesticides which are applied
directly to rice fields for mosquito control and which
are infiltrated into nearby water bodies.

Journal of the Water Pollution Control Federation. Water
Pollution Control Federation. Monthly, known as Sewage
Works Journal until 1950, and Sewage & Industrial
Wastes until 1960.

Consists of compilations of references to and bibliographic citations of the major technical literature
relevant to water pollution control topics that have
been published worldwide in the previous year. The
June issue of the Journal is the Annual Literature
Review.

Kool, H. J., et al. "Toxicology Assessment of Organic Compounds in Drinking Water." CRC Critical Reviews in
Environmental Control 12 (No. 4 1982): 307-353.

Investigates the carcinogenic and mutagenic effects of
organic compounds in drinking water. Discussion on the
identification of organic mutagens and the health significance of the presence of these compounds in drinking water is given.

Meyers, Theodore R. and Jerry D. Hendricks. "A Summary of
Tissue Lesions in Aquatic Exposures to Environmental
Contaminants, Chemotherapeutic Agents, and Potential
Carcinogens." Marine Fisheries Reviews 44 (No. 12
1982): 1-17.

Consolidates the pre-1981 literature regarding pathological changes caused by particular compounds in the
aquatic environment. This review is a useful reference
for aquatic toxicologists and aquatic pathologists.

Schmitt, Christopher J., et al. "National Pesticide Monitor-
ing Program: Residues of Organochlorine Chemicals in
Freshwater Fish, 1980-81." Archives of Environmental
Contamination and Toxicology 14 (1985): 225-260.

Analyzes residues of organochlorine chemicals in 315
composite samples of whole fish collected in 1980-81
from 107 stations nationwide as part of the National
Pesticide Monitoring Program.

Southward, A. J. "An Ecologist's View of the Implications
    of the Observed Physiological and Biochemical Effects
    of Petroleum Compounds on Marine Organisms and Ecosys-
    tems." Philosophical Transactions of the Royal Society
    of Britain 297 (No. 1087 1982): 385-399.

    Reviews field studies and experimental data on the
    lethal and sublethal effects of oil to marine animals.

Taylor, Margaret C., et al. "Effects of Zinc on Humans,
    Laboratory and Farm Animals, Terrestrial Plants and
    Fresh Water Aquatic Life." CRC Critical Reviews in
    Environmental Control 12 (No. 2 1982): 113-181.

Woltering, Daniel M. "The Growth Response in Fish Chronic
    and Early Life Stage Toxicity Tests: A Critical
    Review." Aquatic Toxicology 5 (1984): 1-21.

    Examines toxicity data from 25 years of fish life
    cycle, partial chronic and early life stage tests to
    determine the utility of the standard fish chronic
    toxicity endpoints, in particular the growth response,
    to chemical hazard evaluations.

Reports and Documents

Aquatic Plants: Ecology and Environment. 1977-January,
    1983. (Citations from the Selected Water Resource Ab-
    stracts Data Base). Springfield, VA: National Techni-
    cal Information Service, 1983.

    Discusses the effects of domestic sewage, industrial
    wastes, and oil spills on the aquatic environment.

Asbestos in Drinking Water. 1977-June, 1984. (Citations
    from the Selected Water Resources Abstracts Data Base).
    Springfield, VA: National Technical Information Ser-
    vice, 1984.

    Discusses the health effects of asbestos fiber contami-
    nated drinking water.

Bailey, H. C., et al. Toxicity of TNT (Trinitrotoluene)
    Wastewaters to Aquatic Organisms. Volume 4. Chronic
    Toxicity of 2,4-Dinitrotoluene and Condensate Water.
    Menlo Park, CA: SRI International, 1982.

    Presents and discusses the results of early life stage
    and chronic studies on 2,4-dinitrotoluene, condensate
    water, and photolyzed condensate water. Early life

stage studies were conducted on 2,4-DNT and condensate water with rainbow trout, channel catfish and fathead minnows. This report is the last in a series of four reports on the toxicity of 2,4,6-trinitrotoluene (TNT) wastewater to aquatic organisms.

Commercial Shellfish Industry: Water Pollution Effects. 1977-March, 1984. (Citations from the Selected Water Resource Abstracts Data Base). Springfield, VA: National Technical Information Service, 1984.

Concerns the various types of pollutants and their effects on the survival of shellfish, and their effects on human consumers. Considered are bacteria, heavy metals, oil and thermal pollutants.

Connolly, J. P. and R. V. Thomann. WASTOX (Water Quality Analysis Simulation for Toxics), a Framework for Modeling the Fate of Toxic Chemicals in Aquatic Environments. Part 2. Bronx, NY: Manhattan College, Department of Environmental Engineering and Science, 1985.

Describes a generalized model for estimating the uptake and elimination of toxic chemicals by aquatic organisms.

Dredging: Environmental Aspects. 1977-March 1984. (Citations from the NTIS Data Base). Springfield, VA: National Technical Information Service, 1984.

Reports on the environmental aspects of dredging. Studies include pollution control, effects on water quality, and sediment transport. Biological effects are included in a companion bibliography.

Ecology of Pesticide Water Pollution. 1978-May 1984. (Citations from the NTIS Data Base). Springfield, VA: National Technical Information Service, 1984.

Cites literature on pesticides in the aquatic environment. Topics covered are the effects upon plants, animals, and humans. Included are lakes, rivers, oceans, estuaries, streams, and ground water.

Fisheries: Environmental Aspects. 1977-November, 1983. (Citations from the Selected Water Resources Abstracts Data Base). Springfield, VA: National Technical Information Service, 1983.

Emphasizes the effects of temperature, toxins, acidification, oil spills, heavy metals and chemical wastes on fisheries and marine wildlife.

Harrison, F. L.   Review of the Impact of Copper Released
     Into Freshwater Environments.   California: Lawrence
     Livermore National Laboratory, 1984.

     Reviews the concentrations of copper in the abiotic and
     biotic compartments of freshwater ecosystems, and the
     effects on biota of increased amounts of copper in the
     water and sediments.

Koch, L. M. and R. C. Young.   Effects of Selected Inorganic
     Coal-Gasification Constituents on Aquatic Life: An
     Annotated Bibliography.   Muscle Shoals, AL: Tennessee
     Valley Authority, Office of Natural Resources, 1983.

     Concentrates on primary inorganic pollutants of concern
     which result in the aqueous discharges of high-pressure
     coal-gasification technologies.   These pollutants
     include ammonia, cyanide, sulfide, and boron.

Lech, J. J. and M. J. Melancon.   Hazardous Chemicals in Fish:
     Wisconsin Power Plant Impact Study.   Milwaukee, WI:
     Medical College of Wisconsin, Inc., 1984.

     Assesses the role of fish as vectors for organic chemi-
     cal contaminants arising from the operation of a coal-
     fired power plant.   The results indicate that selected
     organic chemicals which could be released by a power
     plant into the aquatic environment will be taken up by
     fish. Humans who might consume such fish may be exposed
     to those chemicals at varying concentrations.

Marine Disposal of Radioactive Wastes.   1978-1983.   (Cita-
     tions from the Life Sciences Collection Data Base).
     Springfield, VA: National Technical Information Ser-
     vice, 1983.

     Emphasizes the disposal of solid radioactive wastes in
     the sea, groundwater, and under the sea bed.   Concern
     for the aquatic life under concentrations of these
     wastes is examined.

Metal Processing Wastes: Water Pollution.   1977-October
     1982.   (Citations from the NTIS Data Base).   Spring-
     field, VA: National Technical Information Service,
     1982.

     Examines water pollution control, abatement planning,
     and health effects from metal processing.

Ocean Waste Disposal.   1977-January, 1984.   (Citations
     from the Selected Water Resources Abstracts Data Base).
     Springfield, VA: National Technical Information Ser-
     vice, 1984.

Concerns the effects and potential impacts of radioactive, industrial and sewage type wastes being dumped
into the oceans.

Pesticide Toxicity in Aquatic Environments. June 1983-
April 1985. (Citations from the Selected Water
Resources Abstracts Data Base). Springfield, VA:
National Technical Information Service, 1985.

Examines the accumulation of pesticides in the aquatic
environment, and the methods to determine their
effects.

Polychlorinated Biphenyls in the Freshwater Environment.
1977-January, 1984 (Citations from the Selected Water
Resources Abstracts Data Base). Springfield, VA:
National Technical Information Service, 1984.

Concerns the occurrence and effects of polychlorinated
biphenyls in the freshwater environment. Topics
include sources, distribution, and accumulation rates
for specific regions, bioaccumulation, pollutant paths,
and introduction into the food chain, and results of
long and short term monitoring of selected areas.
Detection methods for PCB accumulation in fish and
surficial sediments are also considered.

Pulp and Paper Mill Bleaching Effluents: Toxicity and
Control Techniques. 1972-March 1984. (Citations from
the Institute of Paper Chemistry Data Base). Springfield, VA: National Technical Information Service,
1984.

Considers the toxic effects that pulp and paper mill
bleaching processes and materials have on the surrounding environment. Environmental impacts on fishes,
water animals and aquatic plants are discussed.

Rice, S. D., et al. Effects of Petroleum Hydrocarbons on
Alaskan Aquatic Organisms: a Comprehensive Review of
All Oil-Effects Research on Alaskan Fish and Inverte-
brates Conducted by the Auke Bay Laboratory, 1970-81.
Auke Bay, AK: National Marine Fisheries Service, Auke
Bay Laboratory, 1984.

Reviews and summarizes all oil-effects research by the
Auke Bay Laboratory from the beginning of these studies
in 1970 through 1981. Both published and unpublished
results from 62 studies are included. Research is
reviewed according to subject: toxicity, sublethal
effects, etc.

Roy, W. R., et al. <u>Geochemical Properties of Coal Wastes and the Toxicological Effects on Aquatic Life</u>. Champaign, IL: Illinois State Geological Survey Division, 1984.

Assesses the toxicological effects of leachates from solid wastes generated by coal mining, cleaning and gasification on aquatic organisms.

<u>Sewage Effects in Marine and Estuarine Environments</u>. 1977- August 1985. (Citations from the NTIS Data Base). Springfield, VA: National Technical Information Service, 1985.

Covers the effects that sewage effluents and sludge have upon marine plants and animals, and problems due to ocean dumping.

<u>Thermal Pollution: Biological Effects</u>. 1974-August 1982. (Citations from the NTIS Data Base). Springfield, VA: National Technical Information Service, 1982.

Includes abstracts covering the effects of thermal pollution on fish, shellfish, plants, and microorganisms in both fresh and salt water. Emphasis is on the effects on growth, ecology, metabolism, and heat tolerance.

<u>Toxicity Bioassays: Water Pollution Effects on Aquatic Animals and Plants</u>. 1977-May 1985. (Citations from the Selected Water Resources Abstracts). Springfield, VA: National Technical Information Service, 1985.

Concerns the toxicity bioassay studies of water pollution effects on growth, reproduction, and mortality of aquatic animals and plants. Metals, chemicals, pesticides and herbicides from industrial and agricultural wastes are analyzed.

Van der Schalie, W. H. <u>Toxicity of Nitroguanidine and Photolyzed Nitroguanidine to Freshwater Aquatic Organisms</u>. Fort Detrick, MD: Army Medical Bioengineering Research and Development Laboratory, 1985.

Determines the acute toxicity of nitroguanidine to ten species of freshwater aquatic organisms.

<u>Water Pollution Effects of Metals on Fresh Water Fish</u>. 1964-March, 1983. (Citations from the NTIS Data Base). Springfield, VA: National Technical Information Service, 1983.

Includes studies on water quality as a result of pollution by sewage effluents and industrial wastes. Discussed are the bio-accumulation and toxic effects of
copper, zinc, mercury, chromium, cadmium, lead and
iron.

Water Pollution: Uptake of Heavy Metals by Shellfish and
    Marine Plants. 1974-1983. (Citations from Oceanic
    Abstracts). Springfield, VA: National Technical Infor-
    mation Service, 1984.

Focuses on heavy metal contamination of marine plants
and shellfish. Evaluates the toxicity levels and long
term effects on the ecology of the marine environment,
as well as the growth rate, and the long term effects
on the food chain.

Data Bases

Aqualine
Producer: Water Research Centre
Time Span: 1974 to present
Coverage: International
Updated: Monthly, 4500 records per year

Contents: Contains citations, with abstracts, to the
worldwide literature on every aspect of water, wastewater, and the aquatic environment. Topics covered
include groundwater, wastewater treatment, drinking
water quality, sludge utilization, groundwater pollution, sewerage systems, and other water related topics.
Corresponds to printed WRC Information.

Aquatic Sciences and Fisheries Abstracts (ASFA)
Formerly: Aquatic Biology Abstracts; Current Bibliography
            for Aquatic Sciences and Fisheries
Producer: Food and Agriculture Organization; International
            Oceanographic Government Commission of UNESCO
Time Span: 1978 to present
Coverage: International
Updated: Monthly, 2000 records per month

Contents: Covers all aspects of the aquatic environment, both marine and fresh water, emphasizing oceanography, marine biology, ecology, marine pollution, and
fisheries.

Canadian Environment (CENV)
Formerly: Canada Water (CWA)
Producer: WATDOC
Time Span: 1970 to present
Coverage: Canada
Updated: Monthly, 500 records per month

> Contents: Includes literature on water quality, pollu-
> tion, aquatic life, and marine technology.

Centre National Pour l'Exploitation des Oceans (CNEXO)
Producer: Centre National pour l'Exploitation des Oceans,
            Bureau National des Donnees Oceaniques, Centre
            Oceanologique de Bretagne
Time Span: 1972 to present
Coverage: International
Updated: 3000 records per year

> Contents: Covers English and French citations on aqua-
> culture, marine corrosion, marine pollution, and the
> impact on the environment including pathology of aqua-
> tic animals. Designed to complement the other files
> merged into DOCOCEAN, this file consists primarily of
> references to non-periodical literature.

DATAGRAF
Producer: Sigma Data Services Corp., under contract to the
            Council on Environmental Quality
Time Span: Varies by data base
Coverage: United States
Updated: Varies by data base

> Contents: Provides capabilities for studying combin-
> ations of environmental, natural resource, public
> health, and demographic data. IDB (Integrated Database)
> collects data from various sources that includes water
> quality data from studies on the concentrations of
> dissolved metals and pesticides in drinking water for
> 312 U.S. counties. STORET contains approximately 60
> million observations of water quality indicators mea-
> sured at about 200,000 monitoring sites in the U.S.
> Includes information on laboratory analysis of the
> concentrations of substances in water. NASQAN (National
> Stream Quality Accounting Network) contains daily,
> monthly, and quarterly hydrological measurements col-
> lected by over 500 monitoring stations in U.S. geologic
> sub-basins. Contains 153 variables including metals,
> biological water quality indicators, pesticides, and
> radioactive elements in water and sediment.

Delft Hydro
Formerly: Delft
Producer: Delft Hydraulics Laboratory Library
Time Span: 1976 to present
Coverage: International
Updated: 7500 records per year

>Contents: Includes water management, coastal morpholo-
>gy, coastal and offshore engineering, dredging, aquatic
>pollution, and aquatic ecosystems.

Distribution Register of Organic Pollutants in Water (WDROP)
    (also known as WaterDROP)
Producer: Environmental Protection Agency
Time Span: 1970 to present
Coverage: United States
Updated: Irregular

>Contents: Contains citations to identifications of
>organic water pollutants. Most observations come from
>the United States, but those made elsewhere are also
>included. In addition to bibliographic data, detail on
>experimental procedures used in identifying the organic
>pollutants are included.

DOCOCEAN
Producer: Centre National pour l'Exploitation de Oceans,
           Bureau National des Donnees Oceaniques,
           Centre Oceanologique de Bretagne
Time Span: 1964 to present
Coverage: International
Updated: Quarterly

>Contents: Contains references to the worldwide litera-
>ture on these 4 data bases: Aquatic Sciences and Fish-
>eries Abstracts, Centre National pour l'Exploitation de
>Oceans, Oceanic Abstracts, and Pascal. Subjects include
>oceanography, ecological, biological, physical, chemi-
>cal and geological aspects, and exploitation of oceans.

Instructional Resources Information System (IRIS)
Producer: Environmental Protection Agency
Time Span: 1979 to present
Coverage: United States
Updated: 1000 records every 6 months

>Contents: Consists of a specialized file of abstracts
>of educational and instructional materials on water
>quality and water resource materials. It also includes
>other environmental related instructional materials.
>The file includes journal articles and references to
>films, filmstrips, slides, brochures, pamphlets and
>books. Corresponds to the printed Water Quality
>Instructional Resources Information System.

National Ground Water Information Center
Producer: National Water Well Association
Time Span: Earliest data from 1900
Coverage: United States
Updated: Biweekly

> Contents: Contains approximately 26,000 citations to
> the worldwide literature on groundwater and hydrology.
> Covers books, journals, technical reports, conference
> proceedings, and citations from Selected Water Resour-
> ces Abstracts.

National Stream Quality Accounting Network
Producer: Sigma Data Services Corp. (in conjunction with
            the Council on Environmental Quality)
Time Span: Unknown
Coverage: United States
Updated: Unspecified

> Contents: Contains daily, monthly, and quarterly data
> covering water quality. Data include physical proper-
> ties, measures of pesticides, sediments and radioactive
> elements present, and biological properties.

Oceanic Abstracts
Producer: Cambridge Scientific Abstracts
Time Span: 1964 to present
Coverage: International
Updated: Monthly, 6000 records per year

> Contents: Contains citations on all major ocean
> related topics including marine biology, ocean pollu-
> tion, fisheries, and offshore mining. Corresponds to
> coverage of Oceanic Abstracts.

Pascal 885-Pollution
Producer: Centre National de la Recherche Scientific,
            Centre de Documentation Scientifique et Technique
            (France)
Time Span: 1975 to present
Coverage: International
Updated: 1600 records per month

> Contents: Contains literature on water, water pollu-
> tion, and solid wastes.

Scientific Parameters for Health and the Environment,
    Retrieval and Estimation (SPHERE)
Producer: Environmental Protection Agency, Office of
            Toxic Substances
Time Span: 1971-1980
Coverage: International
Updated: Periodically, data base to eventually contain
            information on approximately 4000 chemicals

Contents: Contains 2 files of information on the
health effects of chemical substances. One file,
Aquatic Information Retrieval Data Base (AQUIRE),
contains data on the chronic, bioaccumulative, and
sublethal effects of over 2500 chemical substances on
freshwater and saltwater organisms. Each record covers
a single experiment and includes chemical substance
information, description of test organism, study proto-
col, results, and bibliographic reference.

Selected Water Resources Abstracts
Producer: Water Resources Scientific Information Center
Time Span: 1968 to present
Coverage: International
Updated: Monthly, 7000 items per year

Contents: Covers the water-related aspects of
conservation, use and management including hydrology,
water resources, water supply, quality, planning and
engineering.

Sludge Newsletter.
Producer: Business Publishers, Inc.
Time Span: January 1983 to present
Coverage: United States
Updated: Bimonthly

Contents: Includes literature on pollution control
residuals management, ocean dumping, cadmium uptake and
other controversies.

Waterlit
Producer: South African Water Information Centre
Time Span: 1976 to present
Coverage: International
Updated: Monthly, 10,000 records per year

Contents: Includes literature on sludge disposal,
wastewater and treatment, sewage and solid waste management, health aspects of water, pollution of water,
water ecology and marine biology.

Waternet
Producer: American Water Works Association
Time Span: 1971 to present
Coverage: Primarily United States, Canada, Mexico, and Latin
           America, with some coverage of Europe and Asia
Updated: Quarterly, 3000 records per year

Contents: Contains citations, with abstracts, to literature on water quality, water and wastewater reuse, and
environmental issues related to water which include:
the drinking water industry, water pollution, health
effects, and toxicology.

Audiovisual Materials

Battling Sea Pollution. Tokyo, Japan: NHK Films, 198_.
3/4" U-matic cassette, 30 min., color.

Examines the early stages of pollution of the sea. The
containment and disposal of oil slicks is emphasized.

Biological Studies of River Pollution. Portmouth, Hants,
England: Focal Point Audiovisual Ltd, n.d. 2 x 2 slide
with cassette, 50 frames, color.

Focuses on the biological effects of pollution on rivers. Demonstrates how various aquatic organisms react
to water quality deterioration.

In Our Water. Brooklyn, NY: Foresight Films, n.d. 16mm
film, 60 min., color.

Documents the lives of the Kaler family who live next
to a chemical dumpsite in New Jersey. This film reports
on the waste problem and chemical pollution of the
nation's drinking water.

Protecting Our Water. Portmouth, Hants, England: Focal
Point Audiovisual Ltd, n.d. 2 x 2 slide, 40 frames,
color.

Shows the effects of pollution on inland and coastal
waters from industries, mining and quarrying, oil, sewage and thermal pollution.

Splash. New York: National Film Board of Canada, 1982.
16mm film, optical sound, 12 min., color.

Uses water droplets to demonstrate the importance of
water in urban living and the impact of pollution on
this precious natural resource. (Animated.)

Water - A Clear and Present Danger. New York: ABC News,
1983. 16mm film, optical sound, 26 min., color.

Looks at the health effects of groundwater contamination. Shows that despite efforts to regulate drinking
water, 700 man-made chemicals have been found in water,
some of which are harmful.

Water, More Precious Than Oil. Washington, D.C.: PBS Video,
1982. Beta, VHS, 3/4" U-matic cassette, 60 min.,
color.

Examines the quality of current water supplies, water-related diseases, and water pollution.

When There Are No More Fish in the Sea. Tokyo, Japan: NHK Films, 198_. 3/4" U-matic cassette, 33 min., color.

Conceptualizes man's disregard for nature.

Associations

Columbia National Fisheries Research Laboratory
Fish and Wildlife Service
Department of the Interior
Route No. 1
Columbia, MD 65201   (314) 875-5399

Examines the effects of pesticides on fish. Maintains data on toxicity of pesticides to fish.

Great Lakes United
c/o Robert Boice
R.D. #2
Archer Road
Watertown, NY 13601   Phone #: not available

Promotes the conservation and enhancement of the Great Lakes ecosystem. Target issues include: hazardous and toxic substances, fish and wildlife management and habitat protection.

International Association on Water Pollution Research and Control
Alliance House
29/30 High Holborn
London WCl
England   Phone #: not available

Contributes to the advancement of research in water pollution control and encourages communication and a better understanding among those engaged in the solution of water pollution problems.

International Ocean Disposal Symposium
c/o Dr. Iver W. Duedall
Dept. of Oceanography and Ocean Engineering
Florida Institue of Technology
Melbourne, FL 32901   (305) 768-8008

Exchanges ideas and information among investigators involved in research on the problems associated with

disposing of wastes in the ocean. Organizes information resulting from symposia presentations into peer-
reviewed books prepared for worldwide distribution.

Lake Michigan Federation
53 W. Jackson Boulevard
Chicago, IL 60604   (312) 427-5121

Promotes citizen participation in public policy decisions pertaining to Lake Michigan water quality, coastal planning and hazardous waste containment.

Water Pollution Control Federation
2626 Pennsylvania Avenue
Washington, D.C. 20037   (202) 337-2500

Advances fundamental and practical knowledge concerning
the nature, collection, treatment and disposal of
domestic and industrial wastewaters, and the design,
construction, operation, and management of facilities
for these purposes.

Government Organizations

Aquatic Environmental Research Program
Tennessee Valley Authority
Box 2000
Brown's Ferry Nuclear Plant
Decatur, AL 35602   (205) 386-2030

Provides facilities for studying thermally modified
aquatic environments. The Aquatic Environmental
Research Program is concerned with the examination of
the effects of elevated thermal regimes on the aquatic
ecosystems in large outdoor channels. Program's primary research interest is the effects on freshwater
fish, invertebrates, and algae ecology. Program is
operated at the Biothermal Research Station, which was
placed in operation in 1974 by the TVA in collaboration
with the EPA.

Environmental Protection Agency
Environmental Research Laboratory
Sabine Island
Gulf Breeze, FL 32561   (904) 932-5311

Examines the impact of insecticides and other organic
compounds on marine species and communities and
includes studies on the toxicity of organic pollutants
to estuarine organisms.

Environmental Protection Agency
Environmental Research Laboratory
6201 Congdon Boulevard
Duluth, MN 55804   (218) 727-6692

Conducts research in aquatic toxicology methods; evaluates the predictive capability of these methods in
natural surface waters; provides toxicology data for
Agency use; and develops methods to predict effects of
pollutants in the Great Lakes. Specifically, the Laboratory develops toxicity test methods for aquatic life
for EPA's regulation development; provides primary
Agency consultation on freshwater toxicology problems;
provides toxicity data to confirm effects of problem
chemicals for court cases; and evalutes toxicity of
complex effluents treated by various control technologies.

Environmental Protection Agency
Environmental Research Laboratory
S. Ferry Road
Narrogansett, RI 02882   (401) 789-1071

Conducts research on the effects of pollutants on
marine ecosystems including studies on the chemical and
physical behavior of pollutants in marine systems;
investigations of significant responses of organisms to
pollutant stress; characterization of marine ecosystems
and their responses to stress; development of systems
to quantitate response to specific pollutants; and
development of methods for determining the impact of
specific pollution incidents.

Fish and Wildlife Service
Department of the Interior
C Street Between 18th and 19th Streets, NW
Washington, D.C. 20240   (202) 343-1100

Includes such activities as biological monitoring
through scientific research; surveillance of pesticides, heavy metals, and thermal pollution; studies of
fish and wildlife populations; and ecological studies.

Georgia Environmental Resources Center
Georgia Institute of Technology
Atlanta, GA 30332   (301) 344-3637

Includes studies on surface water pollutants, environmental impact, groundwater pollution from landfills,
and environmental radiation.

Monitoring and Data Support Division
Office of Water Regulations and Standards
Environmental Protection Agency
401 M Street, SW
Washington, D.C. 20460   (202) 426-7760

    Examines the chemical, physical, and biological charac-
    teristics of water, sources and effects of pollutants
    on streams, lakes, and coastal waters, municipal and
    industrial waste discharges, pollution-caused fish kill
    statistics, and related water pollution control infor-
    mation.

Newtown Fish Toxicology Station
Environmental Protection Agency
3411 Church Street
Cincinnati, OH 45244   (513) 684-8601

    Conducts research and development studies related to
    water quality criteria necessary to protect aquatic
    life, and methods of biological evolution of the toxi-
    city of municipal and industrial water - water dischar-
    ges.

Libraries and Information Centers

Coast Guard, Office of Marine Environment and Systems,
    Pollution Incident Reporting System
G-WEP-1/12
2100 Second Street, SW
Washington, D.C. 20593   (202) 426-9571

    Subject Matter: Discharge of oil or hazardous substan-
    ces into the navigable waters of the United States.

Environmental Protection Agency
Environmental Research Laboratory, Corvallis. Library
200 SW, 35th Street
Corvallis, OR 97330   (503) 757-4731

    Subject Matter: Effects in water, air, and soil pollu-
    tants on the ecosystem, marine and freshwater ecosys-
    tems, toxic substances, fish toxicity, and hazardous
    waste.

Environmental Protection Agency
Environmental Research Laboratory. Library
Sabine Island, 6
Gulf Breeze, FL 32561   (904) 932-5311

Subject Matter: Pesticides, marine ecology, fishery biology, and carcinogens.

Gulf Coast Research Laboratory, Gordon Gunter Library
Ocean Springs, MA 39564   (601) 875-2244

Subject Matter: Marine sciences, including biology, physiology, parasitology, chemistry, ecology, oceanography, toxicology, fisheries research and management.

International Joint Commission, Great Lakes Regional Office Library
100 Ouellette Avenue, 8th Floor
Windsor, Ontario N9A 6T3
Canada   (519) 256-7821

Subject Matter: Great Lakes water quality, toxic substances, limnology, wastewater treatment.

James M. Montgomery, Consulting Engineers, Library
555 E. Walnut Street
Pasadena, CA 91101   (213) 796-9141

Subject Matter: Water quality, effluents, drinking water, wastewater.

John Van Oosten Great Lakes Fishery Research Library
Great Lakes Fishery Laboratory
Fish and Wildlife Service
1451 Green Road
Ann Arbor, MI 48105   (313) 994-3331

Subject Matter: Fishery biology, aquatic ecology, pesticide, mercury, Great Lakes, and water pollution.

National Water Data Storage & Retrieval System
Geological Survey. Water Resources Division
National Center
Mail Stop 437
Reston, VA 22092   (703) 860-6879

Subject Matter: Surface water stage and discharge, chemical quality parameters, pesticide and biological concentrations in water, ground and surface water levels.

New Hampshire Water Supply and Pollution Control Commission. Library
Health and Welfare Building, Hazen Drive
Box 95
Concord, NH 03301   (603) 271-3503

   Subject Matter: Water - pollution, biology and quality,
   industrial wastes, waste treatment, lakes, New Hampshire, pesticides, chemistry.

Oil and Hazardous Materials Technical Assistance Data System
Environmental Protection Agency
WH-548-B
401 M Street, SW
Washington, D.C. 20460   (313) 675-5000

   Subject Matter: Physical, chemical and toxicological
   properties of oil and hazardous materials, including
   pesticides, industrial chemicals and oil spill treating
   agents. Emphasis is on water pollution hazards posed
   by these substances.

University of Minnesota, St. Paul. Entomology, Fisheries
   and Wildlife Library
1980 Folwell Avenue
375 Hodson Hall
St. Paul, MN 55108   (612) 373-1741

   Subject Matter: Fisheries, pesticides, water pollution,
   limnology, aquatic biology.

University of Wisconsin, Madison. Water Resources Center
   Library
1975 Willow Drive
Madison, WI 53706   (608) 262-3069

   Subject Matter: Water resources, pollution sources,
   wastewater treatment, and limnology.

Virginia. State Water Control Board. Library
2111 N. Hamilton Street
Box 11143
Richmond, VA 23230   (804) 257-6340

   Subject Matter: Water, water pollution, wastewater,
   groundwater, toxins in water.

Water Pollution Control Federation. Library
2626 Pennsylvania Avenue, NW
Washington, D.C. 20037   (202) 337-2500

   Subject Matter: Water pollution control, wastewater
   treatment and disposal.

Water Research Centre (Great Britain)
Information Service on Toxicity and Biodegradability
Stevenage Laboratory
Elder Way
Stevenage, Herts SG1 1TH
England   0438-2444

>   Subject Matter: Effects of chemicals on sewage treat-
>   ment processes, toxicity of chemicals to aquatic organ-
>   isms, and environmental effects of wastes disposed of
>   in landfills. Service is primarily concerned with
>   freshwater environment.

Water Research Centre (Great Britain)
Library and Information Services
Stevenage Laboratory
Elder Way
Stevenage, Herts SG1 1TH
England   0438-2444

>   Subject Matter: Water research and resources, includ-
>   ing treatment, quality and health, sewage and indus-
>   trial waste water treatment, sludge disposal, pollution
>   and fish studies.

Research Centers and Industrial Laboratories

Academy of Natural Sciences of Philadelphia
Division of Environmental Research, Philadelphia Laboratory
19th Street and the Parkway
Philadelphia, PA 19103
Founded: 1947   (215) 299-1080

>   Research Activities: Basic and applied research and
>   consulting related to fresh and estuarine waters, in-
>   cluding bioassays on physiologic effects of sublethal
>   contaminants; effects of chemical, mechanical and ther-
>   mal stresses from industry on the aquatic environment;
>   and acidification studies.

Archibold Biological Station
P.O. Box 2057
Lake Placid, FL 33852
Founded: 1971   (813) 465-2571

>   Research Activities: Marine biology and biological
>   oceanography. Specialty areas of research in environ-
>   mental chemistry includes studies in priority pollutant
>   analysis in sea water sediments and biological tissue,
>   analysis of toxins and carcinogens in wastewater and
>   drinking water, and effects of open ocean disposal of
>   wastes.

Battelle Marine Research Laboratory
439 West Sequim Bay Road
Sequim, WA 98382
Founded: 1967   (206) 683-4151

   Research Activites: Effects and fate of chemical pollutants in coastal marine ecosystems and identification
   and control of fish and shellfish diseases encountered
   in mariculture. Sponsored programs include research on
   the effects of energy technology development on marine
   biological systems, particularly physiological/biochem-
   ical adaptation to chemical stress and the fate and
   effects of sediment bound pollutants.

Cornell Fishery Research Laboratory
Cornell University
112 Fernow Hall
Ithaca, NY 14853
Founded: 1924   (607) 256-2298

   Research Activities: Fisheries biology and aquatic ecology, including studies on response of fish and fish
   populations to toxic substances, especially heavy
   metals.

Freshwater Institute
501 University Crescent
Winnipeg, MB R3T 2N6
Canada
Founded: 1966   (204) 269-7379

   Research Activities: Biological, technological, and
   environmental aspects of freshwater fisheries, including studies on problems of pollution of lakes, effects
   of acidification and radionuclides, effects of in-
   creases in nutrient levels in lakes, general ecology of
   fish species in natural and polluted waters, quality of
   fishery products, general ecology of Arctic marine
   mammals, and studies on effects of northern industrial
   developments on aquatic life.

Halliburton Company
Laboratory Services Division
900 Gemini Avenue
Clear Lake City
Houston, TX 77058
Founded: Unknown   (713) 486-5700

   Research Activities: Environmental laboratory services
   for terrestrial, aquatic, and fish toxicity; toxicity
   bioassays; and water analyses.

Indiana University, Water Resources Research Center
1005 East 10th Street
Bloomington, IN 47401
Founded: 1964   (812) 337-3848

    Research Activities: Physical, biological, and social
science aspects of water problems under all types of
environments, including stream pollution. Conducts
studies on surface mining, mine drainage, thermal pollution, effects of sanitary landfills on ground- and
surface-water movement and water quality, and contamination effects of water moving through the soil.

Institute for Fisheries Research
212 Museums Annex
Ann Arbor, MI 48109
Founded: 1930   (313) 663-3554

    Research Activities: Sport and commercial fisheries
problems in all waters of the state, including studies
on the effects of pesticides on fishes.

Kappe Associates Inc.
(Scientific Research Division)
14211 Travilah Road
P.O. Box 1036
Rockville, MD 20850
Founded: Unknown   (301) 762-7797

    Research Activities: Industrial wastewater treatabil-
ity studies, chemical analysis and bacteriological
examinations of water and wastewater, and fish toxicity
studies.

Kent State University, Institute of Limnology
Biological Sciences
Kent, OH 44242
Founded: 1967   (216) 672-3613

    Research Activities: Prepares land and lake use models
for major cities, evaluates lake restoration methodologies, and the impact of toxic substances.

Michigan State University, Inland Lakes Research and Study
    Center (ILRSC)
334 Natural Resources Building
East Lansing, MI 48824
Founded: 1982   (517) 353-3742

    Research Activities: Specific topics include effects
of toxic and hazardous materials which enter the water
as waste products, nonpoint pollution sources, and
atmospheric deposition.

Oak Creek Laboratory of Biology
Oregon State University
Department of Fisheries and Wildlife
Corvallis, OR 97331
Founded: 1953   (503) 754-3508

   Research Activities: Aquatic biology, including problems relating to biological effects and control of
   water pollution and experimental studies of physiology,
   toxicology, and ecology of freshwater and marine fish,
   shellfish, and other aquatic organisms, their environmental requirements, and their responses to water quality changes. Conducts training program for graduate
   students in aquatic biology with special emphasis on
   fisheries, water pollution biology, and toxicology.

Oklahoma State University. Center for Water Research
203 Whitehurst Hall
Stillwater, OK 74074
Founded: 1965   (405) 624-6995

   Research Activities: Aquatic ecology, limnology, ecological effects of contaminants, fate of pollutants in
   groundwater, and identification and detection of
   groundwater pollutants.

Oklahoma State University. Water Quality Research Laboratory
Stillwater, OK 74074
Founded: 1967   (405) 624-5551

   Research Activites: Biological aspects of water pollution, including transport, fate, and effect of water
   pollutants, ecotoxicology, analysis of pollutants, and
   bioassays with aquatic organisms.

Pacific Bio-Marine Laboratories, Inc.
P.O. Box 536
Venice, CA 90291
Founded: Unknown   (213) 822-5757

   Research Activities: Basic research includes the environmental effects of water disposal in marine environment.

Rutgers University. Center for Coastal and Environmental
   Studies
104 Doolittle Building
New Brunswick, NJ 08903
Founded: 1971   (201) 932-3738

   Research Activities: Marine sciences and coastal environmental assessments, environmental impacts of nuclear

power plants, effects of thermal enrichment, nonpoint
source pollutants, and effects of toxic contaminants in
surface waters.

Rutgers University. Division of Water Resources
Doolittle Hall
New Brunswick, NJ 08903
Founded: 1965   (201) 932-3596

   Research Activities: Studies on urbanization and its
   effect upon water resources, urban runoff, nonpoint
   pollution sources, and fate of and effects on toxic
   substances in water.

State University College at Buffalo. Great Lakes Laboratory
1300 Elmwood Avenue
Buffalo, NY 14222
Founded: 1966   (716) 878-5422

   Research Activities: Environmental problems of the
   Great Lakes with particular emphasis on the environmental toxicology and chemistry of pollutants and their
   socioeconomic impact. Maintains a library of 30,000
   volumes on the Great Lakes, water pollution, pesticides, toxic chemicals, and aquatic organisms.

State University of New York at Stony Brook
Marine Sciences Research Center
Stony Brook, NY 11794
Founded: 1968   (516) 246-7715

   Research Activities: Coastal marine sciences including
   pollution in coastal waters, waste disposal in the sea
   and ocean policy. The Center operates a comprehensive
   research program that delves into the biology, chemistry and physics of the coastal ocean and the effects
   that society has had on the coastal ocean.

State University Research Center at Oswego
State University College at Oswego
King Hall
Oswego, NY 13126
Founded: 1978   (315) 341-3639

   Research Activities: Environmental studies of Great
   Lakes system, including analytical chemistry, geochemistry, thermal effluent studies, sublethal effects
   of microcontaminants on fish and mammals, groundwater
   contamination, hazardous wastes and environmental communication.

Tulane University. Environmental Health Sciences
    Research Laboratory
F. Edward Hebert Research Center
New Orleans, LA 70118
Founded: 1969   (504) 394-2233

   Research Activities: Water pollution abatement, water
   quality, evaluation and control, toxicology. Interdisciplinary studies include toxic effects of herbicides and pesticides on aquatic organisms, biological
   and physical/chemical treatment of industrial waste.

University of Alaska, Fairbanks. Institute of
    Water Resources
306 Tanana Drive
315 Dickering Building
Fairbanks, AK 99701
Founded: 1965   (907) 474-7775

   Research Activities: Emphasis on problems dealing with
   the water environment of Alaska, including biological
   effects of pollution, thermal pollution, physical,
   biological, and chemical waste treatment, effects of
   thermal discharges into arctic streams, and environmental effects of developments of lakes and streams.

University of Connecticut. Marine Sciences Institute
Avery Point
Groton, CT 06340
Founded: 1968   (203) 446-1020

   Research Activities: Research projects include studies
   on pollution microbiology, microbiology of captive
   marine animals, pathology of marine invertebrates
   caused by parasitism and environmental impact.

University of Hawaii. Hawaii Undersea Research Laboratory
Marine Sciences Building
1000 Pope Road
Honolulu, HI 96822
Founded: 1980   (808) 948-6335

   Research Activities: Fisheries, pollution; physical
   effects of waste disposal at sea; behavioral, biochemical and physiological responses of marine organisms to
   pollutants.

University of Kentucky. Water Resources Research Institute
762 Anderson Hall
Lexington, KY 40506
Founded: 1964   (606) 257-1832

Research Activities: Studies on strip mine-polluted water, impact of surface mining, and bacterial contamination of groundwater.

University of Maryland, College Park. Water Resources Research Center
Symons Hall
College Park, MD 20742
Founded: 1965   (301) 454-6405

Research Activities: Includes studies on the control of wastewater quality of crab processing plant, physical characterization of extrachromosomal deoxyribonucleic acid (DNA) in Chesapeake Bay bacteria from polluted and nonpolluted sites. Also studies sublethal effects of pollutants, including usefulness of life tables as a means for evaluating the impact of chronic toxicity and instream flow requirements for fishes and fisheries in Maryland.

University of Wisconsin-Madison Water Chemistry Program
660 North Park Street
Madison, WI 53706
Founded: 1961   (608) 262-2470

Research Activities: Chemistry of hazardous organic compounds in lakes, atmospheric input of chemicals into lakes, trace metal chemistry in lakes, and adsorption reactions at partical surfaces in aqueous systems.

University of Wisconsin Sea Grant Institute
1800 University Avenue
Madison, WI 53706
Founded: 1968   (608) 262-0905

Research Activities: Studies sources, fates, and effect of microcontaminants, such as PCB, dioxins, phosphorous, heavy metals, and oil in the Great Lakes.

University of Wisconsin-Superior. Center for Lake Superior Environmental Studies
1800 Grand Avenue
Superior, WI 54880
Founded: 1969   (715) 392-8101

Research Activities: Multidisciplinary research of environmental problems confronting the southwestern Lake Superior region, including toxicity of chemicals toward aquatic life.

University of Wyoming
Red Buttes Environmental Biology Laboratory
Box 3166, University Station
Laramie, WY 82071
Founded: 1970   (307) 745-8504

Research Activities: Aquatic toxicology, fish physiology, studies on acid deposition effects on fish, acclimation of fish to organic and metal pollutants.

4.

The Effects of Toxic and Hazardous Materials on Land

Donna Serafin

Introduction

The uncovering of landfills containing accumulated toxic wastes near residential sections of cities, towns, and villages has caused many people much stress. Incidents of new sites are reported periodically, thus keeping this problem before the public.

The migration of contaminants from unsecured landfills often consists of runoff and seepage of leachate, a liquid solution formed by the joining of precipitation and the contents of a landfill. Common contaminants may be heavy metals and minerals, synthetic chemicals, and/or pathogenic organisms. If not pumped out or contained, this leachate joins the groundwater and ultimately interconnecting rivers, lakes or oceans.

Of equal polluting capability are toxic or explosive gases and volatile organic chemicals which leak from landfills and escape either into the atmosphere or underground through porous soils. When these pollutants are synthetic chemicals, they resist natural degradation and accumulate in wildlife, fish, and human beings, thus greatly increasing their toxic effects.

While this chapter deals primarily with the migration of contaminants, particularly toxic wastes, from unsecured landfills and its effect on land, it also includes information sources on groundwater contamination and the effect of pesticide residues in soils. The references listed are not exhaustive, but cite the more important sources concerned with toxic wastes and their effect on land.

Books

Brown, Kirk W., et al., eds. Hazardous Waste Land Treatment.
    Boston: Butterworth, 1983.

    A practical guide which clearly describes the proce-
    dures for selection, design, operation, and monitoring
    of a hazardous waste land treatment facility that meets
    or goes beyond EPA regulations. The July 26, 1982
    regulations are included.

Cheremisinoff, Paul N. and Kenneth A. Gigliello. Leachate
    from Hazardous Waste Sites. Lancaster, PA.: Technomic,
    1983.

    Presents guidelines on the formation, generation, sam-
    pling methodology, control and treatment of leachate
    from solid waste sites. Processes by which leachate is
    formed at solid waste land disposal sites and methods
    to estimate its quantity at a particular site are
    discussed. Detailed methods of sampling are explained
    including sample hazard classification, sample types,
    safety considerations and actual sampling methodology.
    A broad discussion of the control and treatment of
    leachate from solid waste land disposal facilities is
    presented as are proper techniques and methods used to
    minimize leachate production.

Cope, C. B., et al. The Scientific Management of Hazardous
    Wastes. Cambridge: Cambridge University Press, 1983.

    Provides a clear description of waste disposal prac-
    tices in the United Kingdom. Of particular relevance
    are chapters six, which discusses the characteristic
    properties of landfill leachate; seven, which covers
    leachate management; and eight, which examines the geo-
    chemistry of hazardous waste disposal. Also discussed
    are alternative methods to landfill and factors which
    affect pollutant attentuation.

Fuller, Wallace Hamilton and Arthur W. Warrick. Soils in
    Waste Treatment and Utilization. Volume 1, Land Treat-
    ment. Boca Raton, FL: CRC Press, 1984.

    Provides useful information for the management of land
    treatment facilities including the use of waste as
    fertilizers, biodegradation as a waste treatment system
    and soil attenuation.

Fuller, Wallace Hamilton and Arthur W. Warrick. Soils in
    Waste Treatment and Utilization. Volume 2, Pollutant
    Containment, Monitoring and Closure. Boca Raton, FL:
    CRC Press, 1984.

Concentrates on soil selection, use and manipulation of
soils in trenches, encapsulations and landfills.
Includes discussions on landfill liners and gas, odor,
and aerosol controls. Provides screening protocols for
waste treatment sites, predicts transformations and
mobility of pollutants in soil.

Krueger, Raymond F. Treatment and Disposal of Pesticide
    Wastes. American Chemical Society Symposium Series No.
    259. Washington, D.C.: American Chemical Society,
    1984.

Emphasizes the problems met in the disposing of pesticide waste and the development of better disposal
plans. These include chemical, biological and physical
methods of disposal. The analytical and modeling
approaches which apply to waste disposal conditions are
covered. EPA guidelines on safe disposal and their
regulatory effects are discussed.

National Research Council. Geophysics Study Committee.
    Groundwater Contamination. Washington, D.C.: National
    Academy Press, 1984.

Includes a review of the extent of groundwater contamination throughout the United States. This volume in
the Studies in Geophysics series focuses on the
physical and chemical processes that control the movement of contaminants in groundwater. A number of chapters examine case studies of sites of groundwater contamination.

Overcash, Michael R., ed. Decomposition of Toxic and Non-
    toxic Organic Compounds in Soils. Lancaster, PA:
    Technomic, 1981.

Consists of 43 reports on chemicals in soils, many of
which are available for the first time in English. Four
areas are emphasized: spills, transfer of pollutants
from air to land, purposeful use of land to treat
wastes and agricultural chemicals.

Parr, James F., et al., eds. Land Treatment of Hazardous
    Wastes. Park Ridge, NJ: Noyes, 1983.

Reviews and assesses current knowledge and management
practices for ultimate disposal of hazardous wastes
through land treatment systems. Identifies areas of
research that would improve the technology of land
treatment. Part one of the book examines processes
that influence the effects of land-applied waste such
as bioassay, effects of toxics on the food chain, plant
uptake of inorganics, interaction with soils, and fate
of pathogens and composting. Part two discusses prob-

lems and potentials of land application for eleven
different industries. Among them are petroleum, pulp
and paper, textile, wood preservative, leather and
munitions manufacturing wastes.

Pye, Veronica I., et al. Groundwater Contamination in the
United States. Philadelphia: University of Pennsylvania Press, 1983.

Explores the sources of groundwater contamination
including hazardous wastes, agricultural runoff, accidental spills, land disposal of sludge, mining, highway
deicing salts, brine disposal association with the
petroleum industry, and radioactive sources. Numerous
references and tables concerning the above are present
as well as the geographical extent of contamination;
monitoring; remedial action and rehabilitation of aqui-
fers. Federal statutes, state and local measures are
described.

Shuckrow, Alan J. and Andrew P. Pajak. Hazardous Waste
Leachate Management Manual. Pollution Technology
Review Series, No. 92. Park Ridge, NJ: Noyes, 1982.

Analyzes the four major options for controlling, treating, and disposing of leachate from hazardous waste
disposal sites: waste generation, hazardous waste
treatment, disposal site management, and leachate
treatment and disposal. Provides emergency, monitoring, and safety contingency plans.

Sims, Ronald, ed. Contaminated Surface Soils In-Place
Treatment Techniques. Pollution Technology Review
Series, No. 132. Park Ridge, NJ: Noyes, 1986.

Sittig, Marshall. Landfill Disposal of Hazardous Wastes and
Sludges. Pollution Technology Review Series, No. 62.
Park Ridge, NJ: Noyes, 1980.

Disposal of hazardous materials in landfills is treated
from all aspects including waste disposal alternatives,
regulatory requirements, waste preparation, site selection, landfill design, construction and operation,
monitoring and final land use. The material presented
is based mainly on reports and guidelines issued by the
Environmental Protection Agency.

Smith, Michael A., ed. Contaminated Land: Reclamation and
Treatment. NATO Challenges of Modern Society, Volume
8. New York: Plenum, 1985.

Discusses the treatment of groundwater contaminated by
leachate, on-site processing of contaminated spoils,

covering systems, pollutant transport by groundwater,
toxic and flammable gases and the long term effectiveness of remedial measures. Includes conclusions and
recommendations.

Wood, Eric F., et al., eds. Groundwater Contamination from
Hazardous Wastes. Englewood Cliffs, NJ: Prentice-Hall,
1984.

Provides an overview of the problem of groundwater
contamination focusing on mechanics, evaluation, and
analysis. Also examines the siting of new hazardous
waste landfills. Includes case studies.

Yaron, B., et al., eds. Pollutants in Porous Media: The
Unsaturated Zone Between Soil Surface and Groundwater.
Ecological Studies, Volume 47. New York: Springer-
Verlag, 1984.

Deals with that land area between soil surface and
groundwater which may act as filter or reservoir for
pollutants. Specifically discusses surface interactions between pollutants and porous media; how biological processes govern pollutants' persistence in the
unsaturated zone; presents new ways of modelling transport of pollutants through porous media. Includes case
studies dealing with pollution sources and management.

Proceedings

Conway, Richard A., ed. Hazardous Solid Waste Testing.
Proceedings of the First Conference on Hazardous Solid
Waste Testing, Ft. Lauderdale, Florida, January 14-15,
1981. ASTM Special Technical Publication, No. 760.
Philadelphia: American Society for Testing and
Materials, 1981.

Includes sections on laboratory extraction and leaching
procedures; analysis of residues, extracts, soils, and
groundwaters; and evaluation of land disposal and
materials sites.

Disposal of Hazardous Waste. Proceedings of the Sixth
Annual Research Symposium, Chicago, Illinois, March 17-
20, 1980. EPA-600/9-80-010 (PB-175086). Cincinnati,
OH: Municipal Environmental Research Laboratory, Office
of Research and Development, Environmental Protection
Agency, 1980.

Describes research projects dealing with seven
broad categories in the management of hazardous wastes

through land disposal: waste sampling and characterization; transport and fate of pollutants; pollutant control; co-disposal; landfill alternatives; remedial actions; and thermal destruction techniques.

Francis, Chester W. and Stanley I. Auerbach, eds. Environment and Solid Wastes: Characterization, Treatment and Disposal. Proceedings of the Fourth Life Sciences Symposium on Environmental and Solid Wastes, Gatlinburg, Tennessee, October 4-8, 1981. Boston: Butterworth, 1983.

Deals with the environment and solid wastes including testing methods prior to landfill disposal; remedial actions at solid waste landfills; land treatment of hazardous wastes; chemical indicators of leachate contamination in groundwater; migration of contaminants; responsible long-term use of agricultural and urban land for solid waste disposal; food chain toxic pathways for toxic materials in wastes.

Jackson, Larry P., et al., eds. Hazardous and Industrial Waste Management and Testing. Proceedings of the Third International Symposium on Industrial and Hazardous Solid Wastes, Philadelphia, Pennsylvania, March 7-10, 1983. ASTM Special Technical Publication, No. 851. Philadelphia: American Society for Testing and Materials, 1984.

Includes such titles as "Vapors, Odors, and Toxic Gases from Landfills"; "Effect of Chemical Treatment of Fly Ash on the Leaching of Metals"; "Hazardous Waste Landfill Research". The book itself is divided into four sections dealing with analysis of wastes and waste disposal sites, amelioration of wastes in the disposal environment, waste as resource and national perspectives in waste management.

Lagrega, M. D. and D. A. Long, eds. Toxic and Hazardous Waste. Proceedings of the Sixteenth Mid-Atlantic Waste Conference on Industrial Waste, State College, Pennsylvania, June 24-26, 1984. Lancaster, PA: Technomic, 1984.

Contains several papers dealing with alternative methods to treatment of groundwater contamination, hazardous sludge and soil detoxification. Also includes papers on a new method of predicating leachate generation, using computer models to locate sources of groundwater pollution and the impact on soil biota of a land treatment method.

Land Disposal of Hazardous Waste. Proceedings of the
    Seventh Annual Research Symposium, Philadelphia,
    Pennsylvania, March 16-18, 1981. EPA-600/9-81-002b
    (PB81-173882). Cincinnati, OH: Municipal Environmental
    Research Laboratory, Office of Research and
    Development, Environmental Protection Agency, 1981.

    Covers technical areas involving hazardous waste such
    as transport and fate of pollutants, containment, land
    treatment of such and what characterizes hazardous
    waste.

Land Disposal of Hazardous Waste. Proceedings of the Eighth
    Annual Research Symposium, Ft. Mitchell, Kentucky,
    March 8-10, 1982. EPA-600/9-82-002 (PB82-173022).
    Cincinnati, OH: Municipal Environmental Research
    Laboratory, Environmental Protection Agency, 1982.

    Provides a compilation of research projects on land
    disposal, incineration and treatment of hazardous
    wastes.

Land Disposal of Hazardous Wastes. Proceedings of the Ninth
    Annual Research Symposium, Ft. Mitchell, Kentucky, May
    2-4, 1983. EPA-600/9-83-018 (PB84-118777).
    Cincinnati, OH: Municipal Environmental Research
    Laboratory, Environmental Protection Agency, 1983.

    Updates seven technical areas of hazardous waste
    disposal on land as follows: research overviews, design
    and operation of landfills, pollutant movement;
    pollutant control liners, waste modification, remedial
    action, and cost economics.

Land Disposal of Hazardous Wastes. Proceedings of the Tenth
    Annual Research Symposium, Ft. Mitchell, Kentucky,
    April 3-5, 1984. EPA-600/9-84-007 (PB84-177799).
    Cincinnati, OH: Municipal Environmental Research
    Laboratory, Environmental Protection Agency, 1984.

    Compilation of current research dealing with hazardous
    waste land disposal including these topics: in-situ
    hazardous waste stabilization by injection grouting;
    reactivity of grouts to leachates; encapsulation of
    drums containing hazardous waste using welded poly-
    ethylene. Also discusses permeability of compacted
    soils to solvent mixtures and petroleum products;
    designing earthen liners for landfills; and methods for
    predicting composition of leachate from hazardous
    waste.

<u>Land Disposal of Hazardous Wastes</u>. Proceedings of the
     Eleventh Annual Research Symposium, Cincinnati, Ohio,
     April 29-May 1, 1985. EPA-600/9-85-013 (PB85-196376).
     Cincinnati, OH: The Laboratory, 1985.

<u>Land Disposal of Hazardous Wastes</u>. Proceedings of the
     Twelfth Annual Research Symposium, Cincinnati, Ohio,
     April 21-23, 1986. Cincinnati, OH: The Laboratory,
     1986.

     Report and order numbers not available at time of
     publication.

Lehman, John P., ed. <u>Hazardous Waste Disposal</u>. Proceedings
     of the NATO Committee on the Challenges of Modern
     Society Series. Symposium on Hazardous Waste Disposal,
     Washington, D.C., October 5-9, 1981. New York: Plenum,
     1983.

     Provides a comprehensive review of the international
     status of hazardous waste disposal. Contains papers on
     landfill research and practice in the United Kingdom,
     United States, Canada, and West Germany.

O'Leary, P. R. and J. T. Quigley, eds. <u>Proceedings of the
     Sixth Annual Madison Conference of Applied Research and
     Practice on Municipal and Industrial Waste</u>. Madison,
     Wisconsin, September 14-15, 1983. Madison: University
     of Wisconsin Extension, 1983.

     Papers dealing with applied research and practice on
     municipal and industrial waste. Topics include struc-
     tural design and retrofitting of landfills; control,
     prevention and prediction of leachate seeps; co-dispo-
     sal of industrial waste; and liners, both flexible
     membrane and soil cement synthetic.

Perry, R., ed. <u>Effects of Waste Disposal on Groundwater and
     Surface Water</u>. Symposium on the Effects of Waste
     Disposal on Groundwater and Surface Water at the First
     International Association of Hydrological Scientific
     General Assembly, Exeter, England, July 19-30, 1982.
     IAHS Publication, No. 139. Reston, VA: International
     Association of Hydrological Sciences, 1982.

     Jointly sponsored by the International Association of
     Hydrological Sciences and UNESCO. Major papers include
     modelling of pollution transport in aquifers, the pol-
     lution potential of sanitary landfills and the effects
     of waste disposal on groundwater and surface water
     leachate attenuation, solute movement.

Proceedings of the National Conference on Management of
  Uncontrolled Hazardous Waste Sites. Washington, D.C.,
  November 29-December 1, 1982. Silver Springs, MD:
  Hazardous Materials Control Research Institute, 1982.

  Contains articles dealing with detection of contaminants in soils. Among them are NDT methods for detecting buried containers in silty clay; electrical-resis-
  tivity techniques for liner leaks; biological sampling
  at abandoned sites; vadose zone monitoring at landfills, impoundments and land treatment disposal areas;
  composting to treat TNT and RDX contaminated soils;
  leachate treatment and mathematical modelling of pollutant migration. Also includes an international study
  of contaminated lands.

Sweeney, Thomas L., et al, eds. Hazardous Waste Management
  for the Eighties. Proceedings of the Second Ohio Environmental Engineering Conference, Columbus, Ohio, March
  1982. Ann Arbor, MI: Ann Arbor Science, 1982.

  A compilation of conference papers providing practical
  and useful information on hazardous waste management,
  specifically the monitoring, recovery and treatment of
  polluted groundwater; seepage; clay liner permeability
  and the practical aspects of land treatment of hazardous materials.

Symposium on Theory and Practice of the Use of Soil Applied
  Herbicides. Proceedings of the Symposium on Theory and
  Practice of the Use of Soil Applied Herbicides,
  Versailles, France, December 8-9, 1981. Wageningen:
  European Weed Research Society, 1981.

  Of particular relevance is the paper on the "Effects of
  Residues of Herbicides in the Soil on Subsequent
  Crops." Other papers discuss the migration of herbicides through soil columns, and the effects of herbicides on soil mechanisms.

van Duijvenbooden, W., et al., eds. Quality of Groundwater.
  Proceedings of an International Symposium, Noordwij-
  kerhout, The Netherlands, March 23-27, 1981. New York:
  Elsevier, 1981.

  Major papers presented include "Effect of Industrial-
  Urbanized Landscapes on Underground Waters," "Ground-
  water Pollution by Volatile Halogenated Hydrocarbons:
  Sources of Pollution and Methods to Estimate Their
  Relevance," and "The Impact of Point Source Pollution
  on Groundwater Quality."

Abstracts and Indexes

The majority of the literature on hazardous and toxic materials and its effects on land is reported in the primary
journals dealing with environmental pollution.   However, a
significant portion of the information is found in the
proceedings of conferences and symposia and/or technical
reports.

The accessing of journal articles and conference papers is
available through abstracting and indexing journals and
recently has been simplified through the use of on-line data
bases.   Several, including <u>Pollution Abstracts</u> and
<u>Environment Abstracts</u> are described in Chapter 1.

Periodicals

<u>Archives of Environmental Contamination and Toxicology</u>.
    Springer-Verlag.   Bimonthly, published since 1972.

    An international, interdisciplinary repository of full
    length articles which describe original experimental or
    theoretical research work dealing with the scientific
    views of contaminants in the environment.

<u>Bulletin of Environmental Contamination and Toxicology</u>.
    Springer-Verlag.   Monthly, published since 1966.

    Reports monthly on the latest advances and discoveries
    regarding the contamination and pollution of soil,
    water, food and air.

<u>Environmental Pollution. Series B: Chemical and Physical: An
    International Journal</u>.   Elsevier.   Bimonthly, published
    since 1980.

    Original research papers deal with chemical and physi-
    cal pollutant distribution and new techniques for their
    study and measurement.   Disposal problems, landscape
    pollution, pesticides, sewage disposal and soil protec-
    tion are among many subjects discussed.   Includes book
    reviews.

<u>Ground Water</u>.   Water Well Journal Publishing Co.   Bi-
    monthly, published since 1963.

    Publication of the Association of Ground-Water Scien-
    tists and Engineers, this journal provides research
    reports on groundwater-related information.   Special

features includes: "Reader's Forum," "Ground Water in the News," "New Publications," and "Discussion of Papers".

Ground Water Age. National Trade Publications, Inc. Monthly, published since 1966.

Trade journal to serve the pump installation, water well drilling and geohydrologist fields. Includes product literature.

Ground Water Monitoring Review. Water Well Journal Publishing Co. Quarterly, published since 1981.

Reports on the latest technology in monitoring ground water resources.

Ground Water Newsletter. Water Information Center, Inc. Bimonthly, published since 1972.

Provides up-to-date coverage on all ground water related information. Includes book reviews.

Journal of Environmental Quality. Published cooperatively by the American Society of Agronomy, Crop Science Society of America, and Soil Science Society of America. Quarterly, published since 1972.

Reports original research articles, reviews and analyses, and book reviews. Each article is preceded by an abstract.

Journal of Environmental Science and Health. Part B: Pesticides, Food Contaminants, and Agricultural Wastes. Dekker. Bimonthly, published since 1985.

Reports on original research concerned with improvements on existing methods of dealing with pesticide residue; chemical contaminants and their metabolic fate; contamination of the biosphere, detoxification methods and toxicological consequences.

Land Pollution Reporter. Freed Publishing Co. Bimonthly, published since 1969.

Provides information on soil erosion, strip mines and other land pollutants and discusses possible solutions to these problems.

Pesticides and Toxic Substances Monitoring Report.
Environmental Protection Agency, Office of Pesticide
and Toxic Substances. Survey and Analysis Division.
Irregular, published since 1980.

Reports on the Survey and Analysis Division's findings
of evidences of human and environmental exposure to
pesticides and other toxic chemicals.

Water, Air and Soil Pollution: An International Journal of
Environmental Pollution. D. Reidel Publishing Co.
Published 8 times a year, since 1971.

Audiovisual Resources

Hazardous Waste Disposal. Fullerton, CA: James L. Ruhle &
Associates, 1983. Filmstrip with cassette, color, 50
frames.

In Our Own Backyard - The First Love Canal. Produced by
Buffalo Documentary Group. Oley, PA: Bullfrog Films,
1982. Videocassette (issued as U-matic 3/4 inch, or
Beta 1/2 inch, or VHS 1/2 inch), color, 59 min.

The Killing Ground. Produced by ABC News. New York: ABC
Wide World of Learning, 1979. Videocassette (issued
as U-matic 3/4 inch or Beta 1/2 inch or VHS 1/2 inch),
sound, color, 52 min. (Issued also as a motion picture,
2 reels, 16mm., color, 48 min.).

Toxic Wastes - Disposing or Dumping? Produced by New York
Times Co. Westminster, MD: Random House School
Division, 1983. Filmstrip (83 frames, 35mm.), sound
cassette, color, 17 min.

Research Centers and Industrial Laboratories

Bio-Dynamics
P.O. Box 43
East Millstone, NJ 08873   (201) 873-2550

Fields of research & development: Analytical chemistry
on soil pollutants; toxicological and metabolic studies
in foods, pesticides and agricultural chemicals.

Bio-Search and Development Co., Inc.
12700 Prospect Avenue
Route 30
Kansas City, MO 64146   (816) 942-3421

Fields of research & development: Agricultural chemicals, including pesticide residue application, and crop monitoring.

Controls for Environmental Pollution, Inc.
1925 Rosina
P.O. Box 5351
Sante Fe, NM 87502   (505) 982-9841

Fields of research & development: Environmental studies, pesticide and herbicide analysis in air, water, soil and biological specimens.

Forest-Ag Environmental Protection Service
3483 Golden Gate Way, Suite 219
P.O. Box 745
Lafayette, CA 94549   (415) 284-5212

Fields of research & development: Pesticides, soils and nutrition, pesticide residues in the environment.

Kentucky Testing Laboratory Corp.
968 Swan Street
Louisville, KY 40204   (502) 583-5256

Fields of research & development: Pesticide residues; chemical and bacteriological testing and analysis.

North Carolina State University
Pesticide Residue Research Laboratory
3709 Hillsborough Street
Raleigh, NC 27607   (919) 737-3391

Fields of research & development: Effects of pesticides on plants, animals, soils, water and air with particular emphasis on residues on food and feed crops. Also studies behavior of herbicides and pesticides in soils.

Libraries and Information Centers

Documentation and Information Service
Environmental Toxicology Library
University of California
Davis, CA 95616   (916) 752-2562

Subjects: Industrial contaminants, hazardous materials,
hazardous waste management, pesticides, food toxicants;
food and feed additives.

Publications: Pesticides: A Selected Bibliography

Services: Maintains pesticide data bank and provides
statistical summaries from data bank on fee basis;
answers inquiries; on-site use of periodical, book and
technical report collection; access to TOXLINE.

Robert S. Kerr Environmental Research Center Library
Environmental Protection Agency
Box 1198
Ada, OK 74820   (405) 332-8800

Subjects: Agricultural pollution, soil treatment,
ground water quality and pollution.

Services: Interlibrary loan, open to the public for
reference use only.

Rachel Carson Council, Inc. - Library
8940 Jones Mill Road
Chevy Chase, MD 20815   (301) 652-1877

Subjects: Pesticides, pest management programs, toxic
substances.

Special Collections: EPA Pesticide Product Information

Publications: Publications on toxic substances and
pesticides and alternatives to their use

Services: Open to the public by appointment

Residuals Management Technology, Inc. - Library
1406 E. Washington Avenue, Suite 124
Madison, WI 53703   (608) 255-2134

Subjects: Solid and hazardous waste management, hydrogeology, environmental engineering

Services: Library open to public for reference use only
and by appointment; interlibrary loan

Information on toxic and hazardous waste management, pesticides, industrial and agricultural pollution is available in many of the Environmental Protection Agency's libraries. See Chapter 1.

Associations & Organizations

Governmental Refuse Collection and Disposal Association
8401 Dixon Avenue, Suite 4
P.O. Box 7219
Silver Spring, MD 20910   (301) 585-2898

> Areas of interest: Improvement of solid waste management

> Publications: Newsletter (monthly)

> Services: Annual seminar and equipment show; sponsors seven training programs including hazardous waste technologies and practices, principles of landfill operations

National Agricultural Chemicals Association
1155 15th Street, NW
Washington, D.C. 20005   (202) 296-1585

> Areas of interest: Pesticides, toxicology, agricultural chemicals

> Publications: Actionews (biweekly), restricted distribution

> Services: Provides educational and safety literature, brochures, and audiovisual materials; answers inquiries

Soil Society of America
677 S. Segoe Road
Madison, WI 53711   (608) 274-1212

> Areas of interest: Soil mineralogy, soil testing and plant analysis, fertilizer technology and use, soil chemistry, soil fertility and plant nutrition

> Publications: Soil Society of America Journal, (bimonthly), Journal of Environmental Quality (quarterly)

> Services: Annual convention in conjunction with the American Society of Agronomy and the Crop Science Society of America

Government Agencies

Municipal Environmental Research Laboratory
Office of Research and Development
Environmental Protection Agency
26 W. St. Clair
Cincinnati, OH 45268   (513) 684-7951

   Areas of interest: Waste water treatment; solid and
   hazardous waste management

Office of Pesticides and Toxic Substances
Environmental Protection Agency
401 M Street, SW
Washington, D.C. 20460   (202) 382-2090

   Areas of interest: Pesticides use and their effects;
   pesticides regulation; toxic chemicals regulation,
   monitoring of pesticide residue levels in food, humans
   and wildlife.

   Publications: Pesticides & Toxic Substances Monitoring
   Report

Office of Solid Waste and Emergency Response
Environmental Protection Agency
401 M. Street, SW
Washington, D.C. 20460   (202) 382-4610

   Areas of interest: Guidelines and standards for land
   disposal of hazardous wastes

Robert S. Kerr Environmental Research Laboratory
Environmental Protection Agency
P.O. Box 1198
Ada, OK 74820   (405) 332-8800

   Areas of interest: Effects of pollutants on soil; agricultural wastes, soil treatment systems, ground water
   quality and pollution

   Services: On-site use of the collection; answers
   inquiries; lends materials

Soil Conservation Service
Department of Agriculture
Fourteenth Street and Independence Ave., SW
Washington, D.C. 20250   (202) 447-4543

   Areas of interest: Soil and water conservation, agricultural pollution control; rural community development

Publications: Soil surveys

Services: Technical assistance to landowners, soil maps and other resource data; soil surveys

5.
Effects of Hazardous Substances in the Air

Deborah Husted

Introduction

Toxic and hazardous waste pollutants are formed and find
their way into our atmospheric environment from a diverse
array of sources: combustion processes (fuel oil, gasoline/
diesel fuel, wood, coal, and natural gas); industrial/
manufacturing processes (e.g., smelting, electroplating,
welding, and petroleum refining); electrical energy generation; mining; incineration of municipal and industrial
wastes; and emissions from toxic waste dumps and accidental
spills of hazardous, vaporous materials. Categories of
atmospheric pollutants range from organic and inorganic
gases to particulate materials. Types of atmospheric
contaminants can include: polycyclic aromatic hydrocarbons;
aldehydes; nitrogen, carbon, and sulfur dioxides; and liquid
and solid particulates. Finally, the list of individual
names of substances that have carcinogenic, mutagenic, and
toxicological properties is distressingly long as well. For
example: arsenic, carbon monoxide, lead, formaldehyde,
dioxins, ozone, acetone, and asbestos may all be included as
highly problematic substances that can be found in indoor
and outdoor air environments, where they pose risks for the
health and well being of humans and vegetation. Hence, it
is clear that the nature and process of atmospheric contamination is neither easily categorized nor resolved. However, through increased understanding and knowledge of
atmospheric pollutants, it will be possible to plan for
longrange solutions to this highly complex, difficult
situation.

The scope of this chapter is the selective examination of
print information sources (generally appearing after 1979)
pertaining to the sources, formation, chemistry, properties,
behavior, modelling, and control of toxic, hazardous atmospheric pollutants. Some information sources concerning
direct effects of such pollutants on biological systems have
been included as well, although the overwhelming majority of
sources pertain to the nature and activities of these
contaminants in the air.

Several items bear further delineation:

-- A comprehensive source of information, which predates the scope of this chapter, should not be overlooked. The Environmental Protection Agency's Air Pollution Technical Information Center [Environmental Protection Agency, Library Services Office, Mail Drop 35, Research Triangle Park, NC 27711 (919) 541-2777] issues both the APTIC data base and microfilm copies (limited availability) of technical literature on air pollution (indexed and abstracted) covering the time span of 1966-1976. Although the data base is available through DIALOG, it has not been updated.

-- The reviews section of this chapter contains selected review and research articles with lengthy, timely bibliographies on hazardous atmospheric pollution. The report section generally comprises bibliographies generated from online data bases. These may be ordered from the National Technical Information Service, and the NTIS numbers are included.

-- In reflecting on several discussions found in sources contained in this chapter, a few research trends stand out in their importance with regard to the presence of heavy metal and other dangerous pollutants in the air. One trend concerns the need to investigate not only the various ramifications of individual pollutants in the air, but also the need to investigate the synergistic effects of many such pollutants in the atmosphere. Another trend concerns the investigation of long-term, continued low-level exposure to pollutants. Clearly, this field of environmental research has many important contributions to make toward the understanding and control of atmospheric pollutants.

Books

Benarie, Michel M. Urban Air Pollution Modelling. Cambridge, MA: MIT Press, 1980.

Consists of a compilation of methodologies, statistical formulae and models which may be applied toward air pollution prediction estimation. Describes multi-source Gaussian plume and short-time models and concepts. Discusses mass transport balance, multibox models, statistical relationships, long-term plume modelling, and air pollution forecasting.

Bower, Frank A. and Richard B. Ward, eds. Stratospheric Ozone and Man. 2 volumes. Boca Raton, FL: CRC Press, 1982.

Analyzes the nature, measurement, modelling, and distribution of stratospheric ozone; processes affect-

ing ozone; and social/governmental response to the
problems associated with ozone depletions. Specifically, both atmospheric and human generated ozone
perturbations are examined. Includes discussions of
trends in government and industry ozone research.

Brunner, Calvin R. _Hazardous Air Emissions from Incinera-
tion_. New York: Chapman and Hall, 1985.

Delineates the connection between air quality and the
incineration of hazardous wastes. Discusses
fundamentals of incineration processes, estimates of
emission rates, in addition to air pollution control
processes (such as wet gas scrubbing, electrostatic
precipitation, gas cleaning, and filters). Examines
the influence of incineration of hazardous wastes on
air quality by analyzing dioxins, emission of odors,
and gaseous and particulate atmospheric pollutants.

Calvert, Seymour and Harold M. Englund, eds. _Handbook of
Air Pollution Technology_. New York: Wiley, 1984.

Contains thirty-eight articles concerning effects of
toxic air pollutants on the atmosphere and methods and
devices used to control atmospheric pollution. Two
articles analyze the characteristics of gaseous and
particulate pollutants. Contains an informative listing of hazardous air pollutants, their chemical and
toxicity characteristics and concentration levels.

Cohn, Louis F. and Gary R. McVoy. _Environmental Analysis of
Transportation Systems_. New York: Wiley, 1982.

Chapter six, "Transportation Air Quality," pertains to
the analysis of the effects of transportation pollution
on overall air quality levels. Identifies and
discusses most commonly found air pollutant species
(carbon monoxide, hydrocarbons, photochemical oxidants,
nitrogen oxides, and lead) associated with transportation facilities. Considers emission rates, dispersion
of these air pollutants, and air pollution models.

Connell, Des W. and Gregory J. Miller. _Chemistry and
Ecotoxicology of Pollution_. New York: Wiley, 1984.

Provides a cohesive conceptualization of chemical
interactions of pollutants and atmospheric, aquatic,
and terrestrial environments and their toxicological
and ecotoxicological ramifications. Chapter one
includes a discussion of the chemodynamics of atmospheric pollutants. Chapter eleven, "Atmospheric
Pollutants," examines several aspects of dangerous air
pollutants including: sources, atmospheric metals,

aerosols, toxic effects, photochemical behavior, and global effects of atmospheric pollutants.

Constance, John D. Controlling In-Plant Airborne Contaminants: Systems Design and Calculations. Mechanical Engineering, 21. New York: Dekker, 1983.

Assesses the characteristics and control of airborne hazardous substances in indoor working environments. Summarizes the basics of gas/vapor chemistry, physics, and behavior. Reviews control of indoor airborne contaminants via various methods of ventilation, moisture control, and contaminated air cleaning systems. Appendix A.2: "Control of Explosive or Toxic Air-gas Mixtures."

Cullis, C. F. and J. G. Firth, eds. Detection and Measurement of Hazardous Gases. London: Heinemann, 1981.

Examines the nature, measurement, analysis, and sampling of hazardous gases with special reference to industrial environments and new developments in sampling instrumentation.

Dix, H. M. Environmental Pollution: Atmosphere, Land, Water, and Noise. New York: Wiley, 1981.

Chapter six contains a discussion of the physical effects of pollution in the atmosphere and stratosphere. It contains specific descriptions of carbon monoxide, sulphur dioxide, ozone, nitrogen dioxide, and particulates. Source material cited is generally dated (1970-1976).

Ehrenfeld, John R., et al. Controlling Volatile Emissions at Hazardous Waste Sites. Pollution Technology Review, 126. Park Ridge, NJ: Noyes, 1986.

Analyzes presently available control technologies for emissions released during hazardous waste treatment, storage, and disposal processes. For each major type of facility for hazardous waste management examined, sources of air emissions are provided and assesses various methods used for control of emissions. Estimation techniques for air emissions from storage tanks, wastewater treatment processes, drum storage and handling facilities, and waste piles are examined.

Grimmer, Gernot. Environmental Carcinogens: Polycyclic Aromatic Hydrocarbons: Chemistry, Occurrence, Biochemistry, Carcinogenicity. Boca Raton, FL: CRC Press, 1983.

Reviews the effects of polycyclic aromatic hydrocarbons (PAH) in the environment. Contains current knowledge about the occurrence and chemistry of atmospheric pollution of PAH and the biologic, carcinogenic, and epidemiologic effects on human populations.

Hesketh, Howard E. and Frank L. Cross. Fugitive Emissions and Controls. The Environment and Energy Handbook Series. Ann Arbor, MI: Ann Arbor Science, 1983.

Discusses the definition, sources, data, control and measurement of hazardous emissions. Provides ground level concentration data for a number of toxic pollutants.

Jennings, M. S., et al. Catalytic Incineration for Control of Volatile Organic Compound Emissions. Pollution Technology Review, No. 121. Park Ridge, NJ: Noyes, 1985.

Concerns the technology and processes whereby the volatile organic compounds from industrial wastes can be controlled. (Incineration at a temperature range of 1300-1500 degrees F (or 700-900 degrees F with the use of catalysts) can be used to control these compounds.) Reviews current state of catalytic incineration technology in some industries that desire to control volatile organic compounds and various topics related to the catalytic incineration process, testing and analysis, and suitability factors.

Manahan, Stanley E. Environmental Chemistry. 4th ed. Boston: Willard Grant Press, 1984.

Focuses upon environmental chemistry with regard to the properties and effects resulting from the presence of hazardous and toxic substances in various environments. Several chapters pertain to the atmospheric presence of toxic pollutants and wastes: "The Nature and Composition of the Atmosphere," "Gaseous Inorganic Pollutants and Oxides in the Atmosphere," "Particulate Matter in the Atmosphere," "Photochemical Smog," "Organic Pollutants in the Atmosphere," and "Atmospheric Monitoring."

Meszaros, E. Atmospheric Chemistry: Fundamental Aspects. Studies in Environmental Science, No. 11. New York: Elsevier, 1981.

Original Hungarian text (1977) is revised and enlarged to reflect changes in knowledge since 1977. Explains basics of atmospheric structure, chemistry. Provides detailed description of atmospheric cycles of several toxic trace constituents, such as carbon compounds (including carbon monoxide), ozone, nitrogen compounds (nitrous oxide, nitrogen oxide), and sulphur compounds. Describes airborne activity of these chemicals, aerosols, trace substance atmospheric removal, and influence of climatic variation on atmospheric chemical composition. Complemented by many text figures and an extensive bibliography.

Moreton, Janet and N. A. R. Falla. Analysis of Airborne Pollutants in Working Atmospheres: The Welding and Surface Coatings Industries. Analytical Sciences Monographs. London: Chemical Society, 1980.

Covers sampling and analysis of levels of particulate and gaseous toxic pollutants found in indoor environments in the welding industry, with some discussion of external air pollution as well (primarily in the United Kingdom).

National Research Council, Board on Atmospheric Sciences and Climate, Global Tropospheric Chemistry Panel. Global Tropospheric Chemistry: A Plan for Action. Washington, D.C.: National Academy Press, 1984.

Provides a general overview of tropospheric chemistry; recommendations for enhancing the understanding of the topic; and discussion of present knowledge of atmospheric cycles, procedures for modelling, and measuring tropospheric dynamics.

National Research Council, CO_2/Climate Review Panel. Carbon Dioxide and Climate: A Second Assessment. Washington, D.C.: National Academy Press, 1982.

Assesses studies that minimize the projected effect of carbon dioxide on surface temperatures and diverges from these analyses. Noted are the significant effects that increasing concentrations of airborne carbon dioxide; trace gases (methane, nitrous oxide, chlorofluormethanes, and ozone); and atmospheric aerosols can have on climatic alteration.

National Research Council, Committee on Indoor Pollutants. Indoor Air Pollutants. Washington, D.C.: National Academy Press, 1981.

This 537 page volume examines the nature, character, and sources of indoor air pollution. Some of the more toxic substances identified are asbestos, formaldehyde, and radioactivity. Major emphasis is directed to health effects of various indoor air contaminants. Basic components of such pollution are described with some emphasis on possible synergistic effects within contained environments.

Neely, W. Brock. <u>Chemicals in the Environment: Distribution,</u>
<u>Transport, Fate, Analysis</u>. Pollution Engineering and
Technology, Volume 13. New York: Dekker, 1980.

Examines environmental pathways taken by toxic,
hazardous chemicals and various models (mathematical,
computer, compartmental) which may be used to predict
and understand pollutant properties and behavior.
Discusses the various compartments of the atmosphere,
hydrosphere, lithosphere and their chemical/physical
properties. Provides data for atmospheric nitrogen
emissions and background levels, constituents (major
and minor) of chemical species in the troposphere, and
atmospheric modelling parameters.

Organisation for Economic Co-Operation and Development.
<u>Control Technology for Nitrogen Oxide Emissions from</u>
<u>Stationary Sources</u>. Washington, D.C.: OECD Publica-
tions and Information Center (distributor), 1983.

Reviews nitrogen oxide emission pollution effects,
characteristics, and sources (environmental and anthropogenic). Delineates control mechanisms and processes
such as catalytic and non-catalytic reduction and dry
process controls.

Organisation for Economic Co-Operation and Development.
<u>Photochemical Smog: Contribution of Volatile Organic</u>
<u>Compounds</u>. Washington, D.C.: OECD Publications and
Information Center (distributor), 1982.

Provides emissions data for volatile organic compounds
in OECD countries and their contributions to photochemical smog. Enumerates the toxic components of smog
and their natural and anthropogenic sources.

Organisation for Economic Co-Operation and Development,
Group on Energy and Environment. <u>Coal: Environment</u>
<u>Issues and Remedies</u>. Washington, D.C.: OECD Publica-
tions and Information Center (distributor), 1983.

The ultimate aim of this book is to recommend policies
on the environmentally responsible use of coal. It
also provides an overview of the nature of coal and its
contamination of the environment through mining, combustion, and disposal. Reviews atmospheric effects of
sulfur oxides, nitrogen oxides, acid deposition, trace
elements, and particulates.

Orlemann, J. A., et al. <u>Fugitive Dust Control Technology</u>.
Pollution Technology Review, No. 96. Park Ridge, NJ:
Noyes, 1983.

Focuses on the source identification and control
measures for fugitive dust emissions. Presents related
data for thirty industrial categories providing information about sources, process descriptions, characteristics of particles, emission factors, and the availability of control techniques.

Pucknat, A. W., ed. Health Impacts of Polynuclear Aromatic
    Hydrocarbons. Environmental Health Review, No. 5.
    Park Ridge, NJ: Noyes, 1981.

    Explores polynuclear aromatic compounds (PNA) and their
    atmospheric chemistry. Describes PNA characteristics
    in the environment. Concentrations, fate, and
    chemical/physical reactivity in the atmosphere are
    reviewed. Compiles health effects (exposure, toxicity,
    carcinogenicity, teratogenicity, and mutagenicity).

Purves, David. Trace-Element Contamination of the Environ-
    ment. Rev. ed. Fundamental Aspects of Pollution
    Control and Environmental Science, Volume 7. New York:
    Elsevier, 1985.

    Summarizes the sources, means of dispersal, deposi-
    tional sites and effects on plants and animals of toxic
    metals (primarily lead, mercury, and cadmium) and non-
    metals (boron, fluorine, and others).

Rau, John G. and David C. Wooten, eds. Environmental Impact
    Analysis Handbook. New York: McGraw-Hill, 1980.

    Written in response to the U.S. National Environmental
    Policy Act and covers environmental impact analyses
    (socioeconomic, air, noise, energy, water, vegetation,
    and wildlife). Chapter three examines air quality
    impact analysis. Covers several fundamentals of air
    pollution: the nature of air pollution sources (point,
    line, area, industry types, solid waste, external com-
    bustion, etc.); types of pollutants are enumerated, and
    considerable attention is devoted to the mechanics and
    significance of air pollution modelling. This 165 page
    chapter has an 118 item bibliography.

Rodhe, H., et al. Tropospheric Chemistry and Air Pollution.
    WMO, No. 583. Technical Note/World Meteorological
    Organization, No. 176. Geneva: Secretariat of the
    World Meteorological Organization, 1982.

    Examines the atmospheric pollutants and transport,
    dispersion, rainfall, and stagnation; various modes of
    chemical atmospheric transformations and processes
    (such as gas phase and liquid phase transformation,
    hydroxyl chemistry, formation of ozone, sulphate and
    nitrate formation); types of meteorological dispersion

models; and budgets (global and regional) of trace
constituents in the atmosphere.

Schwartz, Stephen E., ed. Trace Atmospheric Constituents:
    Properties, Transformations, and Fates. Advances in
    Environmental Science and Technology, Volume 12. New
    York: Wiley, 1983.

Analyzes the presence, concentrations, metamorphosis,
chemistry, and transport of atmospheric trace
constituents. Directs attention to the presence of
sulphur and nitrogen compounds in the atmosphere from
their aqueous-phase chemistry, including their
conversion to nitric and sulfuric acids. Also examines
dispersion of these compounds, carbon particulates, and
ozone together with their effects over a range of time-
scales and distances. Includes both theoretical and
field investigations of the properties and abundance of
trace constituents in the atmosphere.

Stern, Arthur C., et al. Fundamentals of Air Pollution.
    2nd ed. Orlando, FL: Academic, 1984.

Describes the nature of air pollution (scope, sources,
primary/secondary pollutants) and its effects on
health; vegetation; and the atmospheric, soil, and
water environments. It contains useful information on
atmospheric chemistry and categorization of hazardous
wastes (toxic; flammable; explosive; irritating; corrosive; radioactive; bioaccumulative; and carcinogenic,
teratogenic, and mutagenic substances), and the need to
avoid atmospheric contamination by such wastes.

Thain, William. Monitoring Toxic Gases in the Atmosphere
    for Hygiene and Pollution Control. New York: Pergamon,
    1980.

Discusses the nature of toxic chemical gases in the
atmosphere and methods and techniques available for
detecting, monitoring, and sampling these hazardous
gases. Reviews sampling/collection methods and tools,
types of monitoring, and techniques of measurement
(electrochemical reactors, spectral absorption, solid
state sensors, etc.)

Wadden, Richard A. and Peter A. Scheff. Indoor Air Pollu-
    tion: Characterization, Prediction and Control.
    Environmental Science and Technology Series of Mono-
    graphs, Textbooks, and Advances. New York: Wiley,
    1983.

Examines indoor/outdoor sources of indoor pollution.
Presents models for prediction (mixing factor, empiri-
cal, source and sink term models) of indoor air

quality. Describes a number of hazardous air pollutants (carbon monoxide, nitrogen oxide, asbestos,
formaldehyde) and covers health effects. Also examines
indoor air quality standards and methods for improvement.

Weber, Erich, ed. Air Pollution: Assessment Methodology and
    Modeling. NATO Challenges of Modern Society, Volume 2.
    New York: Plenum, 1982.

    Discusses technical, conceptual, and historical aspects
    of air pollution modelling and air quality management
    systems. Especial attention is devoted to Gaussian
    plume models, air pollution emission inventory and
    projection. Includes a glossary of air pollution assess-
    ment methodology and modelling terms.

Whitten, Robert C. and Sheo S. Prasad, eds. Ozone in the
    Free Atmosphere. New York: Van Nostrand Reinhold,
    1985.

    Consists of a baseline study of ozone including the
    importance, distribution, photochemistry, transport,
    and tropospheric presence of ozone. Concluding chap-
    ters examine the perturbations of stratospheric ozone
    including anthropogenic sources and the biological,
    climatic effects of such perturbations and reductions.

Periodicals

Atmospheric Environment: an International Journal. Pergamon.
    Monthly, published since 1967.

    Contains research and review articles, short communica-
    tions, letters, book reviews, and discussions of pre-
    liminary research results concerning the effects of
    chemical and particulate substances on the atmospheric
    environment. Although some articles concern the
    effects of airborne pollutants on terrestrial and
    floral ecosystems, many articles pertain to the
    description and analysis of source pollutants on the
    atmosphere, atmospheric dispersion and transport of
    pollutants, and modelling and measurement of atmos-
    pheric pollutants.

Environmental Pollution. Series A: Ecological and
    Biological. Elsevier. Monthly, published since 1970.

    Contains research articles and book reviews pertaining
    to the biological and ecological effects of pollutants.

Air pollution articles and reviews pertain to the ecological ramifications of atmospheric pollution.

Environmental Pollution. Series B: Chemical and Physical. Elsevier. Eight times per year, published since 1980.

Contains research articles by researchers (predominately from outside of the United States) covering the chemical, physical aspects of environmental pollution. The articles describe the effects of dangerous pollutants on various portions of the environment with a fair portion devoted to the effects of these pollutants on the air. Air pollution articles cover topics such as the analysis, assessment, monitoring, and modelling of atmospheric pollutants.

Environmental Spectrum. Cooperative Extension Service, Cook College, Rutgers University. Bimonthly, published since 1968. Formerly **Air Pollution Advisory**.

This newsletter monitors air pollution information. Provides current listing of publications pertinent to air pollution concerns.

Journal of Environmental Science and Health. Part A: Environmental Science and Engineering. Dekker. Eight times per year, published since 1968.

The focus is on providing information needed to make engineering solutions to various environmental problems largely caused by pollution. Air pollution articles concern the monitoring, effects, and analysis of hazardous substances. Also contains announcements and book reviews.

Journal of Hazardous Materials. Elsevier. Quarterly, published since 1975.

Contains review and research articles, brief communications, book reviews and meeting reports on the subjects of the management, handling, disposal and risk assessment of hazardous substances and materials. Although the primary emphasis of the articles is on these responses to the environmental hazards, emphasis is also directed toward the nature and properties of dangerous pollutants. Atmospheric topics are well represented.

Journal of the Air Pollution Control Association. Air Pollution Control Association. Monthly, published since 1951.

Presents technical and review papers, summary articles
and current research pertaining to various air pollution subjects: control, monitoring, measurement, modelling, sources, analysis, behavior, toxicity, and
trends. Especial attention is directed toward toxic/
hazardous waste management and treatment. "Current
literature" department contains very timely book
reviews and listings of articles from journals of
related interest.

Proceedings

Berg, G. G. and H. D. Maillie, eds. Measurement of Risks.
Proceedings of the 13th Rochester International
Conference on Environmental Toxicity, University of
Rochester, New York, June 2-4, 1980. Environmental
Science Research, Volume 21. New York: Plenum, 1981.

Includes several papers on the biological risks
associated with exposure to dangerous atmospheric
pollutants: "Risks of contaminated air and drinking-
water," "Problems and possibilities of determining the
carcinogenic potency of inhalable fibrous dusts and
polycyclic aromatic-hydrocarbons," and "Use of tests of
pulmonary-function to measure effects of air-pollu-
tants."

Bonsignore, G. and G. Cumming, eds. The Lung and its
Environment. Proceedings of the Symposium, Erice,
Sicily, June 16-21, 1980. Ettore Majorana Inter-
national Science Series: Life Sciences, Volume 6. New
York: Plenum, 1982.

Contains papers that study various health problems
connected with the lung and some environmental causes
associated with lung and respiratory disorders. Papers
pertaining to air related causes: "The particulates in
the atmosphere and their intra-pulmonary deposition,"
"Pollution in contained environments," and "The
metabolism of chemicals by the lung."

Cooke, M. and A. J. Dennis, eds. Polynuclear Aromatic
Hydrocarbons: Formation, Metabolism and Measurement.
Proceedings of the 7th International Symposium, Colum-
bus, Ohio, October 26-28, 1982. Columbus: Batelle
Press, 1983.

Pertains to the sources, measurement, analysis,
formation and distribution of polynuclear aromatic
hydrocarbons and their mutagenic, carcinogenic and
biologic aspects. Some papers discussing atmospheric
aspects include: "Formation mechanisms of PAH in soot

flames," "Polynuclear aromatic compounds in fluidized-bed combustion of bituminous coals, subbituminous coal and oil shale," and "A generalized model for estimating the concentration of PAH in urban air."

Dunn, C. D. R., et al., eds. Environmental Pollution and Man. Proceedings, Cape Town, South Africa, April 30-May 4, 1984. Immunology and Hematology Research Monographs, No. 3. Garden City, NY: Image Books, 1984.

Contains several papers specifically referring to the effects of air pollutants: "Environmental pollution and the lung," "Airway reactivity and environmental pollutants," and "Background and non-particulate pollutants." Also includes examinations of ethylene-oxide, agrochemicals, and health hazards resulting from employment in the chemical industry.

Engel, A. J., et al., eds. Emission Control from Stationary Power Sources: Technical, Economic and Environmental Assessments. Proceedings of the 87th National Meeting of the American Institute of Chemical Engineers, Houston, Texas, and Boston, Massachusetts, 1979. AICHE Symposium Series, No. 201, 1980, Volume 76. New York: American Institute of Chemical Engineers, 1980.

Attention, in these selected papers, is directed toward the effects and control of toxic pollutants released into the atmosphere by production of power. Various approaches covered: urban air particulate organic-carbon concentrations, possible health effects of coal fly-ash, trace-metal chemical analysis of coal fly-ash, and studies of coal combustion and the resulting mineral matter vaporization.

Englund, H. M., ed. Hazardous Waste Incineration: Selected Papers from an APCA Annual Meeting. Proceedings of the 77th Annual Meeting of the Air Pollution Control Association, San Francisco, California, June 24-29, 1984. Pittsburgh: Air Pollution Control Association, 1984.

Pertains to the control of hazardous substances released into the atmosphere via hazardous waste incineration. Papers concerning the nature of these releases into the atmosphere: "Modifications to decrease fine particulate-emissions from a hazardous-waste incinerator," and "Sampling and analytical methods for assessing toxic and hazardous organic emissions from stationary sources.

Frederick, E. R., ed. <u>Specialty Conference on Measurement</u>
<u>and Monitoring of Non-Criteria (Toxic) Contaminants in</u>
<u>Air</u>. Proceedings of the Conference, Chicago, Illinois,
March 22-24, 1983. Pittsburgh: Air Pollution Control
Association, 1983.

Concerns the character, analysis, measurement, sampling
and monitoring of toxic air pollutants (such as
volatile organic compounds, particulates, benzo-a-
pyrene, and PCBs). Papers about the nature of such
airborne contaminants include: "Toxic organic air
pollutants in the Arctic," "Characterization of ambient
air around abandoned chemical waste dump sites,"
"Coastal impacts of PCB incineration operations in the
Gulf of Mexico," and "Population exposure to hazardous
air pollutants from waste combustion in industrial
boilers and RCRA regulated boilers."

Goldberg, E. D., ed. <u>Atmospheric Chemistry</u>. Proceedings of
the Dahlem Workshop on Atmospheric Chemistry, Berlin,
Federal Republic of Germany, May 2-7, 1982. Physical
and Chemical Sciences Research Reports, Volume 4.
Dahlem Workshop Report. New York: Springer-Verlag,
1982.

Several papers from this workshop on atmospheric chemistry focus on the effects of hazardous elements on the
atmosphere: "The production and fate of reduced volatile species from toxic environments," "The production
and fate of reduced volatile molecular-species in the
environment: metals and metalloids," and "Tropospheric
gases, aerosols and photochemical reactions."

Grace, J., et al., eds. <u>Plants and Their Atmospheric</u>
<u>Environment</u>. Proceedings of the 21st Symposium of the
British Ecological Society, University of Edinburgh,
Edinburgh, Scotland, March 26-30, 1979. Boston: Black-
well Scientific, 1981.

Two papers pertain to the effects of air pollutants on
plant biology: "The exchange of carbon-dioxide and air-
pollutants between vegetation and the atmosphere," and
"Turbulent transfer of sulfur-dioxide to cereals: a
case study."

Harrison, Roy M., ed. <u>Pollution: Causes, Effects, and</u>
<u>Control</u>. Papers presented and organized by the Contin-
uing Education Committee of the Royal Society Chemistry,
University of Lancaster, September 13-15, 1982.
London: Royal Society of Chemistry, 1983.

Six papers pertain to air pollution (with emphasis on
toxic, hazardous pollutants) concerning chemical analysis of key atmospheric pollutants, air pollution

routes, atmospheric dispersion and modelling of toxic air pollutants, and control systems for pollution emissions.

Hartwig, S., ed. Heavy Gas and Risk Assessment. Proceedings of the Symposium on Heavy Gas, Frankfurt, Federal Republic of Germany, September 3-4, 1979. Boston: Kluwer Boston, 1980.

Examines the subject of environmental risks from heavy gas release and dispersion into the atmosphere by studying modelling and risk analysis approaches. Papers highlighting this topic: "Models for evaluating the probabilities and consequences of possible outcomes following the releases of toxic or flammable gases," "Dispersion of gasoline in road tunnels after release of major quantities," and "Analysis of the potential explosion effects of flammable gases during short time release into the atmosphere."

Hartwig, S., ed. Heavy Gas and Risk Assessment II. Proceedings of the 2nd Symposium, Frankfurt, Federal Republic of Germany, May, 1982. Boston: Kluwer Boston, 1983.

Analyzes the process and substance of environmental risk assessment with regard to heavy gas atmospheric release and dispersion. Examples of papers discussing risks such gases pose to the atmosphere include: "Further analysis of catastrophic LNG spill vapor dispersion," "Entrainment mechanisms of air in heavy gas clouds," and "Examples of analyses of gas cloud explosion hazards."

Hazardous Materials Controls Research Institute. National Conference on Management of Uncontrolled Hazardous Waste Sites. Proceedings of the National Conference, Washington, D.C., November 29-December 1, 1982. Silver Spring, MD: Hazardous Materials Controls Research Institute, 1982.

Primarily concerned with hazardous sites monitoring and control, several papers pertain specifically to atmospheric considerations: "Hazardous waste incineration," "Air-pollution problems on uncontrolled hazardous waste sites," and "Air modeling and monitoring for site evacuation."

Heavy Metals in the Environment. Proceedings of the 4th International Conference, Heidelberg, September, 1983. Edinburgh: C.E.P. Consultants, 1984.

Contains several papers concerning the characteristics, behavior, analysis, and concentrations of atmospheric heavy metals. For example: "Mercury levels in the air of a Mediterranean area," "Emission of lead by highway traffic," "The atmospheric emission of trace elements from anthropogenic sources in Europe," and "Bromine to lead ratios in atmospheric particles at urban and rural sites."

Institution of Chemical Engineers. _Assessment and Control of Major Hazards_. Proceedings of the Symposium, Manchester, England, April 22-24, 1985. Institution of Chemical Engineers Symposium Series, No. 93. EFCE Publication Series, No. 322. Rugby, England: Pergamon, 1985.

Examines hazards inherent in the handling, transport, and storage of toxic, dangerous substances and materials. Risk assessment and remedial procedures in accidents are emphasized. Papers of pertinence to hazards and the air: "The HSE program of research and model development of heavy gas dispersion," "An acoustic model for predicting the overpressures caused by the deflagration of a ground lying vapor cloud," and "The atmospheric dispersion of heavy gases--an update."

Institution of Chemical Engineers. _Assessment of Major Hazards_. Proceedings, University of Manchester, Institute of Science and Technology, Manchester, England, April 14, 1982. Institution of Chemical Engineers Symposium Series, No. 71. EFCE Publication Series, No. 25. Rugby, England: Institution of Chemical Engineers, 1982.

Papers concern the definition and assessment of toxic hazards. Of especial relation to the effects of such hazards to the air are the topics of heavy, dense gas dispersion in the atmosphere; superheated fluid atmospheric discharge; hazardous material evaporation; and the effects and assessments of fireball and explosions.

International Colloquium on Atmospheric Pollution. 15th, Paris, France, May 4-7, 1982. In: _Science of the Total Environment_ 23 (1982).

Examines air pollution from a number of viewpoints such as forecasting, modelling, dispersion rates, ozone concentrations, and transport. Examples of papers related to harmful effects: "A study of physicochemical characteristics of respirable dust in an Indian coal-mine," "Detection and impact prediction of hazardous substances released to the atmosphere," and "Atmospheric contamination of archeological monuments in the Agra region (India)."

Keith, L. H., et al., eds. <u>Chlorinated Dioxins and Dibenzo-furans in the Total Environment II</u>. Proceedings of the Symposium, at the 186th Meeting of the American Chemical Society, Kansas City, Missouri, August 28-September 2, 1983. Boston: Butterworth, 1985.

Contains papers examining disparate approaches to analyzing the presence of chlorinated dioxins and dibenzofurans in the environment. Papers pertaining to the atmospheric environment: "Emissions of polychlor-inated dibenzo-para-dioxins and polychlorinated dibenzofurans from a resource recovery municipal incinerator," "Formation of polychlorinated dibenzo-furans and other potentially toxic chlorinated pyroly-sis products in PCB fires," "Levels of chlorinated organics in a municipal incinerator," and "Overview of dioxin formation in gas and solid-phases under munici-pal incinerator conditions."

Kneip, T. J. and P. J. Lioy, eds. <u>Aerosols, Anthropogenic and Natural, Sources and Transport</u>. Proceedings of the Conference, New York, NY, January 9-12, 1979. Annals of the New York Academy of Sciences, Volume 338. New York: New York Academy of Sciences, 1980.

Examines sources, formation, size, distribution, chemistry, transport, and effects of atmospheric aerosols. Papers showing a variety of approaches to this topic: "Trace-element abundances and chemistry of atmospheric aerosols," "Mesoscale and synoptic scale transport," "Arctic haze perturbation of the polar radiation budget," and "A review of urban plume studies."

Koziol, M. J. and F. R. Whatley, eds. <u>Gaseous Air Pollutants and Plant Metabolism</u>. Proceedings of the 1st International Symposium, Oxford, England, August 25, 1982. Boston: Butterworth, 1984.

Examines the direct biological effects of pollutant gases on plant metabolism, respiration, membranes, and cell ultrastructure. Some papers representative of these topics: "Effects of gaseous air pollutants on stromal reactions," "Emissions of volatiles from plants under air pollution stress," and "Defining gaseous pollution problems in North America."

Liu, Benjamin Y. H., et al., eds. <u>Aerosols: Science, Technology, and Industrial Applications of Airborne Particles</u>. Proceedings of the 1st International Aerosol Conference, September 17-21, 1984, Minneapolis, Minnesota. New York: Elsevier, 1984.

Discusses aerosols with reference to deposition, measurement, sampling, characterization, generation, and indoor characteristics. Some papers pertaining to atmospheric aerosols include: "Mutagenic activity and composition of organic particulates from domestic heating and automobile traffic," "Effect of diesel vehicles on visibility in California," "The optical effect of fine-particle carbon on urban atmospheres," and "Particle emissions and concentrations of trace elements by a coal fired power plant."

London, J., ed. <u>Quadrennial International Ozone Symposium</u>. Proceedings of the Symposium, Boulder, Colorado, August 4-9, 1980. Boulder, CO: National Center for Atmospheric Research, 1981.

Analyzes the subject of ozone with reference to measurement, estimation, effects, assessment, analysis, and meteorology. Some papers concerning atmospheric pollution: "The atmospheric photo-oxidation cycle and the influence of troposphere pollution on ozone," "The observation of long-lived trace gases in the stratosphere," and "Ozone effects on climate."

Long, L. A. and Glenn E. Schweitzer, eds. <u>Risk Assessment at Hazardous Waste Sites</u>. ACS Symposium Series, No. 204. Washington, D.C.: American Chemical Society, 1982.

Based on a symposium sponsored by the American Chemical Society Committee on Environmental Improvement at the 183rd Meeting of the American Chemical Society, Las Vegas, Nevada, March-April 1982. Focuses on evaluating the hazardous waste site risk with emphasis on the effect on groundwater in the United States. However, the evaluation of effects to the air are also discussed. Stresses risk areas, evaluation, and location of sites.

<u>Long-range Transport of Airborne Pollutants and Acid Rain Conference</u>. Papers Presented. Albany, New York, April 27-30, 1981. In: <u>Water, Air and Soil Pollution</u> 18, (1982).

Concerns the effects and nature of atmospheric pollution and acid precipitation/deposition. Papers pertaining to air pollutants: "Source regions of summertime ozone and haze episodes in the eastern United States," "Estimates for the long-range transport of air pollution," and "Impact of New York State emission sources on class-1 areas."

Mackay, D., et al., eds. <u>Physical Behavior of PCBs in the Great Lakes</u>. Proceedings of the Conference, Toronto, Ontario, Canada, December, 1981. Ann Arbor, MI: Ann Arbor Science, 1983.

Although most papers pertain to the topic of PCB poisoning of water, several papers concern their effects in
the air: "Evidence for the atmospheric flux of polychlorinated-biphenyls to Lake-Superior," "PCBs in the
Lake-Superior atmosphere 1978-1980," and "Survey of
polychlorinated-biphenyls in ambient air across the
Province of Ontario."

Moo-Young, M., et al., eds. <u>Waste Treatment and Utilization:
Theory and Practice of Waste Management</u>. Proceedings
of the 2nd International Symposium, Waterloo, Ontario,
Canada, June 18-20, 1980. Elmsford, NY: Pergamon,
1982.

Provides analysis of the management, toxicity, effects,
and analysis of hazardous wastes. Papers pertain
primarily to water related topics, but several apply to
air concerns: "Air pollution control at a tar sands
plant," "Long-range transport implications for control
of ambient air-quality," and "On the deposition of
aerosols: a review and recent theoretical and experimental results."

Murray, Frank, ed. <u>Fluoride Emissions: Their Monitoring and
Effects on Vegetation and Ecosystems</u>. First Austral-
asian Fluoride Workshop, Sydney, Australia, August 31-
September 1, 1981. New York: Academic, 1982.

Provides analyses of fluoride atmospheric emissions,
concentrations, modelling predictions, monitoring
techniques, and effects from coal fired plants and
smelting processes. Also examines the topic of
problems associated with the types of toxicity and
growth influencing effects of atmospheric fluoride
deposition on plants and its introduction into
ecosystems.

<u>National Conference on Management of Uncontrolled Hazardous
Waste Sites</u>. Proceedings of the National Conference,
Washington, D.C., November 29-December 1, 1982. Silver
Spring, MD: Hazardous Materials Control Research
Institute, 1982.

Examines a multiplicity of approaches to the analysis,
management, handling, and risk assessment of hazardous
wastes. Although the papers concern many aspects of
such wastes, several papers relate to atmospheric
topics, including: "Air pollution problems of
uncontrolled hazardous waste sites," "Air emission
monitoring of hazardous waste sites," and "Hazardous
waste incineration."

National Waste Processing Conference, 1982. <u>Meeting the
Challenge</u>. 10th Biennial Conference, New York, NY, May
2-5, 1982. New York: American Society of Mechanical
Engineers, 1982.

Contains papers examining waste processing and handling
such as energy recovery, air pollution control, incineration, and sampling and analysis. Some of the papers
pertaining to atmospheric aspects of hazardous wastes
are "Refactories for hazardous waste incineration: an
overview," "Emissions and emission control in modern
municipal incinerators," and "The formation of nitrogen
oxides in hazardous waste incineration."

National Waste Processing Conference, 1984. <u>Engineering,
the Solution</u>. 11th Biennial Conference, Orlando,
Florida, June 3-6, 1984. New York: American Society of
Mechanical Engineers, 1985.

Includes papers which discuss hazardous waste
processing and air pollution topics: "Considerations
regarding incineration of industrial plastic and hazardous wastes," "Combustion of scrap oil for steam generation," and "Overview of hazardous waste incineration
technology."

Ooms, Gijsbert and Hendirk Tennekes, eds. <u>Atmospheric
Dispersion of Heavy Gases and Small Particles</u>. International Union of Theoretical and Applied Mechanics.
Symposium, Delft, The Netherlands, August 29-September
2, 1983. New York: Springer-Verlag, 1984.

Includes papers concerning the mechanisms, modelling,
entrainment and dispersion of heavy, dense pollutant
gases with some attention devoted to particle
dispersion.

Robinson, F. A., ed. <u>Environmental Effects of Utilising
More Coal</u>. Proceedings of a Conference Organized by
the Council of Environmental Science and Engineering,
Royal Geographical Society, London, December 11-12,
1979. London: Royal Society of Chemistry, 1980.

Papers cover the effects of toxic air pollutants from
coal burning on the atmosphere including the carcinogenic, meteorologic, and biologic effects.

Schneider, T. and L. Grant, eds. <u>Air Pollution by Nitrogen
Oxides</u>. Proceedings of the U.S.-Dutch International
Symposium, Maastricht, The Netherlands, May 24-28,
1982, organized by the Ministry of Health and Environmental Protection, The Netherlands, and the Environmental Protection Agency. Studies in Environmental
Science, 21. New York: Elsevier, 1982.

This 1100 page volume of conference papers covers many
diverse aspects of current research and information on
nitrogen oxide pollution. Of relevance to its effects
in the air are papers concerning the global cycle,
formation, inventory of sources, transport, transformation, physical atmospheric processes of nitrogen
oxide, and its relation to acidic deposition. The
papers are generally well referenced with exceptionally
current bibliographies.

Schryer, David R., ed. Heterogeneous Atmospheric Chemistry.
    Geophysical Monograph, 26. Washington, D.C.: American
    Geophysical Union, 1982.

    Contains papers presented at (or submitted to) a con-
    ference, "Multiphase Processes, Including Heterogeneous
    Catalysis: Its Importance to Atmospheric Chemistry,"
    Albany, New York, June 29-July 3, 1981. Emphasis is on
    multiphasic, heterogeneous processes on air quality,
    an area now beginning to be more fully researched and
    appreciated. Papers of particular relevance: "Sulphate
    in the atmospheric boundary layer: concentration and
    mechanisms of formation," "The relative importance of
    various urban sulfate aerosol production mechanisms-a
    theoretical comparison," and "Soot catalyzed atmospher-
    ic reactions."

Shriner, D. S., et al., eds. Atmospheric Sulfur Deposition:
    Environmental and Health Effects. Proceedings of the
    2nd Life Sciences Symposium on Potential Environmental
    and Health Consequences of Atmospheric Sulfur
    Decomposition, Gatlinburg, Tennessee, October 14-18,
    1979. Ann Arbor, MI: Ann Arbor Science, 1980.

    Provides coverage of topics of significance to the
    subject of sulfurous content of environmental atmos-
    pheres: health effects; emission rates, characteris-
    tics, and sources; ecological effects on forests,
    waters, and soils; atmospheric interactions; and atmos-
    pheric interfaces with terrestrial boundaries.

Sorsa, Marja and Harri Vainio, eds. Mutagens in Our
    Environment. Proceedings of the Twelfth Annual Meeting
    of the European Environmental Mutagen Society, Espoo,
    Finland, June 20-24, 1982. Progress in Clinical and
    Biological Research, Volume 109. New York: Alan R.
    Liss, 1982.

    Papers primarily concerned with the biological effects
    of toxic mutagens. Three papers concern specifically
    the effects and sources of genotoxic, mutagenic meter-
    ials in the air.

Stober, W. and D. Hochrainer, eds. <u>Aerosols in Science,</u>
<u>Medicine, and Technology: Aerosols in and from Indus-</u>
<u>trial Processes</u>. Proceedings of the 9th Conference,
Duisburg, Federal Republic of Germany, September 23-25,
1981. Schmallenberg, Germany: Gesellschaft für
Aerosolforschung, 1981.

Includes papers concerning the nature, sources,
measurement, chemistry, and biological and effects of
aerosols. Some papers concerning atmospheric aspects:
"Comments on the implications for health of the
physical and chemical characteristics of airborne
particles," "Dispersion and deposition of lead from
motor exhausts and its effect on vegetables grown along
a highway," and "Aerosols generated at an open-pit
coal-mine."

Swann, Robert L. and Alan Eschenroeder, eds. <u>Fate of</u>
<u>Chemicals in the Environment: Compartmental and Multi-</u>
<u>media Models for Predictions</u>. ACS Symposium Series,
No. 225. Washington, D.C.: American Chemical Society,
1983.

Based on a symposium sponsored by the ACS Division of
Pesticide Chemistry, American Chemical Society, Kansas
City, Missouri, September 12-17, 1982. Papers of especial relevance to this topic: "Environmental fate and
transport at the terrestrial-atmospheric interface,"
and "Modeling of human exposure to airborne toxic
materials."

<u>Toxic Materials in the Atmosphere: Sampling and Analysis</u>. A
Symposium Sponsored by ASTM Committee D-22 on Sampling
and Analysis of Atmospheres, Boulder, Colorado, August
2-5, 1981. ASTM Special Technical Publication, 786.
Philadelphia: ASTM, 1982.

These conference papers explore the topic of ambient
air sampling and analysis in the workplace with
specific attention to air monitoring instrumentation,
passive monitors, and techniques for chemical analysis
of toxic, hazardous substances.

Tucker, Richard E., et al., eds. <u>Human and Environmental</u>
<u>Risks of Chlorinated Dioxins and Related Compounds</u>.
Proceedings of the International Symposium on
Chlorinated Dioxins and Related Compounds, Arlington,
Virginia, October 25-29, 1981. Environmental Science
Research, Volume 26. New York: Plenum, 1983.

Approaches the topic of polychlorinated dibenzo-p-
dioxins (PCDDs) and polychlorinate dibenaofurans (PCDFs)
from several perspectives. Discusses their toxicity and
resulting effects in human, laboratory, natural and biologic environments and their risk, assessment and man-

agement. Several articles concern monitoring and analysis of airborne PCDDs and PCDFs from furnace, wood fuel,
incineration, and other combustion sources.

Turoski, V., ed. Formaldehyde: Analytical Chemistry and
    Toxicology. Proceedings of the Symposium of the 187th
    Meeting of the American Chemical Society, St. Louis,
    Missouri, April 8-13, 1984. Advances in Chemistry
    Series, Volume 210. Washington, D.C.: American
    Chemical Society, 1985.

Versino, B. and G. Angeletti, eds. Physico-chemical
    Behaviour of Atmospheric Pollutants. Proceedings of
    the 3rd European Symposium, Varese, Italy, April 10-12,
    1984. Boston: Kluwer Boston, 1984.

    Contains over 60 papers concerning air pollutant compo-
    sition and activities. Various topics are covered
    including acidic deposition; sampling, analysis, and
    measurement of air pollutants; and toxic effects as
    measured in plants, snow and glaciers. Specific papers
    of relevance to atmospheric sources and effects:
    "Photochemical air pollution in Denmark: weekday
    effects and evidence of large-scale formation,"
    "Characterization of suspended particulate matter in a
    lead smeltery area," and "Identification of the sources
    of particulate polycyclic aromatic-hydrocarbons in the
    urban atmosphere."

Waters, M. D., et al., eds. Short-Term Bioassays in the
    Analysis of Complex Environmental Mixtures III.
    Proceedings of the Symposium, Chapel Hill, North Caro-
    lina, January 25-27, 1982. Environmental Science
    Research, Volume 27. New York: Plenum, 1983.

    Approaching the subject of toxic substance biological
    impact, this symposium presented several papers on
    atmospheric pollution of biologically dangerous sub-
    stances: "Mutagens in airborne particulate pollutants
    and nitro-derivatives produced by exposure of aromatic-
    compounds to gaseous-pollutants," "Application of muta-
    genicity tests for detection and source assessment of
    genotoxic agents in the rubber work atmosphere," and
    "Some aspects of mutagenicity testing of the particu-
    late phase and the gas-phase of diluted and undiluted
    automobile exhaust."

Waters, M. D., et al., eds. Short-Term Bioassays in the
    Analysis of Complex Environmental Mixtures IV.
    Proceedings of the Symposium, Chapel Hill, North Caro-
    lina, March 27-29, 1984. Environmental Science
    Research, Volume 32. New York: Plenum, 1985.

Includes papers on testing and analysis of air
pollutant matter with respect to toxic, mutagenic
properties. In regard to such pollutants in the air,
several papers apply: "Mutagenicity studies of New
Jersey amibient air particulate extracts," "The effect
of atmospheric transformation upon the bacterial
mutagenicity of airborne organics," and "Generation and
characterization of complex gas and particle mixtures
for inhalation toxicologic studies."

Wispelaere, C. De., ed. Air Pollution Modeling and its
    Application. Proceedings of the 11th International
    Technical Meeting. NATO Challenges of Modern Society,
    Volume 1. New York: Plenum, 1981.

    Papers of direct pertinence to this subject: "Physical
    modeling of 40 cubic meter long spills at China Lake,
    California," "Application of a photochemical dispersion
    model to the Netherlands and its surroundings," and
    "Entrainment through the top of a heavy gas cloud."

Wispelaere, C. De., ed. Air Pollution Modeling and its
    Application III. Proceedings of the 13th International
    Technical Meeting. NATO Challenges of Modern Society,
    Volume 5. New York: Plenum, 1984.

    Papers concerning varying approaches to air pollution
    modelling include: "Simulation of atmospheric effects
    of industrial heat releases," "Plume modeling from
    meteorological Doppler radar data," "A new Monte Carlo
    scheme for simulating Lagrangian particle diffusion
    with wind shear effects," and "Numerical simulations of
    atmospheric releases of heavy gases over variable
    terrain."

Wispelaere, C. De., ed. Air Pollution Modeling and its
    Application IV. Proceedings of the 14th International
    Technical Meeting. NATO Challenges of Modern Society,
    Volume 7. New York: Plenum, 1985.

    Over 40 papers pertain to a diverse array of air
    pollution models: numerical-simulation, long-range
    transport, wet deposition, mesoscale lagrangian puff-
    model, tracer release system, and mass-flow simulation.
    Of especial interest to this topic: "A literature study
    on tracer experiments for atmospheric dispersion
    study," "The relation of urban model performance to
    stability," "Downwind hazard distances for pollutants
    over land and sea," and "Methodologies to validate
    multiple source models."

Reports and Documents

Air Pollution Effects on Plants. April, 1976–June 1984.
(Citations from the NTIS Data Base). (PB85-866564)
Springfield, VA: National Technical Information
Service, 1984.

Citations describe the effects of industrial/automotive
emissions and combustion products on various plants
(forests, vegetation, and agricultural crops). Also
covers the ramifications of acid rain on plants.

Air Pollution Emission Factors. 1970–July 1985. (Citations
from the NTIS Data Base). (PB85-863827) Springfield,
VA: National Technical Information Service, 1985.

References pertain to emission inventories originating
from mobile, stationary and industrial sources reviewed
by kind of air pollution source and particular chemical
constituents.

Air Pollution Tracer Studies in the Lower Atmosphere. 1970–
April 1985. (Citations from the NTIS Data Base).
(PB85-858322) Springfield, VA: National Technical
Information Service, 1985.

Bibliographic citations cover numerous types of tracers
used in the detection of air pollutant pathways in the
lower atmosphere. Reviews air pollution tracer motions
(based on project analyses and materials and techniques
used) emanating from urban, nuclear power, and industrial sources.

Atmospheric Modeling of Air Pollution. November 1981–August
1984. (Citations from the NTIS Data Base).
(PB84-873488) Springfield, VA: National Technical
Information Service, 1984.

Citations cover a variety of air pollution models based
on numerous criteria (atmospheric diffusion, wind,
photochemistry, and stability; topographic features;
wake impacts; depositional features; and heat islands
in urban areas). Includes studies of the modelling of
air pollutants deriving from stationary and mobile
sources.

Automobile Air Pollution: Automotive Fuels. 1970–September
1984. (Citations from the NTIS Data Base).
(PB84-875632) Springfield, VA: National Technical
Information Service, 1984.

References pertain to additives and alternative fuels (gasahol, hydrogen, methane, and natural gas) used to abate automotive air contamination. Studies the nature of gasoline air pollution and air improvements made by additives.

Benzene Toxicity. 1978-November 1985. (Citations from the NTIS Data Base). (PB86-852597) Springfield, VA: National Technical Information Service, 1985.

Citations primarily describe the biological aspects of the toxicity of the substance Benzene. Toxic effects are reviewed from long/short term exposures and methods for the determination of amounts of benzene content in the workplace, atmosphere, and human body, are included.

Cadmium Pollution. November 1980-April 1983. (Citations from the NTIS Data Base). (PB85-861599) Springfield, VA: National Technical Information Service, 1985.

Cadmium Pollution. May 1983-1985. (Citations from the NTIS Data Base). (PB85-861607) Springfield, VA: National Technical Information Service, 1985.

Citations examine cadmium pollution of air and water including analyzing, monitoring, and detecting its presence in the environment. Additionally, it contains citations pertaining to studies of cadmium releases, sources, transport behavior, and toxicity effects.

Carbon Monoxide Toxicity. 1978-1985. (Citations from the NTIS Data Base). (PB86-853132) Springfield, VA: National Technical Information Service, 1985.

References primarily concerned with the biological effects of carbon monoxide. Sources (such as tobacco smoke, air pollution, and occupational environments) of carbon monoxide are studied as well.

Coal Tar Hazards. 1976-February 1985. (Citations from the NTIS Data Base). (PB85-855302) Springfield, VA: National Technical Information Service, 1985.

Citations concern dangers of coal tar in the workplace and environment as well as control technology for releases of coal tar, and industrial coal mining safety topics, and inherent coal dust hazards.

Combustion of Plastics and Elastomers. November 1980–
    November 1985. (Citations from the NTIS Data Base).
    (PB86-852035) Springfield, VA: National Technical
    Information Service, 1985.

    Bibliographic citations pertain to plastic and
    elastomer combustion and the products from their
    combustion. Covers the plastic/elastomer: toxic gases
    resulting from the combustion process, chemistry of
    combustion, and additives used to reduce the associated
    hazards.

Coon, D. Indoor Air Quality in Tight Houses: A Literature
    Review. (DE85900574) Toronto: Ecology House, 1984.

    With the inclusion of Canadian literature sources, this
    review focuses on the overall current understanding of
    housing sealing on indoor air quality, and highlights
    areas in need of further research. The types and
    sources of indoor air pollutants are reviewed in light
    of the ramifications for health and suitability. Also
    includes a summary of measurements of indoor air
    contaminants which have been found in conventional
    houses.

Formaldehyde. June 1970–1985. (Citations from the NTIS Data
    Base). (PB86-853009) Springfield, VA: National Tech-
    nical Information Service, 1985.

    Bibliographic references focus primarily on the health
    complications of formaldehyde which enters the body
    through its inhalation. Analytical, sampling, and
    workplace quantification topics are included.

Hartman, M. W. and G. D. Rives. Literature Review and
    Survey of Emissions from Residential Wood Combustion
    and Their Impact. (PB85-197820) Research Triangle
    Park, NC: Radian Corp., 1985.

    Based on a 53 report literature survey on woodstove
    combustion, ambient air effects, design, and emissions.
    Contains descriptions of woodstove emissions of carbon
    monoxide, hydrocarbons, nitrogen oxides, particulate
    matter, and polycyclic organic material, notably benzo-
    a-pyrene. Also contains U.S. surveys of ambient air
    impacts.

Hunt, W. F., Jr., et al. Compilation of Air Toxic and Trace
    Metal Summary Statistics. (PB84-245273) Research
    Triangle Park, NC: Environmental Protection Agency,
    Office of Air Quality Planning and Standards, 1984.

    Summarizes toxic air data from several sources (Dr. H.
    B. Singh of SRI International, trace metal and benzo-a-

pyrene data from the Storage and Retrieval of Aerometric Data, National Aerometric Data Bank). Contains
tables of Type A pollutants -- potentially carcinogenic
priority chemicals and Type B pollutants -- chemicals
also under consideration. Also incudes an attachment
summarizing benzene and arsenic data. In order to
address health questions of long-term exposure to these
dangerous pollutants, compilations of data on some of
these pollutants focus upon long-term averaging time.

Indoor Air Pollution. June 1976-1982. (Citations from the
    NTIS Data Base). (PB84-851583) Springfield, VA:
    National Technical Information Service, 1983.

Indoor Air Pollution. January 1983-October 1983. (Citations
    from the NTIS Data Base). (PB84-876812) Springfield,
    VA: National Technical Information Service, 1984.

Indoor Air Pollution. November 1983-October 1984.
    (Citations from the NTIS Data Base). (PB85-870459)
    Springfield, VA: National Technical Information
    Service, 1985.

Indoor Air Pollution. November 1984-October 1985.
    (Citations from the NTIS Data Base). (PB85-870467)
    Springfield, VA: National Technical Information
    Service, 1985.

    Citations pertain to possible sources and kinds of
    dangerous indoor air pollutants as: asbestos, carbon
    monoxide, formaldehyde, natural gas, nitrogen oxide,
    particulates, and radon/daughters, as well as water,
    insulation material, and air recirculation.

Industrial Health Hazards Due to Atmospheric Factors.
    October 1978-March 1982. (Citations from the NTIS Data
    Base). (PB83-807438) Springfield, VA: National
    Technical Information Service, 1983.

Industrial Health Hazards Due to Atmospheric Factors. April
    1982-May 1983. (Citations from the NTIS Data Base).
    (PB83-807446) Springfield, VA: National Technical
    Information Service, 1983.

    Abstracts of research reports concerning various
    industrial health concerns based on occupational
    exposure to potentially toxic air pollutants. Some of
    these pollutants include: asbestos, radon, platinum,
    nickel, chromium, manganese, pesticides, polyvinal
    chloride, and silicon dioxide.

Miles, A. J. and J. A. Williams. <u>National Dioxin Study Tier
    4 -- Combustion Sources: Initial Literature Review and
    Testing Options</u>. (PB85-216166) Research Triangle
    Park, NC: Radian Corp., 1984.

Tier 4, National Dioxin Study examines whether
substantial atmospheric releases of dioxin are made by
combustion sources. In preparation for the Tier 4
dioxin testing program, this literature review was
conducted in order to compile a summary of combustion
source dioxin emission data and provide potential
categories of combustion origins for the program.
Appendices contain tables of dioxin emission data base
summarizations and an extensive listing of references.

<u>Pollution Caused by Ammunition Manufacturing</u>. 1970-January
    1983. (Citations from the NTIS Data Base).
    (PB85-867505) Springfield, VA: National Technical
    Information Service, 1985.

<u>Pollution Caused by Ammunition Manufacturing</u>. February
    1983-September 1985. (Citations from the NTIS Data
    Base). (PB85-867513) Springfield, VA: National
    Technical Information Service, 1985.

Various manifestations of pollution (air, solid, and
water) emanating from munition plant operations are
covered in the citations. Includes toxicity reviews as
well as studies of chemicals and approaches to pollution monitoring and control.

<u>Pollution Effects of Beryllium and Beryllium Compounds</u>.
    1970-March 1985. (Citations from the NTIS Data Base).
    (PB85-858694) Springfield, VA: National Technical
    Information Data Base, 1985.

Includes citations pertaining to beryllium and
beryllium derived compounds largely released into the
environment by the burning and transformation
processing of coal. Included are examinations of
toxological aspects, detection of sources, and
pollution control techniques.

<u>Polychlorinated Biphenyls in the Environment</u>. September
    1980-April 1985. (Citations from the NTIS Data Base).
    (PB85-858413) Springfield, VA: National Technical
    Information Service, 1985.

References describe air and water pollution sources of
polychlorinated biphenyls as well as their points of
origin, dispersion methods, and distribution. Toxic
aspects of PCBs, their impact on the environment, and
examinations of the extent of their effects in certain
places are included also.

<u>Toxic and Potentially Toxic Substances: Effects on the
   Integumentary System</u>. 1974-May 1983. (Citations from
   the NTIS Data Base). (PB83-864504) Springfield, VA:
   National Technical Information Service, 1984.

   Citations pertain to toxic/potentially toxic substances
   that affect humans via inhalation or dermal absorption.
   Hazardous substances concerned include: workplace
   chemicals, heavy metals, and pesticides.

<u>Toxicity of Zinc</u>. 1970-June 1985. (Citations from the NTIS
   Data Base). (PB85-862191) Springfield, VA: National
   Technical Information Service, 1985.

   Bibliographic references concern the presence of zinc
   in the environment, the emission sources of atmospheric
   zinc trace elements, health effects from zinc pollution, and approaches to monitoring zinc in the environment.

<u>Toxicology of Ozone</u>. 1970-November 1985. (PB86-850559)
   (Citations from the NTIS Data Base). Springfield, VA:
   National Technical Information Service, 1985.

   Citations cover the subject of the toxic impact of
   ozone in a variety of spheres: human, agricultural, and
   forestal. Additionally, the combinational (or synergistic) effect of ozone and nitrogen/sulphur oxides and
   workplace effects are inlcuded also.

Data Bases

<u>Air/Water Pollution Report</u>
Producer: Business Publishers, Inc.
Time Span: 1983 to present
Coverage: International
Updated: Weekly
Vendor: NewsNet

   Although primarily concerned with legal aspects of air
   and water enviornmental matters, coverage is also
   devoted to news for the air/water pollution control
   industry and reports about research in this field as
   well. File is based upon reporters' verification,
   interviews and statements.

Asbestos Information (ASB)

Producer: Programme de Recherche sur L'Amiante
           de l'Université de Sherbrooke, et al.
Time Span: 1870 to present
Coverage: International
Updated: Quarterly
Vendor: QL Systems

Consists of a file (based upon journal articles, books, government reports, conference proceedings, theses, and bibliographies) pertaining to asbestos, including its effects in the environment, ramifications of asbestos for health and safety, asbestos production and mining, and the legal aspects of asbestos related issues.

Environmental Assessment Data Systems (EADS)

Comprised of four component data bases:

Fine Particle Emissions Information System (FPEIS)
Solid Discharge Data System (SDDS)
Gaseous Emissions Data System (GEDS)
Liquid Effluents Data System (LEDS)

Producer: Environmental Protection Agency. Industrial
           Environmental Research Laboratory. Environmental
           Assessment Data Systems.
Time Span: Unknown
Coverage: United States
Updated: Unknown
Vendor: Environmental Assessment Data Systems;
           limitations and restrictions apply.

These waste stream data bases contain data pertaining to properties, origins, formation, and analysis of these four types of pollutant forms (gaseous, fine particle, liquid, and solid) of discharges resulting from manufacturing, waste managment, and energy production processes.

National Emissions Data Systems (NEDS)

Producer: National Air Data Branch, Environmental
           Protection Agency
Time Span: Unknown
Coverage: United States
Updated: Unknown
Vendor: National Air Data Branch; limitations and
           restrictions apply

Contains atmospheric emissions data from 140,000 point sources from 3300 areas within fifty-five U.S. states and territories.

National Emissions Inventory System
Producer: Environment Canada, Environmental Protection
            Service, Technology Transfer Division, Waste
            Management Branch
Time Span: 1970 to present
Coverage: Canada
Updated: Biannually

Consists of an inventory file of 600 point sources of
human generated point and source air pollutants derived
from over eighty sectors of the Canadian economy.

Storage and Retrieval of Aerometric Data (SAROAD)
Producer: Environmental Protection Agency, National Air
            Data Branch
Time Span: 1958 to present
Coverage: United States
Updated: Unknown
Vendor: National Air Branch; limitations and restrictions
            apply

Consists of air monitoring and air quality data which
is submitted by network of stations that monitor air
samples throughout the United States.

Reviews

Altshulter, A. P.   "Review paper: The Role of Nitrogen
    Oxides in Nonurban Ozone Formation in the Planetary
    Boundary Layer Over N. America, W. Europe and Adjacent
    Areas of Oceans." Atmospheric Environment 20 (1986):
    245-268.

    Reviews current knowledge about the behavior, levels,
    and distribution of nonurban ozone and nitrogen oxides;
    lifetimes of nitrogen oxides; and ozone and the
    generation of ozone within plumes with respect to
    nitrogen oxides. Assesses modelling techniques for
    ozone generation in plumes in urban areas, plumes
    emanating power plants, and ozone transport over long
    distances.

Batifol, Francoise M. and Claude F. Boutron.   "Atmospheric
    Heavy Metals in High Altitude Surface Snows from Mont
    Blanc, French Alps." Atmospheric Environment 18
    (1984): 2507-2515.

    Presents an analysis of heavy metal deposition levels
    in the snows removed from the Alps with the intention
    of determining present and historical levels of heavy
    metal concentrations in the atmosphere and comparing
    data with that reported by other investigators.

Cawse, P. A.   "Inorganic Particulate Matter in the
     Atmosphere." Environmental Chemistry: A Review of the
     Literature Published up to mid-1980. Volume 2.
     Specialist Periodical Report. London: Royal Society of
     Chemistry, 1982.

     Reviews the literature concerning methods used in the
     analysis, measurement, remote sensing, and filtering of
     atmospheric inorganic particulates. Directs attention
     to particulate characteristics and properties of
     background aerosols, marine aerosols, gas-phase and
     photochemical reactions, and trends in particulate
     levels in the atmosphere. Identifies and reviews
     sources of air particulates, transport, modelling of
     particulate dispersion, global inventory cycles of
     elements, and particulate atmospheric removal and
     deposition.

Colbeck, I. and R. M. Harrison. "Tropospheric Ozone." In:
     Bowen, H. J. M., et al., Environmental Chemistry: A
     Review of the Literature Published up to the End 1982.
     Specialist Periodical Report. London: Royal Society of
     Chemistry, 1984.

     Reviews the history and understanding of atmospheric
     chemistry of ozone. Assesses ozone sources, photo-
     chemistry, sinks, and distribution in unpolluted tropo-
     spheric areas. Evaluates the understanding of the
     formation and destruction of ozone in polluted air,
     rural and urban ozone, meteorological factors connected
     with ozone pollution, and biological effects of ozone
     on humans and vegetation.

McLaughlin, S. B.   "Effects of Air Pollution on Forests: A
     Critical Review." Journal of the Air Pollution Control
     Association 35 (1985): 512-534.

     Reviews the historical, thematic background to the
     direct effects of air pollutants on forest environments
     and places present conditions into prespective. Exam-
     ines variance in concentrations of major air pollutants
     in the United States and Europe. Presents a summary of
     recent evidence attributing forest declines to air
     pollutants as a causative agent and discusses the
     mechanisms of these changes. Additionally, addresses
     the issue of prospective research on this subject.

Polcyn, Andrew J. and Howard E. Hesketh. "A Review of
     Current Sampling and Analytical Methods for Assessing
     Toxic and Hazardous Organic Emissions from Stationary
     Sources." Journal of the Air Pollution Control
     Association 35 (1985): 54-60.

     Assesses the necessary criteria for determining the
     choice of methodology in the sampling and analysis of

air pollutants with attention directed toward the difficult aspects of measurement of toxic organic compounds in the atmosphere.

Roberts, James M., et al. "Measurements of Aromatic Hydrocarbon Ratios and NOx Concentrations in the Rural Troposhere: Observation of Air Mass Photochemical Aging and NOx Removal." Atmospheric Environment 18 (1984): 2421-2432.

Studies of tropospheric photochemistry concerning the removal of aromatic hydrocarbons (toluene, ethylbenzene, benzene, and othro(o)xylene) and nitrogen oxides through the photochemical aging process. Establishes correlation between rural air quality, hydroxyl abundance, and proximity to urban sources.

Rosenkranz, Herbert S., et al. "The Genotoxicity, Metabolism and Carcinogenicity of Nitrated Polycyclic Aromatic Hydrocarbons." Journal of Environmental Science and Health C3 (1985): 221-272.

Places into context the current understanding of polycyclic aromatic hydrocarbons in the atmosphere and their mutagenic, genotoxic characteristics. Additionally, reviews their sources (incomplete combustion processes--notes that nitroarenes have been detected in emissions from airplane, gasoline and diesel engines, incineration processes, and residential and kerosene heaters; cigarette smoke; particles in fly ash; urban atmopheres; and effluent from service stations), chemical compositions, and formation in combustion processes.

Sterling, T. D. and A. Arundel. "Possible Carcinogenic Components of Indoor Air: Combustion Byproducts, Formaldehyde, Mineral Fibers, Radiation, and Tobacco Smoke." Journal of Environmental Science and Health C2 (1984): 185-230.

Reviews the properties, sources, types, and toxicity of indoor air pollutants. Focuses upon the levels of concentrations of hazardous, genotoxic indoor air pollution and the potential health risks posed by such substances in non-industrial indoor environments.

Wilkins, E. S. and M. G. Wilkins. "Review of Toxicity of Gases Emitted from Combustion, Pryolysis of Municipal and Industrial Wastes." Journal of Environmental Health A20 (1985): 149-175.

Reviews the types of toxic and potentially toxic gases resulting from the incineration of wastes. Underscores the dangerous nature of carbon chain compounds with low molecular weight (for example, the aldehydes) and the need for analyses of long-term exposure to noxious gases and their toxic concentrations in and directly near waste incineration sites. Also discusses the presence and sources of viral and bacterial releases from the combustion of wastes.

Wolff, Eric W. and David A. Peel. "The Record of Global Pollution in Polar Snow and Ice." Nature 313 (1985): 535-540.

Evaluates reports of the levels of airborne particulates (sulphates and heavy metals), trapped atmospheric gases, and stable isotope ratios of snow from polar regions as an index of global pollution levels. Discusses the significance of geographical and historical patterns of the global transport of air pollutants.

Associations, Research Centers, and Government Organizations

Air Pollution Control Association
P.O. Box 2861
Pittsburgh, PA 15230   (412) 621-1090

With a broadly based membership of 8,000 (in government, research, meteorology, control equipment manufacture, education and other areas), this organization aspires to promote research and inquiry into the abatement of air pollution and attendant problems. Actively promotes seminars, meetings, continuing education, specialized committee work, and publications.

Air Pollution Modeling and Monitoring Laboratory
University of Michigan
2213 Space Research Building
Ann Arbor, MI 48109   (313) 764-3335

Directs efforts toward the creation of models and sampling equipment for studies of air pollution and rain. Specifically researches new methods of modelling the behavior, transport, and nature of air pollutants to be used in the areas of air quality concerns. Also researches in the field of instrumentation for the analysis of air pollution and quality.

Air Pollution Research Laboratory
University of Florida
Gainesville, FL 32611   (904) 392-0846

Directs research activities toward air pollution origins, analysis, and control. Provides educational training in the area of air pollution. Additionally places emphasis on atmospheric contaminant chemistry and behavior.

Air Resources Laboratory
University of Nevada
P.O. Box 60220
Reno, NV 89506   (702) 972-1676

Directs activities toward the monitoring and assessment of Great Basin and Mohave Valley air quality, with attention aimed toward air pollution simulation, measurements, and effects. Develops and implements networks for the monitoring of air quality in the Great Basin.

Association of Local Air Pollution Control Officials
444 N. Capitol Street, NW, Suite 306
Washington, D.C. 20001   (202) 624-7864

Serves as a mechanism for directors of air pollution control agencies to exchange information in the area of air pollution control administration. Additionally, facilitates interagency interaction on local, state, and federal levels. Issues publications and sponsors meetings.

Atmospheric Sciences Research Center
State University of New York
Earth Sciences Building, Room 324
1400 Washington Avenue
Albany, NY 12222   (518) 457-7584

Emphasizes research in atmospheric physics, aerosols, and meteorology; directs research in such areas as air pollution, conservation of energy, fogs, and solar energy.

Bowdoin College
Marine Station
Brunswick, ME 04011   (207) 725-8731

Conducts research into the biological and chemical ramifications of air pollutants with attention directed toward investigations of oil spill contaminations in temperate/tropical zones.

California Air Resources Board
Research Division
P.O. Box 2815
Sacramento, CA 95812   (916) 445-0753

Analyzes and investigates air pollution topics from
origin to control, with research efforts directed
toward acid rain as well. Also measures emissions and
pollution levels in the air. Has 20,000 volume air
pollution library; sustains California Emission
Inventory and California Air Quality Data data bases;
issues publications; and sponsors meetings, conferences, and seminars.

Center for Air Pollution Impact and Trend Analysis
Washington University
Lindell and Skinker Boulevards
St. Louis, MO 63130   (314) 889-6099

Researches the effects of anthropogenic nitrogen,
hydrogen, and sulfur pollutants on the atmosphere.
Additionally, investigates the nature, dispersement,
and sources of atmospheric haze. Supports a library on
topics related to the Center's research activities.

Centre for Research on Environmental Quality
York University
4700 Keele Street
Downsview, Ontario M3J 1P3
Canada   (416) 667-3326

Examines the toxicological effects of environmental
pollutants with especial emphasis on atmospheric contaminants. Researches the physical, biological, and
carcinogenic effects of particulates and mutagenic
substances on humans and the atmosphere.

Environmental Engineering Research Center
University of Florida
College of Engineering
Gainesville, FL 32611   (904) 392-0834

Investigates a number of environmental, ecological
areas in need of engineering solutions. Includes
attention to the topics of the chemistry of the
atmosphere, control of air pollution, and acid rain.

Environmental Protection Agency
Environmental Sciences Research Laboratory
Mail Drop 59
Research Triangle Park, NC 27711   (919) 541-2191

Conducts research into air pollutant behavior and
resulting effects on atmospheric, terrestrial, and
aquatic environments. The laboratory supplies the EPA
with the ability to establish source emission standards, sponsors/reviews research into air pollution via

grant activity in the United States, and provides
assistance in related technical areas to the federal
government.

Environmental Protection Agency
Fluid Modeling Facility
Research Triangle Park, NC 27711   (919) 541-2811

Serves as a mechanism for the delineation and prediction of the influences of atmospheric processes on
source emissions. Performs modelling of air pollutant
emissions in order to explore the relationship between
atmospheric pollution and meteorology.

Environmental Protection Agency
Health Effects Research Laboratory
Research Triangle Park, NC 27711   (919) 541-2281

Organizes research around the question of the influence
of environmental pollution on biological systems. Air
pollutants stand as a central focus of the research as
well as carcinogens in the environment, chemical
contaminants, and toxic studies of pesticides.

Environmental Protection Agency
Industrial Environmental Research Laboratory
Highway 54 at Alexander Drive
Research Triangle Park, NC 27711   (919) 541-2821

Primarily concerned with the methodology and technology
required for control of pollution resulting from industrial activity and energy usage. Reducing emissions of
nitrogen oxide, and the control or cessation of
emissions of nitrogen oxide, stationary source sulfur
oxides, and particulates serve as primary directions in
research.

Environmental Protection Agency
National Air Data Branch
Mail Drop 14
Research Triangle Park, NC 27711   (919) 541-5694

Analyzes, collects, and distributes U.S. air quality
data. Focuses on data pertaining to atmospheric
pollution, sources of air pollution, and air quality.
The EPA and its affiliated contractors and other
governmental agencies throughout the country serve as
the sources of this data. Maintains data bases and
issues data compilations on air quality.

Environmental Studies Laboratory
University of Utah
University of Utah Research Institute
391 Chipeta Way, Suite C
Salt Lake City, UT 84108   (517) 524-3463

    Investigates air pollution, with an emphasis on direct
effects of air pollutants on air visibility, vegetation, composition of lake chemistry, acid deposition,
and ecological systems.

Gordon Environmental Studies Laboratory
University of Montana
Botany Department
Missoula, MT 59812   (406) 243-2671

    Involved in a wide array of activities pertaining to
air pollution research including: the impact of air
pollution on plants, modelling of air pollutant
transport, quality of air, atmospheric repercussions of
coal combustion and aluminum smelting, and ecological/
biological results of dangerous contaminants and wastes
in the environment.

Industrial Gas Cleaning Institute
700 N. Fairfax Street, Suite 304
Alexandria, VA 22314   (703) 836-0480

    Promotes the advancement of air pollution control
equipment development and improvement for stationary
air pollutant sources. Data and information on the
effects of industrial cleaning of gas, air pollution,
and industry concerns are provided. Issues publications, sponsors meetings, and operates clearinghouse
for pollution control.

Inhalation Toxicology Research Institute
P.O. Box 5890
Albuquerque, NM 87123   (505) 844-2203

    Operated by the Lovelace Biomedical and Environmental
Institute for the Department of Energy, this facility
investigates the toxic properties of materials emitted
into the air by the combustion and gasification of
coal, diesel fuel, nuclear energy, and other energy
technologies. Analyzes not only the behavior, nature,
and chemical properties of various toxic airborne
materials but also analyzes the health effects of such
materials on the body once they have been inhaled.

Laboratory for Atmospheric Research
Washington State University
College of Engineering
Pullman, WA 99164   (509) 335-1526

Researches air pollutant behavior in several areas: air contaminant transport and dispersion, atmospheric chemistry of gases and pollutants, and consequences for vegetation of air contaminants. Conducts measurements of air pollutants in the United States and in locations throughout the world.

Louisiana State University
Hazardous Waste Research Center
3418 CEBA Building
Baton Rouge, LA 70803   (504) 388-6770

Researches the nature of hazardous wastes. Covers the process of incineration of such wastes as well as the containment, management, behavior, and elimination of toxic wastes.

National Oceanic and Atmospheric Administration
Air Resources Laboratories
8060 13th Street
Silver Spring, MD 20910   (301) 427-7684

Serves as a part of the Environmental Research Laboratories and, in turn, are comprised of five facilities: Headquarters group (Silver Spring, MD); Atmospheric Turbulence and Diffusion Laboratory (Oak Ridge, TN); Field Research Office (Idaho Falls, ID); Geophysical Monitoring for Climatic Change Laboratory (Boulder, CO); Meteorology Laboratory (Triangle Park, NC); and Solar Radiation Facility (Boulder, CO). Major thrust of this research is to determine and anticipate the environmental effects of anthropogenic activities. Headquarters group works on the development of computer simulation models for the atmospheric transport of pollutants on the local, regional and global levels.

National Oceanic and Atmospheric Administration
Atmospheric Turbulence and Diffusion Laboratory
P.O. Box E
Oak Ridge, TN 37830   (615) 576-1232

Conducts research in a number of areas (with specific reference to the influence on the atmosphere of energy production) including: simulation modelling of the transport and dispersion of air pollutants, nature of plumes resulting from the release of effluents, regional impacts from the production of energy, and the analysis of climate and forest meteorology.

Pennsylvania State University
Center for Air Environment Studies
226 Fenske Laboratory
University Park, PA 16802   (814) 865-1415

Investigates a comprehensive range of air pollution
research topics including: air pollutant chemistry;
nature and control of gases and aerosols; air pollutant
contamination and impact on various types of environments; and the toxicological composition of toxic
atmospheric pollution. Publishes Air Pollution Titles.
Produces an internally available Air Pollution Titles
data base.

State and Territorial Air Pollution
  Program Administrators
444 N. Capitol Street, NW, Suite 306
Washington, D.C. 20001   (202) 624-7864

Comprised of administrators and their staff members for
state and territorial air pollution programs. Serves
as a forum for exchange, communication, cooperation,
and assistance for those governmental administrators
who effect programs for air pollution control.

Statewide Air Pollution Research Center
University of California, Riverside
Riverside, CA 92521   (714) 787-5124

Research activities directed toward a wide range of air
pollutant and atmospheric chemistry topics including:
air pollution control; physical nature, mutagenicity,
and behavior of atmospheric pollutants; direct effects
of air pollutants on vegetation; monitoring, modelling,
and analysis of various types of air contaminants and
their atmospheric chemistry. Additional activities
include the sponsorship of conferences and the collection of bibliographic materials in the area of air
pollution research.

Tennessee Valley Authority
Air Resources Program
River Oaks Building
Muscle Shoals, AL 35660   (205) 386-2555

Collects and analyzes data; performs field and
laboratory research in the following areas: atmospheric
dispersion, transport and chemistry of pollutants, and
the effects on terrestrial environments of atmospheric
pollution.

Libraries

Air Pollution Control Association
Library
P.O. Box 2861
Pittsburgh, PA 15230   (412) 621-1090

  Emphasizes collections in air pollution and air pollution control subject areas.

California Air Resources Board
Library
1131 S Street
P.O. Box 2815
Sacramento, CA 95812   (916) 323-8377

  Collects in the areas of air pollution, air quality,
  and environmental topics. Has extensive air pollution
  microfiche holdings.

Envirodyne Engineers, Inc.
Library
12161 Lackland Road
St. Louis, MO 63141   (314) 434-6960

  Includes the subject areas of air pollution, water
  pollution, management of toxic wastes, and analytic
  chemistry topics.

Environmental Education Group
Library
5762 Firebird Court
Camarillo, CA 93010   (213) 340-7309

  Maintains collections in environmental pollution topics
  including air pollution and control technologies.

Environmental Protection Agency
Environmental Research Laboratory, Corvallis
Library
200 SW 35th Street
Corvallis, OR 97330   (503) 757-4731

  Contains subject areas which include air, water, soil
  pollution effects; toxic materials; and dangerous
  wastes. Maintains significant document holdings
  pertaining to the effects of air pollutants and acid
  rain.

Environmental Protection Agency
Motor Vehicle Emission Laboratory
Library
2565 Plymouth Road
Ann Arbor, MI 48105   (313) 668-4311

Includes subject coverage for automotive engineering
topics and includes mobile source air pollution topics
as well.

Environmental Protection Agency
National Enforcement Investigations
Library
Denver Federal Center, Building 53
Denver, CO 80225   (303) 234-5765

Maintains holdings in a variety of pollution areas
including air pollution, pollution control activities
in industry and agriculture, toxic wastes, and water
quality.

Environmental Research and Technology, Inc.
Information Center
696 Virginia Road
Concord, MA 01742   (617) 369-8910

Organizes collections around environmental science
subjects including pollution of air and water, management of toxic and hazardous wastes, and chemistry.

Environmental Research and Technology, Inc.
Western Regional Office
Library
975 Business Center Drive
Newbury Park, CA 91320   (805) 499-6582

Collects in the subject areas of environmental air
pollution, chemistry of the environment, and environmental impact statements.

Franklin Institute
Franklin Research Center
Information Management Department
20th & Race Streets
Philadelphia, PA 19103   (215) 448-1227

Contains coverage in environmental areas including air
and water pollution; ozone technological developments;
and toxicologic, carcinogenic, and workplace hazard
topics.

**Galson and Galson, P.C., and Galson Technical
   Services, Inc.**
Information Center
6601 Kirkville Road
East Syracuse, NY 13057   (315) 437-7181

   Includes subject areas pertaining to air pollution,
   pollution dispersion models, toxic waste, and meteorological topics.

Iowa State Department of Water, Air and Waste Management
Technical Library
Henry A. Wallace Building
Des Moines, IA 50319   (515) 281-8895

   Contains collections on the quality of air and water,
   technological developments in chemistry, radiation, and
   solid wastes.

Minnesota State Pollution Control Agency
Library
1935 W. County Road, B-2
Roseville, MN 55113   (612) 296-7719

   Maintains subject collections in the areas of air and
   water pollution control, hazardous wastes, and
   pollution resulting from solid wastes.

National Oceanic and Atmospheric Administration
Atmospheric Turbulence and Diffusion Laboratory
Library
Box E
Oak Ridge, TN 37830   (615) 576-1236

   Has collections directed toward the subject areas of
   air pollution, energy, and forest meteorology.

Pennsylvania State University
Center for Air Environment Studies
CAES Information Services
225 Fenske Laboratory
University Park, PA 16802   (814) 865-1415

   Organizes subject collections around air pollution,
   impact of air pollutants on health, acid rain
   investigations, and emissions from fossil fuels.

Texas State Air Control Board
Library
6330 Highway 290 E.
Austin, TX 78723   (512) 451-5711

Includes holdings concerning to air pollution, chemistry, engineering, and meteorology. Contains extensive holdings about technical aspects of air pollution matters on microfiche.

6.
Acid Rain

P. J. Koshy

Introduction

Acid rain is a problem of progress, a result of industrial
revolution. The emission of sulfur dioxide and nitrogen
oxides occurs mainly from the combustion of fossil fuels.
Natural sources such as volcanoes also contribute sulfur
dioxide to the atmosphere. These gases transform into sulfuric acid and nitric acid particles in the atmosphere and
are carried long distances by the wind and deposited on the
earth's surface. The particles in turn are washed down into
rivers and lakes, increasing the acidity of the waters.
During this process the land is leached, thus affecting
plant life. Fish and other aquatic life, intolerant of
higher acidity levels, die and disappear.

This problem was recognized earlier by the Scandinavian
countries. In the United Nations acid rain was first raised
as an international issue in 1972 at the Stockholm
Conference and has since been a major topic of discussion by
governments, politicians, scientists, economists, environmentalists and the public in general. Research work in
North America gained momentum in the '70s. A great deal of
literature has been published discussing the sources of acid
rain, the atmospheric chemistry of this transformation,
transportation across political boundaries, the many forms
of deposition, mitigation methods, scientific modeling, and
the short and long term effects of acid rain in our environment. Perhaps the flood of information on acid rain influ-
enced Robert E. Trumbule and Marilyn Tedeschi to title
their article for <u>Science and Technology Libraries</u> 4, No. 2,
(Winter 1983) as "Acid Rain Information: Knee Deep and
Rising".

During the last decade an increasing number of scientific
papers, periodical articles and books were published under
many headings, such as acid rain, acid precipitation, wet
deposition, dry deposition, acid snow, acid sleet, acid
hail, acid frost, acid fog, etc. Perhaps acid precipitation
or acid deposition may be more appropriate a term, but for

the purpose of this chapter the most commonly used term "acid rain" is being used. The literature covered is a selection of publications in the English language from 1980-1985, with an emphasis on the subject as it affects the United States and Canada.

Books

Acid Rain Resources Directory. St. Paul, MN: Acid Rain Foundation, April, 1984.

Technical and nontechnical reports available from governmental and nongovernmental sources. Audiovisual material, educational curriculum, conference proceedings etc., are included. A useful source for locating material on this subject.

Bhumralkar, Chandrakant M., ed. Meteorological Aspects of Acid Rain. Acid Precipitation Series, Volume 1. Boston: Butterworth, 1984.

In thirteen chapters the authors discuss the role of the atmosphere in the formation of acid rain, its chemistry, transportation and deposition. Theoretical models are also presented. Includes bibliography, charts, graphs and index. This volume evolved as a result of a symposium held in conjunction with American Chemical Society's Las Vegas meeting on Acid Precipitation in 1982, and is the first in a series of 9 volumes.

Boyle, Robert H. and Alexander R. Boyle. Acid Rain. New York: Schocken Books, 1983.

The authors outline the causes and the various effects of acid rain with emphasis on the impact on aquatic and plant life. The political problems and the industry arguments are dealt with in the last two chapters. Names and addresses of organizations and a bibliography provide sources for additional information.

Bricker, Owen P., ed. Geological Aspects of Acid Deposition. Acid Precipitation Series, Volume 7. Boston: Butterworth, 1984.

These six chapters collectively highlight the problem of acid rain as it interacts with the lithospere. This geological approach deals with the Eastern United States and Canada. An entire chapter is devoted to the Adirondack Region of New York State.

Bubenick, David V., ed. <u>Acid Rain Information Book</u>. 2nd
    ed. Park Ridge, NJ: Noyes, 1984.

    This is a revised and enlarged version of the original
    <u>Acid Rain Information Book</u> published in 1982. It is an
    excellent source for up-to-date information on all
    aspects of acid rain. Besides covering the sources,
    transformation, transportation, deposition and the
    adverse effects of acid rain, it also deals with mon-
    itoring programs, statistical and regional models,
    regulative and mitigative strategies, and current and
    proposed research on acid rain. A summary of issues
    dealt with and bibliography are given at the end of
    each chapter. There is no index but according to the
    editor the arrangement of the table of contents is
    expected to serve as a subject index.

Carroll, John E. <u>Acid Rain: An Issue in Canadian-American</u>
    <u>Relations</u>. Washington, D.C.: Canadian-American
    Committee, 1982.

    Sponsored by the C. D. Howe Institute of Canada and the
    National Planning Association of the USA, the Canadian-
    American Committee seeks to encourage public under-
    standing of the nature of the issues of importance to
    these countries and attempts to develop and disseminate
    ideas consistent with the national goals of these two
    countries. In this book an attempt is made to explore
    the various aspects of this binational problem and to
    suggest ways of dealing with them.

Committee on Atmospheric Transport and Chemical Transfor-
    mation in Acid Precipitation, Environmental Studies
    Board, National Research Council. <u>Acid Deposition:</u>
    <u>Atmospheric Processes in Eastern North America</u>.
    Washington, D.C.: National Academy Press, 1983.

    The Committee on Atmospheric Transport and Chemical
    Transformation in Acid Precipitation was formed in
    January 1982 to study what kind of reduction in
    emissions would be required in order to decrease consi-
    derably the deposition of hydrogen ions in sensitive
    areas. This book gives the results of the study which
    included a review of the current literature and consul-
    tation with experts in the field. Chapters cover atmos-
    pheric processes, deposition, theoretical models,
    empirical observations, and research needs. Includes a
    bibliography at the end of each chapter and biographi-
    cal sketches of committee members in Appendix D.

Cornell University. <u>Acid Precipitation: New York State</u>
    <u>Directory</u>. Ithaca, NY: Center for Environmental
    Research, Cornell University, 1982.

Lists 192 entries of individuals and institutions
engaged in activities involved in acid precipitation
work conducted in New York state. Names, addresses,
telephone numbers and areas of expertise are listed
alphabetically under the headings Aquatic Ecosystems,
Atmospheric Processes, Terrestrial Ecosystems and
Other. A ready reference for quick access to areas of
interest also is given. Available from the Center for
Environmental Research.

Curtis, Carolyn, ed. <u>Before the Rainbow: What We Know About
Acid Rain</u>. Washington, D.C.: Edison Electric Institute,
1980.

Edison Electric Institute, an association of electric
utility companies, exchanges information on matters
relating to the industry among the members, the public,
and the government. Naturally, this publication
represents the viewpoints of the industry and provides
research articles to point out the area of uncertain-
ties associated with acid rain and the need for increased research.

D'Itri, Frank M., ed. <u>Acid Precipitation: Effects on Ecological Systems</u>. Ann Arbor, MI: Ann Arbor Science,
1982.

This is a collection of research papers resulting from
the Conference on Effects of Acid Precipitation of
Ecological Systems convened at Michigan State
University in 1981. Arranged in four parts, it deals
with the history and current status of acid precipitation, a general overview of sources, monitoring, and
effects on aquatic and terrestrial systems. Graphs,
charts, bibliography and an elaborate index are
included.

Durham, Jack L., ed. <u>Chemistry of Particles, Fogs and Rain</u>.
Acid Precipitation Series, Volume 2. Boston: Butterworth, 1984.

This second volume of the Acid Precipitation Series
consists of six chapters by various authors who are the
leaders in their fields. These papers were earlier
presented at the Annual Meeting of the American
Chemical Society at Las Vegas in 1982. Together they
present some of the recent advances in theoretical and
experimental investigations of processes that lead to
the acidification of particles, fogs and rain.

Electric Power Research Institute. <u>Inventory of Acid-Depo-
sition Research Projects Funded by the Private Sector</u>.
Report #EPRI-EA-2889. Palo Alto, CA: Research Reports
Center, 1983.

Contains brief descriptions of over 200 privately
funded research projects on acid deposition and related
research during 1980-82. The information includes the
title, objective, funding and research organization
arranged in four sections - Atmospheric Process, Environmental Effects, Emissions & Monitoring, and Other
Projects. Available from Research Reports Center, Box
50490, Palo Alto, CA 94309. Phone (415) 965-4081.

Environmental Resources Ltd. Acid Rain: A Review of the
    Phenomenon in the EEC and Europe. New York: Unipub,
    1983.

    Summarizes the results of a survey on the problem of
    acid rain conducted by the Environmental Resources
    Ltd., a private consultancy firm in London. Facts
    relating to the effects of dry and wet deposition are
    given to enable the reader to distinguish the differ-
    ence in the argument over the nature of damages.
    Includes a fifteen page bibliography of publications up
    to 1982.

Fraser, J. E. and D. L. Britt. Liming of Acidified Waters:
    A Review of Methods and Effects on Aquatic Ecosystems.
    McLean, VA: General Research Corp., 1982.

    Summarizes current information on liming and how it
    relates to aquatic habitats that are potentially
    impacted by the acidification process. Primary empha-
    sis is on the fishery resource. Introductory informa-
    tion discusses acid-sensitive regions and associated
    aquatic fisheries and the effects of acidification on
    fish. The remaining chapters address preventative and
    mitigative liming, liming techniques, ecological
    responses to liming of aquatic ecosystems, and informa-
    tion needs.

Gay, Kathlyn. Acid Rain. New York: Franklin Watts, 1983.

    After examining the direct and indirect causes and
    effects of acid rain, the author explains what the
    citizen and government can do to alleviate the problem.
    Mentions school projects and available sources for
    project materials. Includes a short reading list and
    index.

Gould, Roy. Going Sour: Science and Politics of Acid Rain.
    Boston: Birkhauser, 1985.

    The author states that the EPA and the industries have
    suppressed or distorted existing research results.
    Gould's arguments are presented in an historical light
    and the scientific explanations are grouped together
    into chapters.

Hicks, Bruce B., ed. Deposition: Both Wet and Dry. Acid
    Precipitation Series, Volume 4. Boston: Butterworth,
    1984.

    This fourth volume in the Acid Precipitation Series
    evaluates wet and dry deposition in North America with
    special chapters on selected areas in Canada, the
    United States and Mexico. These nine chapters are
    contributed by several experts engaged in acid rain
    research.

Howard, Ross and Michael Perley. Acid Rain: The North
    American Forecast. Toronto, Ontario: Anansi Press,
    1980.

    Written lucidly with a minimum of scientific jargon.
    The authors trace the source of acid rain and bring to
    light its extensive impact on plants, and aquatic eco-
    systems. Though they deal with North America as a
    whole, the emphasis is on the economic, political and
    environmental situation in Canada. A useful list on
    page 195 gives the names and addresses of agencies and
    people in the USA and Canada, followed by a short
    bibliography and index.

Levy, Julian A. Acid Deposition in Texas: Technical Summary
    and Perspective. Austin: Texas Energy and Natural
    Resources Advisory Council, 1982.

    Written to serve as a resource document in developing
    an acid deposition work plan for the state of Texas.
    Includes summary of technical information concerning
    critical aspects of the issue as they relate to the
    state of Texas.

Linthurst, Rick A., ed. Direct and Indirect Effects of Acid
    Deposition on Vegetation. Acid Precipitation Series,
    Volume 5. Boston: Butterworth, 1984.

    The impact of acidic deposition on vegetation, an
    extremely complex aspect of acid rain, is dealt with in
    these chapters by the various authors. Both beneficial
    and detrimental impacts on forest vegetation and crop
    plants are presented.

Luoma, Jon R. Troubled Skies, Troubled Waters: The Story
    of Acid Rain. New York: Viking Press, 1984.

    Luoma writes the story of acid rain and the damage it
    has done to the Quetico Provincial Park in Canada and
    the Boundary Waters Canoe Area on the United States
    side. Devoid of technical terms, charts or graphs,
    this book is written for the general public in a story
    teller's style. Explains the damange the lakes have

suffered and expresses the hope that these and tens of thousands of others can be saved. Has no bibliography but an elaborate index.

Ostmann, Robert. Acid Rain: A Plague Upon the Waters. Minneapolis: Dillon Press, 1982.

Robert Ostmann, a distinguished journalist and an award winner for environmental reporting, examines the causes and effects of acid rain in North America and Europe and the social, economic and political ramifications. The short bibliographic essay at the end of the book points out some selected source material essential for the understanding of the problem in general as it is faced in North America and Scandinavia.

Roth, Philip. The American West's Acid Rain Test. Berkeley, CA: World Resources Insitute, 1985.

While most of the books on the subject deal with the northeastern United States and Canada, the authors present the acid rain problem in the west. Chapters deal with resources at risk, the problem itself and a summary of recommendations.

Schnoor, Jerald L., ed.   Modeling of Total Acid Precipitation Impacts. Acid Precipitation Series, Volume 9. Boston: Butterworth, 1984.

Written by leaders in the field, the nine chapters present the state-of-the-art in acid precipitation modeling. These papers resulted from the symposium on Acid Precipitation sponsored by the American Chemical Society at Las Vegas in 1982. These models could be applied to the various stages in acid precipitation and its impact on lakes and streams. Contains bibliographies, charts, graphs and index.

Smith, Oliver F. An Assessment of Acid Rain. Philadelphia: Pennsylvania Environmental Research Foundation, 1982.

Presents an overview of the situation as it was in early 1982, in non-technical terms. Gives a simplified description of the basic chemistry of acid precipitation, transportation of air pollutants, wet and dry deposition and its effects on land and water. Half of this book is devoted to an extended bibliography of periodical articles, newspaper articles, reports and books published through the beginning of 1982. An excellent source, if one is looking for a very concise write-up on the subject.

Vozzo, Steven F., comp. <u>International Directory of Acid Precipitation Researchers</u>. Raleigh, NC: North Carolina State University, 1983.

This edition contains more than 1400 entries, in 149 pages, of researchers engaged in acid precipitation and related subject research arranged alphabetically by Canada, the United States, and European countries. Subject and alphabetical index. Copies available from the Acid Rain Foundation, Inc., 1630 Blackhawk Hills, St. Paul, MN 55122. Also available in microfiche from NTIS (PB84-127158).

Weller, Phil and the Waterloo Public Interest Research Group. <u>Acid Rain: The Silent Crisis</u>. Kitchener, Ontario: Between the Lines, 1980.

Examines the industrial and governmental decisions that led to the persistence of acid rain. It is an effort by the Waterloo, Ontario Public Interest Research Group to provide information for public awareness. Names and addresses of public groups in Canada and the United States are included in the appendix.

Westone, Gregory S. and Armin Rosencranz. <u>Acid Rain in Europe and North America: National Responses to an International Problem</u>. A study for the German Marshall Fund of the United States. Washington, D.C.: Environmental Law Institute, 1983.

Examines in detail the nature and severity of the acid rain problem, the actions taken by the affected nations and a review of the current national and international laws and policies. Appendix B devotes 15 pages to available control technologies. Chapter 15 traces the history of international efforts beginning with the 1972 United Nations Conference on the Human Environment in Stockholm, the UNEP, OECD, the ECE convention up to the 1982 Stockholm Conference.

Zimmer, Michael J. and James A. Thompson, eds. <u>Acid Rain: Planning for the '80s</u>. Rockville, MD: Government Institutes Inc., 1983.

Twelve chapters presenting diverse viewpoints and current status of the acid rain issue by leaders in industry and government officials with a collection of reprints of various articles. Includes selected bibliography, 1980-82, prepared by Library of Congress, Congressional Research Service.

Periodicals

One of the best sources for current literature on any
subject is the periodical article. Most of the research
results appear first as articles in scientific journals.
Regular articles on the various aspects of this topic appear
in general as well as scientific periodicals which are
indexed in the Abstracts and Indexes section of this
chapter.

On page 22 of Acid Precipitation Digest 2, No. 4 (April
1984), two new periodicals on acid rain are mentioned: Acid
Rain Network News, and Acid Rain Intelligence Report, both
available from:

>   North American Water Office
>   1519A E Franklin Avenue
>   Minneapolis, MN 55404

Acid Precipitation Digest. Acid Rain Information Clearinghouse, Center for Environmental Information Inc.
Monthly, published since 1983.

>   Provides the summary of current news, research and
>   events. Includes new periodicals, books, conference
>   proceedings, and published proceedings.

Several periodicals listed in Chapter 1 and other chapters
often contain articles on acid rain:

Atmospheric Environment
Environmental Science and Technology
EPRI Journal
Journal of the Air Pollution Control Association
Journal of Environmental Quality
Journal of Environmental Sciences
Science

Abstracts and Indexes

The term acid rain was first used in 1872 by Robert Angus
Smith, a British chemist, in a book Air and Rain: The Begin-
nings of Chemical Climatology, but it appeared as a subject
heading in most abstracts and indexes in the 1970's, a
hundred years later. The Environment Index used this subject as early as 1973; the Applied Science and Technology
Index in 1982. Public Affairs Information Service, Inc., a
non-profit association of libraries indexing library material in the field of public affairs and public policy used
acid rain as a subject in the PAIS Bulletin in 1980.

(See Chapter 1.) Acid rain as a subject appeared in the
card catalogs in most libraries after the inclusion of this
subject in Library of Congress Subject Headings in its 9th
edition in 1980. Today acid rain is a well accepted subject
in all indexes and abstracts, and thousands of articles are
being published on the subject every year.

Acid Rain Abstracts. EIC/Intelligence. Bimonthly, pub-
    lished since 1984.

    Gives complete bibliographic citations, concise docu-
    ment abstracts, author, subject, source and industry
    indexes. Special sections cover new books, upcoming
    conferences and acid rain trends. Index is cumulated
    in the December issue. Literature coverage is world-
    wide. EIC/Intelligence services include online data
    bases, microfiche document collections, and a full
    range of printed materials in the field of environment
    and energy.

Energy Information Abstracts. EIC/Intelligence. Monthly
    (bimonthly May/June and November/December, published
    since 1980.

    Similar in format and contents organization to the
    Environment Abstracts (see next page) but contains
    fewer references to acid rain since the subject area is
    limited to energy. Cumulated annually with its com-
    panion volume, the Energy Index.

Energy Research Abstracts. Department of Energy, Technical
    Information Center. Biweekly, published since 1976.

    Covers all publications originating from the Department
    of Energy, its laboratories, centers, and contractors.
    Abstracts section is followed by corporate and personal
    author index, subject index, contract number index,
    report number index and order number correlation list.
    Available from NTIS on microfiche and also from the
    Superintendent of Documents, Government Printing
    Office, Washington D.C.

Government Reports Announcements & Index. National Tech-
    nical Information Service. Bimonthly, published since
    1965.

    NTIS is a central source for U.S. Government sponsored
    research, development and engineering reports as well
    as foreign technical reports. Most documents cited are
    available in microform and paper editions. Arranged
    under 22 broad subject catagories and their subcata-
    gories, the entries are arranged alphanumerically.
    Report number, author, title and other bibliographic
    information followed by brief abstracts. Keyword

index, personnel author and corporate author index,
report number and order number index with cumulated
annual index.

Bibliographies

Many research centers, information clearing houses and
libraries publish their own bibliographies. Bibliographic
Index examines 2600 periodicals regularly and gives complete
citations of subject bibliographies published separately or
appearing as part of a publication bearing 50 or more citations. Another source that often cites bibliographies is
the Public Affairs Information Service Bulletin. Confer-
rence papers, books and periodical articles frequently include references as further reading materials which are
valuable sources within their limitations.

NTIS Published Searches, described in Chapter 1, include
several on acid rain. The following are a few taken from
the NTIS Published Searches Master Catalog 1984:

```
PB82-859281/CAC   Acid Precipitation   June 1976-1981
    (380 citations)
PB83-869487/CAC   Acid Precipitation   June 1970-July 1983
    (249 citations)
PB83-872408/CAC   Acid Precipitation   June 1974-Sept 1983
    (149 citations)
PB80-859879/CAC   Acid Precipitation   June 1970-June 1980
    (142 citations)
PB83-869495/CAC   Acid Precipitation   1970-July 1983
    (324 citations)
PB83-857714/CAC   Acid Precipitation   1977-1982
    (169 citations)
PB83-872242/CAC   Acid Precipitation:  Effects on the
    Aquatic Ecosystem   1977-Sept 1983  (242 citations)
PB83-872424/CAC   Acid Precipitation:  Effects on Terres-
    trial Ecosystems   1976-Sept 1983  (274 citations)
PB83-872416/CAC   Acid Precipitation:  Legal, Political,
    and Health Aspects  1976-Sept 1983  (184 citations)
PB83-872259/CAC   Acid Precipitation:  Transfrontier Tran-
    sport of Air Pollutants  1976-Sept 1983 (107 citations)
PB83-801811/CAC   Precipitation Washout  1964-Nov 1982
    (213 citations)
```


Acid Precipitation: A Compilation of World-wide Literature,
    A Bibliography. DOE TIC 3399/STD11. Oak Ridge, TN: DOE
    Technical Information Center, April, 1983.

    Contains 3197 references to journal articles and tech-
    nical reports with abstracts.

<u>Acid Precipitation: A Current Awareness Bulletin</u>. Department
   of Energy, Technical Information Center. Published
   bimonthly since 1983.

   An annotated bibliography based on current information
   entered into the Energy Data Base on acid precipitation
   and closely related subjects from worldwide energy
   R & D information. Each entry is complete with bibliographic information and abstracts. The documents cited
   are not necessarily available from NTIS but may be
   obtained from the publishers of the documents. Available on subscription from NTIS for $40 a year within the
   United States.

<u>Acid Rain: Impacts on Agriculture 1975-1982</u>. NAL-BIBL-83-18.
   Beltsville, MD: USDA, National Agricultural Library,
   1983.

   234 citations of recent resources derived from online
   searches of selected data bases. Copies available from
   Reference Section, USDA National Agricultural Library
   Building, Beltsville, MD 20705.

<u>An Acid Rain Bibliography</u>. PB84-138940. CERL-053d, 2
   volumes. Corvallis, OR: Environmental Protection
   Agency, 1984.

   900 citations covering all aspects of acid rain with
   title, author and subject index.

<u>A Bibliography: The Long-Range Transport of Air Pollutants
   and Acidic Precipitation</u>. Downsview, Ontario: LRTAP
   Program Office, Atmospheric Environment Service,
   Environment Canada, July, 1980.

   Preparation jointly sponsored by the Ontario Ministry
   of the Environment and the Federal Atmospheric Environment Service, Canada. Contains approximately 1300
   citations of periodical articles, conference papers and
   technical reports.

Duensing, Edward. <u>The Environmental Effects of Acid Precip-
   itation</u>. Public Administration Series Bibliography
   No. P624. Monticello, IL: Vance, 1982.

   Based on materials available in the Rutgers University
   Library. Over 140 citations of technical reports,
   periodical articles, and conference proceedings published from 1975-1980.

Dynamic Corporation. <u>Bibliography of Air Pollution and Acid Rain Effects on Fish, Wildlife, and Their Habitats</u>. PB82-246059 Rockville, MD: Fish and Wildlife Service, March, 1982.

Approximately 1900 references are listed alphabetically with full citations focusing specifically on these topics. An unpublished report available through NTIS.

Harrell, Karen Fair. <u>Acid Rain: A Legal and Political Perspective</u>. Public Administration Series Bibliography No. P1319. Monticello, IL: Vance, November, 1983.

Over 170 citations from federal and state documents, proceedings, books, and journal articles published 1979 through 1983.

Koshy, P. J. <u>Acid Rain 1980-1984: A Selected Bibliography</u>. Public Administration Series Bibliography No. P1881. Monticello, IL: Vance, March, 1986.

Over 270 citations of periodical articles, books, conference proceedings, and government reports are arranged by subject for easy use.

Lockerby, Robert W. <u>Acid Rain</u>. Public Administration Series Bibliography No. P928. Monticello, IL: Vance, 1982.

180 citations from 1972 to 1981 covering all aspects of acid rain taken mostly from periodical articles.

Stopp, Harry G., Jr. <u>Acid Rain: A Bibliography of Research Annotated for Easy Access</u>. Metuchen, NJ: Scarecrow Press, 1985.

Contains 886 citations with brief annotations of periodical articles, research reports, books and conference papers published over a 50 year period beginning with 1932 to 1984 and arranged in alphabetical order by author. A subject index provides easy access.

Wiltshire, Denise A. and Margaret L. Evans, comp. <u>Acid Precipitation: An Annotated Bibliography</u>. Geological Survey Circular No. 923. Alexandria, VA: Geological Survey, 1984.

1660 citations on the causes and environmental effects of acidic deposition compiled from computer searches on earth-science and chemistry data bases from late 1800 through 1981. Includes short summaries. Arranged alphabetically by the principal author, with coauthor and subject indexes.

Proceedings

Thousands of scientific meetings take place every year and most of these result in published proceedings of these conferences. Participants being experts in their field of activity, the papers and proceedings represent the state-of-the-art. Some of the sources for locating these proceedings are described in Chapter 1.

Beilke, S. and A. J. Elshout, eds. Acid Deposition. Proceedings of the CEC Workshop organized as part of the Concerted Action "Physico-Chemical Behavior of Atmospheric Pollutants" held in Berlin, September 9, 1982. Dordrecht, Holland: Reidel, 1983.

An attempt is made through these papers to emphasize new scientific ideas, indicate areas of uncertainty and the need for research in the acid deposition problem. Gives an extensive review of the present situation of acid deposition and research in Europe. Includes bibliographies, charts, graphs, and a list of participants with the names of their institutions.

Directory of Published Proceedings: Series PCE.   InterDok Corp. Semiannually, published since 1975.

This series is a compilation of information on pollution control and ecology taken from two other series published by InterDok: Series SEMT - Science, Engineering, Medicine and Technology, and Series SSH - Social Sciences and Humanities.

Gilleland, Diane S., ed. Acid Rain: The Costs of Compliance. Carbondale: Southern Illinois University Press, 1985.

Consists of papers and proceedings of a conference sponsored by the Illinois Energy Resources Commission and the Coal Extraction and Utilization Research Center held in Carbondale, March 18, 1984. This conference, unlike many others, has a geographical focus on the coal producing Illinois State. Participants from different walks of life present their varying viewpoints and discuss several legislative proposals, pending bills and amendments to the Clean Air Act, general aspects of the issue, and the specific impacts of emission reduction strategies.

Gold, Peter S., ed. Acid Rain: A Transjurisdictional Problem in Search of Solution. Proceedings of a Conference, Canadian-American Center, State University of New York at Buffalo, May 1 & 2, 1981. Buffalo, NY: Canadian-American Center, 1982.

This book contains the papers, addresses, and discussions of the conference. Defines the problem from the scientific and regional perspective and examines the economic, political, and legal issues. The appendix contains the laws, agreements, and treaties referred to in the text. An excellent source of material for the background and issues involved between the United States and Canada and the differing viewpoints.

Green, Alex E. S. and Wayne H. Smith, eds. Acid Deposition Causes and Effects: A State Assessment Model. Rockville, MD: Government Institutes, 1983.

Proceedings of the Acid Deposition Science Workshop held March 23-24, 1983 sponsored by Interdisciplinary Center for Aeronomy and Atmospheric Sciences, and Center for Environmental and Natural Resources Programs, University of Florida. Papers cover an overview of the acid deposition situation in the nation in general and in the state of Florida in particular on sources, atmospheric transformation, transportation, and the terrestrial and aquatic effects of deposition. Includes abstracts, charts, models and bibliographic references.

Hutchinson, T. C., ed. Effects of Acid Precipitation on Terrestrial Ecosystems. New York: Plenum, 1980.

This is a collection of papers presented at the NATO Conference on Effects of Acid Precipitation on Vegetation and Soils held at Toronto. Papers by experts in their field cover mostly the effects of acid rain, but the first seven papers in Part I of this book give good coverage of the nature of acid precipitation. Rapporteurs' summary of the sessions and a list of participants with their addresses given at the end of the book provide a good source for further studies.

Jenkins, S. H. and P. Schjodtz Hansen, eds. International Conference on Coal Fired Power Plants and the Aquatic Environment. Copenhagen, Denmark, August 16-18, 1982. New York: Pergamon, 1983.

A collection of ten papers presented at the conference are reproduced in their original form. These papers deal with emission, acidification, transportation, deposition and the effect on aquatic ecosystems. Abstracts, charts, graphs, short bibliography and subject index.

Johnson, R. E., ed. <u>Acid Rain/Fisheries</u>. Proceedings of
the International Symposium on Acidic Precipitation and
Fishery Impact in Northeastern North America, Cornell
University, Ithaca, New York, August 2-5, 1981.
Bethesda, MD: American Fisheries Society, 1982.

Major papers presented were: "The effects of acidification on the chemistry of ground and surface water,"
"Effects of acidification on aquatic primary producers
and decomposers," and "Effects on fish of metals
associated with acidification."

Shriner, David S., ed. <u>Atmospheric Sulfur Deposition:
Environmental Impact and Health Effects</u>. Ann Arbor,
MI: Ann Arbor Science, 1980.

Proceedings of the Second Life Sciences Symposium on
Potential Environmental and Health Consequences of
Atmospheric Sulfur Deposition, Gatlinburg, Tennessee,
October 14-18, 1979 sponsored by Oak Ridge National
Laboratory, Department of Energy, EPA, and Tennessee
Valley Authority. Arranged in eight sections, these
papers deal with the sulfur dioxide emissions, atmos-
pheric transformation, dry and wet deposition and their
effects on environment, animal and plant life. Most
papers have bibliographies.

Toribara, Taft Y., ed. <u>Polluted Rain</u>. New York: Plenum,
1980.

This is the 12th in a series of conferences on environmental toxicity sponsored by the Department of Biology
and Biophysics, University of Rochester. The book is a
collection of conference papers arranged in the order
of the sessions. Bibliographic references are from the
70's. The subjects covered are the chemistry of polluted rain, the mercury problem, effects on plants,
anticipated problems and monitoring systems and legal
aspects. Lists of speakers and their addresses are
included at the end of the book.

Dissertations

Thousands of research students produce theses and dissertations in academic institutions throughout the world. The
main source for locating such valuable work is the <u>Disserta-
tion Abstracts International</u>. <u>Section A: Humanities and
Social Sciences</u>, and <u>Section B: Sciences and Engineering</u>.
Published by University Microfilms International, monthly
since 1938.

University Microfilms International publishes a number of
special subject bibliographies. The latest catalog lists
about a hundred such publications available free. The most
appropriate one for this chapter is <u>Ecology and the Environment: A Catalog of Selected Doctoral Dissertation Research</u>,
which contains citations to 1,616 doctoral dissertations and
175 masters theses published between 1980 and 1984. A few
examples from the list are cited below:

Baker, Joan Patterson. <u>Aluminum Toxicity to Fish as Rela-
    ted to Acid Precipitation and Adirondack Surface Water
    Quality</u>. Ph.D. diss., Cornell University, 1981.

Gottschalk, Marlin Ralph. <u>The Effects of Sulfur-Dioxide and
    Acid Precipitation on Decomposition and Nutrient
    Cycling Process in a Southeastern Deciduous Forest
    Soil</u>. Ph.D. diss., Emory University, 1981.

Tonnessen, Kathy Ann. <u>The Potential Effects of Acid
    Deposition on Aquatic Ecosystems of the Sierra Nevada</u>.
    Ph.D. diss., University of California, Berkeley, 1983.

An online search service offered by UMI called DATRIX DIRECT
and other information on UMI is included in Chapter 1.

Audiovisual Materials

<u>Acid Precipitation in Pennsylvania</u>. University Park, PA:
    Institute of Research on Land and Water Sources, 1983.
    Slides, audiocassettes, 60 min.

    Defines acid precipitation and the extent of research
    done by the Institute. Examines the problem worldwide
    and then concentrates on North America and Pennsylvania
    in particular. Available to be shown to interested
    groups. Contact the Institute at (814) 863-0291.

<u>Acid Rain</u>. Washington, D.C: Environmental Protection
    Agency, 1980. 16mm, color, 20 min.

    Available from the Environmental Protection Agency,
    Film Distribution Center, 1028 Industry Drive, Tukvila,
    Washington, D.C. 98188. Phone (206) 575-1575.

<u>Acid Rain</u>. Friends of the Boundary Waters Wilderness, n.d.
    80 frames, color, 1 cassette, 12 overhead transparen-
    cies and guide.

    Describes the relationship of acid rain to fossil fuel
    consumption and its environmental hazards. Available

from Educational Images Ltd., P.O. Box 3456, West Side
Station, Elmira, NY 14905.

Acid Rain: Requiem or Recovery. Chicago: Film Canada
Center, National Film Board of Canada, 1982. 16mm,
videocassete, color, 27 min.

Examines the causes and impact of acid rain and the
need to increase public awareness. Available from Film
Canada Center, 111E Wacker Drive, Suite 313, Chicago,
IL 60601.

Acid Rain: The Silent Crisis. Bloomington, IN: Indiana
University Audio-Visual Center, n.d. 16mm, color, 25
min.

The Bitter Rain. Toronto, Ontario: Canadian Broadcasting
Service, 1980. Videocassette, color, 22 min.

Examines the causes and effects of acid rain and the
solutions applied to this problem in Scandinavia and
Canada. Distributed by Journal Films, Inc., 930
Pitner, Evanston, IL 60202.

Born on the Wind. Palo Alto, CA: Electric Power Research
Institute, 1983. Video Memo on acid rain. Videocassette, color, 15 min.

Video Memos are capsulized information prepared for the
utility staff. Free to EPRI members. Contact Jeffrey
Lavecchia (415) 855-2928.

Lake Acidification. Palo Alto, CA: Electric Power Research
Institute, n.d. Video Memo on acid rain. Videocassette, color, 20 min.

Examines EPRI study of sources and effects of acid rain
in the Adirondack lakes of New York and the northern
lakes of Wisconsin.

To Catch a Cloud: A Thoughtful Look at Acid Rain.
Petersburg, FL: Modern Talking Picture Service, n.d.
16mm, color, 28 min.

Documentary on acid rain. Animation and on-sight interviews with scientists. Depicts formation of acid rain
and the research currently being conducted.

Free and Inexpensive Materials

Moon, Ilse B. "Free and Inexpensive Materials on Acid
    Precipitation." Collection Building, 5 (2) (Summer
    1983): 67-72.

    A selection of miscellaneous items such as reprint
    articles, newsletters, bibliographies, and pamphlets
    available on request free of cost. This annotated list
    also includes complete bibliographic citations and
    names and addresses of available sources.

Research Centers

A considerable amount of research has been done and is being
done by government agencies, private organizations and academic research institutions. Results are published as
technical reports, journal articles, conference papers and
monographs. The majority of U.S. Government publications
are available from the National Technical Information
Service (NTIS). For a detailed description of Federal
research on acid precipitation, consult chapter 9 of Acid
Rain Information Book, 2nd ed., 1984, pages 359-396.

Electric Power Research Institute (EPRI), a voluntary association of the nations electric utility industry, has under
contract to various research institutions fifty major
programs costing millions of dollars scheduled to be completed between now and 1988. The entire issue of EPRI
Journal, vol. 8, no. 9 (Nov. 1983), devoted to a special
report on acid rain, explores the present state of knowledge
of acid rain and the various research programs EPRI is
pursuing. According to Richard E. Balzhiser, senior vice
president, Research and Development Group, the Institute
plans to spend $175 million over the next five years on
research work directly related to acid rain. In the same
issue, its Washington Report, a regular feature of the
journal, summarizes 200 projects of the Federal Government
research on acid rain.

The following information on research institutions and the
subjects in which they are interested and other information
was provided by those institutions in reply to a questionnaire mailed to them. No effort was made to make the list
or information exhaustive.

Acid Rain Information Clearinghouse
33 South Washington Street
Rochester, NY 14608   (716) 546-3796

The Acid Rain Information Clearinghouse was established
in 1982 by the Center for Environmental Information
Inc., a nonprofit organization, to gather, organize and
disseminate information relevant to the topic of acid
rain. Designated as New York State's acid rain
documentation center, ARIC has access to much of the
world's published literature. Maintains a comprehensive reference library and a computerized retrieval
system on acid rain with access to existing data bases.
Produces specialized bibliographies on acid rain and
related subjects. Organizes seminars, workshops and
conferences for organizations interested in acid rain
information.

Publications: Acid Precipitation Digest. Monthly since
1983. A summary of current news, research and events.
Contains annotated bibliography of current periodical
articles, technical reports, books and conference proceedings.

Air Quality Research Section
Tennessee Valley Authority
Muscle Shoals, AL 35660   (205) 386-2341

Areas of Interest: Atmospheric chemistry; Emission;
Impact on environment; Impact on terrestrial ecosystems; Transportation and deposition.

Publications: Various progress and final research
reports.

Argonne National Laboratory
9700 South Cass Avenue
Argonne, IL 60439   (312) 972-4211

Areas of Interest: Atmospheric chemistry; Control
technology; Economic impact; Emission; Impact on
aquatic ecosystems; Impact on environment; Impact on
terrestrial ecosystems; Transportation and deposition.

Publications: Professional and scientific journal
articles, technical reports, and conference papers.

Atmospheric Sciences Research Center
State University of New York at Albany
1400 Washington Avenue
Albany, NY 12222   (518) 457-7584

Areas of Interest: Atmospheric chemistry; Impact on
environment; Impact on terrestrial ecosystems; Transportation and deposition.

Publications: Project reports and conference proceedings.

Atmospheric Turbulence and Diffusion Laboratory
National Oceanic and Atmospheric Administration
P.O. Box E
Oak Ridge, TN 37831   (615) 576-1233

   Areas of Interest: Atmospheric chemistry; Emission;
   Impact on environment; Transportation and deposition.

   Publications: Scientific papers.

Battelle - Pacific Northwest Laboratories
P.O. Box 999
Richland, WA 99352   (509) 375-2121

   Areas of Interest: Atmospheric chemistry; Economic
   impact; Emission; Impact on aquatic ecosystems; Impact
   on environment; Impact on terrestrial ecosystems;
   Transportation and deposition.

   Publications: Professional and scientific journal
   articles and technical reports.

Center for Northern Studies
Wolcott, VT 05680   (802) 888-4331

   Areas of Interest: Economic impact; Impact on environment; Impact on terrestrial ecosystems; Political
   impact.

   Publications: Northern Ravan (Newsletter). Quarterly,
   free. Other publications appear in appropriate books
   and journals.

Department of Biology
McGill University
422 rue des Moulins,
Mont St. Hilaire,
Montreal J3G 4S6
Canada   (514) 467-1755

   Areas of Interest: Impact on terrestrial ecosystems;
   Transportation and deposition.

   Publications: Research results reported in primary
   literature.

Department of Crop Science
Oregon State University
Corvallis, OR 97331   (503) 754-2964

   Areas of Interest: Impact on agricultural crops; Impact
   on terrestrial ecosystems.

Publications: Research results published in primary journals and reports.

Department of Environmental Engineering Sciences
University of Florida
Gainesville, FL 32611   (904) 392-0838

Areas of Interest: Impact on aquatic ecosystems.

Publications: Professional and scientific journal articles and technical reports.

Department of Soil and Environmental Science
University of California
Riverside, CA 92521   (714) 787-3654

Areas of Interest: Impact on aquatic ecosystems; Impact on environment; Impact on terrestrial ecosystems.

Publications: In professional journals, conference proceedings and research reports.

Electric Power Research Institute
P.O. Box 10412
Palo Alto, CA 94303   (415) 855-2572

Areas of Interest: Atmospheric chemistry; Economic Impact; Emission; Impact on aquatic ecosystems; Impact on environment; Impact on terrestrial ecosystems; Mitigation; Strategic planning; Transportation and deposition.

Publications: EPRI Guide. Quarterly, free. EPRI Research & Development Projects. Annual, $10.00. EPRI Journal. Monthly, free.

Fresh Water Institute
Rensselaer Polytechnic Institute
Troy, NY 12180   (518) 266-6757

Areas of Interest: Impact on aquatic ecosystems; Impact on environment; Transportation and deposition.

Publications: FWI reports and journal articles.

Other activities: Summer courses and seminar series held at the field station in Bolton Landing, New York.

Holcomb Research Institute
Butler University
4600 Sunset Avenue
Indianapolis, IN 46208   (317) 283-9421

Areas of Interest: Impact on aquatic ecosystems; Impact on environment; Impact on terrestrial ecosystems.

Publications: Conference proceedings, reports and journal articles.

Institute for Environmental Studies
University of Toronto
Toronto, Ontario M5S 1A4
Canada   (416) 978-6526

Areas of Interest: Impact on aquatic ecosystems; Impact on environment; Metal cycling related to acidification of terrestrial and aquatic systems.

Publications: In professional and scientific journals, proceedings, monographs and reports.

Other activities: Acid Precipitation Study Group.

Institute for Research on Land and Water Resources
The Pennsylvania State University
University Park, PA 16802   (814) 863-0291

Areas of Interest: Economic impact; Impact on aquatic ecosystems; Impact on environment; Impact on terrestrial ecosystems.

Publications: Institute for Research on Land and Water Resources Newsletter. Quarterly, free (not exclusively on acid rain). Technical reports and reprints.

Other activities: "Acid Precipitation in Pennsylvania" slide show to interested groups, free.

Institute of Ecology
University of Georgia
Athens, GA 30602   (404) 542-2698

Areas of Interest: Atmospheric chemistry; Emission; Impact on environment; Impact on terrestrial ecosystems; Transportation and deposition.

Publications: Research findings in leading scientific journals.

Institute of Environmental Sciences
Miami University
122 Boyd Hall
Oxford, OH 45056   (513) 529-5811

Areas of Interest: Atmospheric chemistry; Economic
Impact; Impact on environment; Political impact; Transportation and deposition.

Publications: In professional and scientific journals
and reports.

Other activities: Operates Multistate Atmospheric Power
Production Pollution Study and US 65 National wind
network precipitation chemistry stations.

Lawrence Livermore National Laboratory
P.O. Box 808 (L-262)
Livermore, CA 94550   (415) 422-1826

Areas of Interest: Atmospheric chemistry; Transportation and deposition.

Publications: Technical reports, occasional reports,
journal articles and conference proceedings.

Massachusetts Department of Environmental
  Quality Engineering
Division of Air Quality Control
One Winter Street, 8th Floor
Boston, MA 02108   (617) 292-5630

Areas of Interest: Economic impact; Emission; Impact
on aquatic ecosystems; Impact on cultural resources;
Impact on environment; Impact on terrestrial ecosystems; Political impact; Transportation and deposition.

Natural Resource Ecology Laboratory
Colorado State University
Fort Collins, CO 80523   (303) 491-1978

Areas of Interest: Deposition monitoring; Impact on
aquatic ecosystems; Impact on environment; Impact on
terrestrial ecosystems; Transportation and deposition.

Publications: NADP Data Report Precipitation Chemistry;
Quarterly. NADP Annual Report. NADP Field Manual. NADP
Siting Manual. (All four publications are currently
free.

Oak Ridge National Laboratory
P.O. Box X, Building 1505
Oak Ridge, TN 37831   (615) 574-7356

Areas of Interest: Atmospheric chemistry; Control technology; Economic impact; Emission; Impact on aquatic
ecosystems; Impact on environment; Impact on terrestrial ecosystems; Mitigation of impacts; Political
impact; Transportation and deposition.

Publications: Technical reports and journal articles.

Other activities: Workshops, symposia and technical meetings are sponsored and hosted by the laboratory.

Ontario Ministry of the Environment
199 Larch Street
Sudbury, Ontario P3E 5P9
Canada   (705) 675-4501

Areas of Interest: Atmospheric chemistry; Impact on aquatic ecosystems; Impact on terrestrial ecosystems; Transportation and deposition.

Publications: In professional journals, proceedings, and technical reports.

Pennsylvania Cooperative Fish and Wildlife Research Unit
Ferguson Building
University Park, PA 16802   (814) 865-4511

Areas of Interest: Impact on aquatic ecosystems; Impact on environment; Mitigation; Political impact.

Publications: Proceedings, technical reports and journal articles.

Other activities: Organizes symposia.

Sandia National Laboratories
P.O. Box 5800
Albuquerque, NM 87185   (505) 844-7328

Areas of Interest: Atmospheric chemistry; Emission; Source apportionment.

Publications: Research reports, proceedings and journal articles.

7.
Health Effects of Hazardous Materials

Sharon A. Keller

Introduction

A vast amount of literature has been published in this
subject area during the past six years. This has been
caused by current research relating to specific incidents
such as Love Canal and Three Mile Island as well as the
continuing discoveries of links between illnesses in humans
and environmental causal agents. Since many of these problems are work-related, primary sources of information on
occupational health hazards and monitoring specific hazards
in the workplace are the Occupational Safety and Health
Administration, the National Institute for Occupational
Safety and Health, the International Labour Organization,
and the World Health Organization.

This chapter does not attempt to be comprehensive. Because
of the volume of extant literature, it is a survey of the
literature in the field, and covers the most important
American publications. Other English-language publications
are included selectively. It should be noted that Scandinavian countries are also significant contributors to the
literature of occupational health.

Harmful effects on humans are emphasized in this chapter.
Effects of alcohol and drugs are excluded, because the
literature in those subject areas is substantial enough to
warrant separate coverage in other books.

Pamphlets and other small publications are excluded, but
many addresses for obtaining this type of material can be
found in the "Associations" section. Important newsletters
are incorporated into the "Periodicals" section.

Many more conferences and their proceedings can be found
in the online data base, Conference Papers Index, or in the
printed Index to Scientific and Technical Proceedings. To
locate information through this index as well as the others
listed, it is necessary to use many terms to identify relevant material. A list of those terms which should be used

include: health hazards, occupational health, occupational
hygiene, occupational exposure, carcinogens, mutagens, teratogens, toxicity, toxicology, reproductive hazards, industrial chemicals, chemical exposure, environmental health, as
well as specific causes of health hazards (radiation, lead,
uranium, asbestos, dust, dioxin, wood).

Numerous medical textbooks have recently been published for
physicians on medical management of specific diseases
(dermatitis, asthma, asbestosis), and many of these texts
have been included.

Libraries have been included only if some public access is
allowed. Permission to use the collection should be
obtained from any library before visiting, however.

Books

Adams, Robert M. <u>Occupational Skin Disease</u>. New York: Grune
& Stratton, 1983.

   Professionals involved with occupational diseases are
   the audience for whom this book was written. There are
   chapters on individual skin diseases, diagnosis, testing, management, prevention and rehabilitation; and
   substances that cause skin diseases. One chapter
   describes various occupations, and lists the irritants
   and standard allergens with which they are associated.

Albert, Adrien. <u>Selective Toxicity: the Physicochemical</u>
   <u>Basis of Therapy</u>. 7th ed. London: Chapman and Hall,
   1985.

   How drugs and agricultural agents affect certain cells
   without harming others is discussed. Included are anti-
   virals, narcotics, chemotherapy, anticancer agents, and
   chelating agents.

American Society of Safety Engineers. <u>Dictionary of Terms</u>
   <u>Used in the Safety Profession</u>. 2nd ed. Park Ridge,
   IL: American Society of Safety Engineers, 1981.

   This dictionary lists over 1600 terms and 150 organizations relating to occupational health and industrial
   hygiene.

Anderson, Kim and Ronald Scott. <u>Fundamentals of Industrial</u>
   <u>Toxicology</u>. Ann Arbor, MI: Ann Arbor Science, 1981.

Written for those in industry who do not have a background in toxicology, this book defines industrial
toxicology and presents a brief history of the field.
The physiology and mode of action of toxicants, types
of exposure, and classification of contaminants are
described, and a chapter is included on information
sources in industrial toxicology.

Barlow, Susan M. and Frank M. Sullivan. Reproductive
     Hazards of Industrial Chemicals: an Evaluation of
     Animal and Human Data. Orlando, FL: Academic, 1982.

This book analyzes the reproductive toxicity of about
50 industrial chemicals from the Threshold Limit Values
list. The first part of the book explains testing
procedures for reproductive hazards. A section on
reproductive hazards associated with different occupational groups is included. The second part lists each
compound, the relevant animal studies, and human studies, a summary and the authors' evaluation of reproductive hazards.

Baselt, Randall C. Biological Monitoring Methods for Indus-
     trial Chemicals. Davis, CA: Biomedical Publications,
     1980.

This book lists chemicals found in the workplace, their
occurrence and usage, concentrations in the blood, their
metabolism and excretion, toxicity, methods of biological monitoring and analysis, and a list of references.
It was written to assist the industrial hygienist in
determining chemical intoxication.

Baselt, Randall C. Disposition of Toxic Drugs and Chemicals
     in Man. 2nd ed. Davis, CA: Biomedical Publications,
     1982.

An alphabetical listing of toxic substances is provided
in this book written for toxicologists, pharmacologists,
and clinical chemists. The occurrence and usage of
each drug and chemical, blood concentrations, metabolism
and excretion, toxicity, and analysis for each entry
are followed by a list of references.

Block, J. Bradford. The Signs and Symptoms of Chemical
     Exposure. Springfield, IL: C. C. Thomas, 1980.

This brief handbook is designed for the physician who
can use it to check a sign or symptom, which is followed by the chemicals which might cause the reaction.
The chemical index lists toxic compounds followed by a
number code for the symptoms which appear with expo-

sure to the compound. It is designed to supplement toxicology textbooks, some of which are listed in the short bibliography.

Bloom, A. D., ed. Guidelines for Studies of Human Populations Exposed to Mutagenic and Reproductive Hazards. New York: March of Dimes Birth Defects Foundation, 1981.

This book results from a conference held after a 1980 cytogenetic study on the residents of Love Canal in New York State. Tables of statistics listing hazardous substances and different outcomes from exposure to them are very useful. References and suggestions for implementation of the guidelines are given.

Blumenthal, Daniel S., ed. Introduction to Environmental Health. New York: Springer-Verlag, 1985.

A textbook designed for college and graduate-level students of environmental health. Topics covered are infectious agents in the environment, chemicals, air pollution, low-level radiation, occupational health, the Occupational Safety and Health Act, hazardous waste management, and environmental health law.

Bowman, Malcolm C. Handbook of Carcinogens and Hazardous Substances: Chemicals and Trace Analysis. New York: Dekker, 1982.

This book is written for researchers who can use it as a guide to analytical methods they can use in the identification and quantitation of carcinogens. The chapters are divided by type of carcinogenic agent. Each chapter has a list of references. A subject index is included.

Bretherick, L., ed. Hazards in the Chemical Laboratory. 3rd ed. London: Royal Society of Chemistry, 1981.

This handbook was written for chemists, as it identifies and discusses hazards in chemical laboratories. The Health and Safety at Work Act of 1974 is discussed in one chapter. Other chapters deal with safety planning and management, fire protection, reactive chemical hazards, chemical hazards and toxicology, health care and first aid, and precautions against radiation. Appendices list the addresses of international labor inspectorates and occupational health authorities, and European and British poison information centers. Most of the book consists of a chapter entitled "Hazardous Chemicals." Four hundred and eight flammable, explosive, corrosive or toxic substances commonly used in

chemical laboratories are described; and their toxic
effects, first aid measures, and spillage disposal are
listed.

Brill, A. Bertrand, ed. Low-level Radiation Effects: a Fact
Book. New York: Society of Nuclear Medicine, 1982.

In this looseleaf handbook, data is presented on radiation effects in animals, humans, and plants. Dosages
received are followed by data on somatic and genetic
effects, and risks. A list of references and recommended reading is provided.

Chenier, Nancy Miller. Reproductive Hazards at Work: Men,
Women, and the Fertility Gamble. Ottawa: Canadian
Advisory Council on the Status of Women, 1982.

There is a Canadian focus to this book, noticeable in
the specific cases which are cited, and legal questions
which are discussed. Specific chemical and biological
hazards are cited along with protective measures which
should be taken. There is a bibliography as well as a
list of 163 references.

Christian, Mildred S., et al., eds. Assessment of Repro-
ductive and Teratogenic Hazards. Advances in Modern
Environmental Toxicology, Volume 3. Princeton, NJ:
Princeton Scientific, 1983.

The first section of this book discusses the problem of
reproductive toxicity in general, and teratogenicity of
drugs, foods, petroleum and radiation, primarily in
animals. The second section contains the proceedings
of two Environmental Protection Agency 1980 conferences: "Assessment of Risks to Human Reproduction and to
Development of the Human Conceptus from Exposure to
Environmental Substances." Female and male reproduction
are discussed in relation to toxicity, estimation of
risk to the fetus, and epidemiology, pharmacokinetics,
and sexual behavior.

Clarke, Stewart W. and Demetri Pavia, eds. Aerosols and the
Lung: Clinical and Experimental Aspects. Boston:
Butterworth, 1984.

Intended for the clinician, the physiology of aerosols
in relation to the lung is described in preparation for
a discussion of therapeutic uses of aerosols and
methods of producing safe aerosols.

Clayton Environmental Consultants, Inc. Medical Management
of Chemical Exposures in the Petroleum Industry. Washington, D.C.: The American Petroleum Institute, 1982.

Physicians and health professionals employed in the
petroleum industry can use the general information in
this publication for evaluating occupational health
hazards. It is a companion to the Industrial Hygiene
Monitoring Manual.

Clayton, George D. and Florence E. Clayton. Patty's
    Industrial Hygiene & Toxicology. 3 vols. 3rd ed. New
    York: Wiley, 1978-1982.

    This group of reference texts are classics in the field
    of occupational health. Chapters in the first volume
    cover many types of potential industrial health pro-
    blems, such as ionizing radiation, agricultural haz-
    ards, heat stress, inhaled dusts and occupational der-
    matoses. Volume 2 deals with toxicology of specific
    substances, such as esters, ethers, metals, epoxy com-
    pounds, phenols, and aldehydes. Volume 3 describes
    monitoring and testing methods, evaluation of exposure
    to several substances, and applications industry.

Deisler, Paul F., Jr., ed. Reducing the Carcinogenic Risks
    in Industry. Occupational Safety and Health, Volume 9.
    New York: Dekker, 1984.

    The point of view of the problem of carcinogenic risk
    in industry in the process of being solved is presented
    by the authors. Relevant federal research is
    described; methods of risk reduction and the role of
    OSHA are discussed.

A Directory of Occupational Health and Safety in Ontario.
    Toronto: Ontario Ministry of Labour, Occupational
    Health and Safety Division, Standards and Programs
    Branch, 1983(?).

    Information on occupational health and safety programs
    and services available in Ontario is provided in this
    directory. Addresses, contacts, objectives, activi-
    ties, testing, capabilities, publications, funds, and
    research are provided for each association, agency or
    laboratory listed.

D'Itri, Frank M. and Michael A. Kamria. PCB's: Human and
    Environmental Hazards. Boston: Butterworth, 1983.

    A comprehensive study presents PCB exposure in the Great
    Lakes, human exposure, the views of the environmenta-
    lists and fishermen, epidemiology, and regulation.

Documentation of the Threshold Limit Values. 4th ed.
    Cincinnati: American Conference of Governmental
    Industrial Hygienists, Inc., 1980.

This is an essential directory in the field of occupational health. It has an alphabetical listing of chemical substances and physical agents present in the workplace. It lists the chemical formula, a chemical name
and describes the effect of the substance in animals
and man according to dosage.

Ecobichon, Donald J. and Robert M. Joy. <u>Pesticides and
Neurological Diseases</u>. Boca Raton, FL: CRC Press,
1982.

This is a comprehensive survey of the toxicological
effects of pesticides on the nervous system. The historical development of the topic, pesticide toxico-
kinetics, an overview of the nervous system, and the
neurotoxicity, pharmacokinetics, and chemistry of
chlorinated hydrocarbon insecticides, cyclodiene and
hexachlorocyclohexane derivatives, organophosphorus
ester insecticides, carbamate ester insecticides, and
mercurial fungicides are thoroughly covered.

Eger, Edmond I. <u>Nitrous Oxide/N_2O</u>. New York: Elsevier,
1985.

The significant sections discuss the mutagenicity,
carcinogenicity, and teratogenicity of nitrous oxide.
Methods of controlling occupational exposure to nitrous
oxide and discussions of whether it should be used make
this an important contribution to the literature of the
field.

Fawcett, Howard H. and William S. Wood. <u>Safety and Accident
Prevention in Chemical Operations</u>. 2nd ed. New York:
Wiley, 1982.

Written with the presumption that all chemical, biological, and radiation sources can be processed, handled,
used, and disposed of safely if precautions are ob-
served, several chapters discuss the health hazards of
working with chemicals. Appendices supply data on
selected hazardous chemicals, the Carcinogen Assessment
Group's List of Carcinogens, and other sources of information.

Foussereau, Jean, et al. <u>Occupational Contact Dermatitis:
Clinical and Chemical Aspects</u>. 1st ed. Copenhagen:
Munksgaard, 1982.

Beginning with a history of skin allergy, a description
of allergy tests, and diagnosis and treatment of contact dermatitis, the authors discuss the disease in
relation to specific occupations. There is a list of
organic pigments and dyes with their chemical descriptions, as well as a section on the principle of

isolation and identification of an allergenic dye. An
alphabetical listing of allergens gives test concentrations and diluents. Allergens in plastic materials,
and an alphabetical list of occupations with which they
may be associated are followed by 71 color plates of
different types of skin reactions.

Frazier, Claude Albee, ed. Occupational Asthma. New York:
    Van Nostrand Reinhold, 1980.

Written for the practicing physician, this book includes chapters on asthma caused by exercise, wrapping
meats in heated plastic, textile work, western red
cedar, wood dust allergy, baking, vinyl chloride, polyvinyl chloride, detergents, cotton dust, silica dust,
asbestos dust, laboratory animals, hoya (sea squirt),
pharmacologic dust, castor-bean dust, porcine trypsin
powder, pine rosin fumes, and epoxy resin systems.
Statistical tables, graphs, and numerous photographs
are included.

Gibson, James E. Formaldehyde Toxicity. New York: Hemis-
    phere, 1983.

This book has been written because of the lack of
printed literature on the toxicity and potential carcinogenicity of formaldehyde. Studies include effects
on humans and animals. Statistical tables are a very
good source of information on morbidity and mortality
of workers exposed to formaldehyde.

Gill, Frank S. and Indira Ashton. Monitoring for Health
    Hazards at Work. Royal Society for the Prevention of
    Accidents, 1982.

Those who are involved with monitoring the workplace
for health hazards, union safety representatives, and
shop stewards will find these instruction sheets, and a
list of units of measurements and abbreviations useful.
Chapters on dust, gases and vapors, heat, ventilation,
noise, light, and other hazards, are introduced with a
description of the health hazards, measurement of the
material, and methods of monitoring. Types of monitoring equipment are discussed. Addresses and telephone numbers of British equipment suppliers and occupational hygiene consultants are given.

Gofman, John W. Radiation and Human Health. San Francisco:
    Sierra Club Books, 1981.

Cancer, leukemia, and chromosome damage are the effects
of ionizing radiation described in this book. It is
designed as a "self-contained" reference book, which
attempts to explain all technical terms used in the

text. It contains many tables of statistics on amount
of exposure in relation to the adverse effect it will
cause, and an extensive bibliography. It is a long and
thorough treatment of this topic.

Gosselin, Robert E., et al. Clinical Toxicology of Commer-
    cial Products. 5th ed. Baltimore: Williams and
    Wilkins, 1984.

    This reference work provides toxicity data for 1642
    substances or classes of substances, over 7000 biblio-
    graphic citations, information on the composition of
    16,000 consumer products, and addresses and phone num-
    bers of 1500 manufacturers. There is a therapeutics
    index, ingredients index, and trade name index.

Grice, H. C., ed. Current Issues in Toxicology. New York:
    Springer-Verlag, 1984.

    The two major sections of this book which is part of a
    series sponsored by the International Life Sciences
    Institute are: "Interpretation and Extrapolation of
    Chemical and Biological Data on Carcinogens to
    Establish Human Safety Standards," and "The Use of
    Short-Term Tests for Mutagenicity and Carcinogenicity
    in Chemical Hazard Evaluation."

Griest, W. H. and M. R. Guevin. Health Effects Investi-
    gation of Oil Shale Development. Lancaster, PA:
    Technomic, 1981.

    A special research project resulted in these 17 studies
    relating to the health effects of shale oil production
    and refining.

Grimmer, Gernot, ed. Environmental Carcinogens: Polycyclic
    Aromatic Hydrocarbons: Chemistry, Occurrence, Biochem-
    istry, Carcinogenicity. Boca Raton, FL: CRC Press,
    1983.

    This is a survey of environmental pollution by polycy-
    clic aromatic hydrocarbons (PAH) and especially their
    biological and carcinogenic effects. The purpose of
    the information is to provide the German government
    with data to mandate standards for air quality. Chap-
    ters are included on the chemistry, occurrence, metabo-
    lism, and epidemiology of PAH.

Gross, Paul and Daniel C. Braun. Toxic and Biomedical
    Effects of Fibers: Asbestos, Talc, Inorganic Fibers,
    Man-made Vitreous Fibers, and Organic Fibers. Park
    Ridge, NJ: Noyes, 1984.

Asbestos and mesothelioma, gastrointestinal cancers,
laryngeal cancer, and the history of asbestos-related
disease are among the subjects covered in the section
of this book on asbestos. Asbestos substitutes which
have come into use since the discovery of health problems due to asbestos exposure are evaluated.

Hamilton, Alice. Hamilton and Hardy's Industrial Toxicology.
4th ed. Littleton, MA: PSG, Inc., 1983.

This standard textbook describes metals and metalloids,
chemical compounds, organic high polymers, pesticides,
radiant energy, and dusts by their uses and industrial
toxicity.

Handbook of Hazardous Materials. 2nd ed. Technical Guide
No. 7. Chicago: Alliance of American Insurers, 1983.

This book contains information not included in the
Handbook of Organic Industrial Solvents or the list of
Threshold Limit Values. Many significant primary
sources are identified in the text. Toxicology,
control of exposure, TLV's, hazard recognition, and
evaluation of exposure are discussed.

Hausen, Bjorn M. Woods Injurious to Human Health: A Manual.
Berlin: de Gruyter, 1981.

Irritation and disease caused by woods are Hodgkin's
disease, adenocarcinoma, asthma, and dermatitis. A
comprehensive table of all woods known to be toxic,
irritating, or sensitivity-inducing gives the use of
the wood, the symptoms it produces, and a list of
references. A botanical, zoological, and subject index, and an index of names of woods provide thorough
access to information in this book.

Hay, Alastair. The Chemical Scythe: Lessons of 2, 4, 5-T
and Dioxin. New York: Plenum, 1982.

Exposure of populations to dioxin are thoroughly discussed. Love Canal, the 1976 overheating of a nuclear
reactor in Seveso, Italy, Agent Orange in the Vietnam
war, the spraying of herbicides in Cambodia and Laos,
and other industrial accidents are described in detail.
Many references to these specific incidents are given.

Hayes, A. Wallace, ed. Toxicology of the Eye, Ear, and Other
Special Senses. New York: Raven, 1985.

Chemical injury to the sense organs is the focus of
this book. Assessing toxic effects using animal test
procedures is emphasized.

Hayes, Wayland J., Jr. Pesticides Studied in Man. Balti-
    more: Williams and Wilkins, 1982.

    This book is divided into 13 chapters by type of pesti-
    cide (inorganic and organometallic, herbicides, fungi-
    cides, synergists, etc.) Each specific pesticide is
    followed by a description of its identity, properties,
    chemical structure, synonyms, physical and chemical
    properties, formulations and uses, and toxicity in
    laboratory animals and humans.

Hazard Assessment of Chemicals. Vol. 1, Current Develop-
    ments. Orlando, FL: Academic, 1981.

    One chapter of this book briefly lists information
    sources in the assessment of toxic substances. The
    other chapters cover health effects in humans, animals,
    laboratory tests, and aquatic environments. Tables and
    lists of references accompany each chapter.

Hemminki, K., et al., eds. Occupational Hazards and Repro-
    duction. New York: Hemisphere, 1985.

    Current knowledge regarding occupational reproductive
    hazards and reproductive biology in men and women is
    presented, with the admission that research in this
    area is still new, so the best-studied subjects such as
    malformation and spontaneous abortion are emphasized.

Hendee, William R., ed. Health Effects of Low Level
    Radiation. East Norwalk, CT: Appleton-Century-Crofts,
    1984.

    The editor emphasizes the book's unbiased approach to
    the uses and health effects of radiation. The purpose
    of the book is to provide an overview of the topic, so
    radiation is defined, and methods of measurement are
    explained. Epidemiologic studies, studies of the
    effects on the fetus, genetic effects, an estimation
    of the risk of cancer from radiation exposure, alterna-
    tives to nuclear power, and management of nuclear waste
    are discussed.

Hill, Rolla and James A. Terzian, eds. Environmental
    Pathology: an Evolving Field. New York: A. R. Liss,
    1982.

    Each chapter describes specific disease states that can
    result from environmental hazards such as radiation,
    silica, asbestos, and drugs. Chapters on "Computer-
    assisted Quantification of Injury to Cells and Tissues"
    and "Concepts in Multivariate Data Analysis" are use-
    ful.

Hook, Jerry B., ed. <u>Toxicology of the Kidney</u>. New York: Raven, 1981.

Assessment of nephrotoxicity, and the effects of environmental chemicals and pesticides are among the reviews written by experts in the field.

International Labour Office. <u>International Classification of Radiographs of Pneumoconioses</u>. Occupational Safety and Health Series, Volume 22. Geneva: International Labour Office, 1980. 22 Radiographs.

This classification includes the types of pheumoconioses which are characterized by regular and irregular opacities. It was drawn up by the ILO, American College of Radiology, and by the Working Group on Radiodiagnosis of the Commission of European Communities. The text accompanying the radiographs is entitled <u>Guidelines for the use of Radiographs of Pneumonoconioses</u>.

International Labour Organization. <u>Occupational Exposure to Airborne Substances Harmful to Health</u>. Geneva: International Labour Organization, 1980.

The ILO code of practice was derived from a meeting of the ILO in coordination with the World Health Organization. A glossary of terms is included.

Kelly, William D., ed. <u>Agricultural Respiratory Hazards</u>. Annals of the American Conference of Government Industrial Hygienists, Volume 2. Cincinnati: American Conference of Government Industrial Hygienists, 1982.

Some of the hazards discussed are infectious and immunologic agents from livestock and farm structures, dusts and allergies, control of grain dust, and the human respiratory response.

Kirsch-Volders, Micheline, ed. <u>Mutagenicity, Carcinogenicity, and Teratogenicity of Industrial Pollutants</u>. New York: Plenum, 1984.

Written for a broad spectrum of readers, information on genetic toxicology in plants, animals, and humans are listed in tables and discussed in detail in chapters which cover the molecular mechanisms of mutagenesis and carcinogenesis and the toxicity, metabolism, mutagenicity, carcinogenicity, teratogenicity, and embryotoxicity of specific metals, insecticides, industrially important monomers, and halogenated hydrocarbon solvents. Each chapter is followed by an extensive list of references.

Langard, Sverre, ed. Biological and Environmental Aspects
    of Chromium. Topics in Environmental Health, Volume 5.
    New York: Elsevier, 1982.

    Production of chromium and occupational exposure, its
    mutagenic and cytogenetic effects, organ toxicity in
    animals, and carcinogenic and dermatologic effects are
    the subjects covered in this volume.

Legator, Marvin S., et al., eds. The Health Detective's
    Handbook: a Guide to the Investigation of Environmental
    Health Hazards by Nonprofessionals. Baltimore: John
    Hopkins University Press, 1985.

    Based on the premise that a high-quality health survey
    can be done at the community level, this manual gives
    background information about identifying health hazards
    and designing an experiment or questionnaire. It also
    provides practical guidelines on how to get information
    and what to do with it.

Mackison, Frank W., et al., eds. Occupational Health Guide-
    lines for Chemical Hazards. DHHS (NIOSH) Publication
    No. 81-123. Washington, D.C.: National Institute for
    Occupational Safety and Health/Occupational Safety and
    Health Administration, 1981.

    This book summarizes information on permissible expo-
    sure limits, chemical and physical properties, and
    health hazards. It provides recommendations for medi-
    cal surveillance, respiratory protection and personal
    protection and sanitation practices for specific chem-
    icals that have federal occupational safety and health
    regulations. A table of NIOSH recommendations for
    occupational health standards has current OSHA Environ-
    mental Standards, NIOSH recommendations for environmen-
    tal exposure limits and health effects. There are sum-
    maries of each substance's permissible exposure limit,
    health hazard information, chemical and physical pro-
    perties, monitoring and measurement procedures, person-
    al protective equipment, sanitation, a listing of com-
    mon operations in which the substance is used and
    control methods which may be effective, emergency first
    aid procedures, spill, leak, and disposal procedures,
    and a list of references.

Magee, Peter N., ed. Nitrosamines and Human Cancer.
    Banbury Report, No. 12. Cold Spring Harbor, NY: Cold
    Spring Harbor Laboratory, 1982.

    In the first section, evidence relating N-nitroso com-
    pounds to human carcinogenesis is presented. Sources
    of exposure are discussed, and epidemiological studies
    are presented.

Maibach, Howard I. and Gerald A. Gellin, eds. Occupational
    and Industrial Dermatology. Chicago: Year Book Medical
    Publishers, Inc., 1982.

    Supplementing Etain Cronin's Contact Dermatitis
    (Churchill Livingstone, 1980) and Jean Foussereau's
    Occupational Dermatology (Munksgaard, 1982), this book
    is written for health professionals who will be able
    to use it to diagnose and treat specific skin diseases
    caused by the work environment. Worker compensation
    and disability are discussed. Dermatitis is discussed
    in relation to beauticians, medical and surgical per-
    sonnel, culinary plants, the wood industry, construc-
    tion work, and silica dust exposure. An appendix
    lists procedures for patch testing.

Marzulli, Francis N. and Howard I. Maibach, eds. Dermato-
    toxicology. 2nd ed. New York: Hemisphere, 1983.

    Covering substances that produce adverse effects to the
    skin, this book includes chapters on skin physiology,
    absorption and light-induced toxicity, photocarcino-
    genesis, cutaneous carcinogenicity, environmental de-
    pigmenting chemicals, chloracne, chemical-induced hair
    loss, heavy-metal-induced hyperpigmentation, and neuro-
    toxic substances. It is written for the physician or
    dermatologist.

Mattison, Donald R. Reproductive Toxicology. Progress in
    Clinical and Biological Research, Volume 117. New York:
    Alan R. Liss, 1983.

    The topics in this book include reproductive toxicology
    in the male and female, and in the fetus and newborn,
    and surveillance of reproductive toxicology. The chap-
    ters in each section cover effects of lead on sperm,
    hypothalamic-pituitary mechanisms, gonadal injury from
    chemotherapy, effect of polycyclic aromatic hydrocarbons
    on oocytes, and occupationally-derived chemicals in
    breast milk.

Mettler, Fred A., Jr. and Robert D. Moseley, Jr. Medical
    Effects of Ionizing Radiation. Orlando, FL: Grune and
    Stratton, 1985.

    Animal studies are not included; sources of exposure
    are identified. Direct effects, carcinogenic effects,
    and genetic effects of ionizing radiation are discussed.

Nater, Johan P. and Anton C. de Groot. Unwanted Effects of
    Cosmetics and Drugs Used in Dermatology. 2nd ed.
    New York: Elsevier, 1985.

Side effects of topical drugs, photochemotherapy, systemic drugs, and cosmetics are presented with references and an index of all compounds mentioned in the text.

National Research Council. Committee on Hazardous Substances in the Laboratory. <u>Prudent Practices for Handling Hazardous Chemicals in Laboratories</u>. Washington, D.C.: National Academic Press, 1980.

Procedures for handling and disposal of hazardous materials in laboratories are discussed in this book which resulted from a study by the National Research Council. It does not include physical or biological hazards in laboratories. Procedures for working with substances that are toxic, corrosive, flammable, or explosive, and with compressed gases are identified. Safety data sheets for a selected group of laboratory chemicals are provided, as well as a selected bibliography of 21 citations on potential hazards of known chemicals. Design of laboratory apparatus and ventilation in the laboratory are discussed. Appendices evaluate published epidemiological studies of chemists and give threshold limit values adopted by ACGIH in 1980.

Needleman, Herbert L. <u>Low Level Lead Exposure: the Clinical Implications of Current Research</u>. New York: Raven, 1980.

The three major sections are divided into papers on population studies of low level lead exposure, animal investigations of low level lead exposure, and public health, economic, and regulatory implications of low level lead exposure.

Nriagu, J. O. <u>Changing Metal Cycles and Human Health</u>. Life Sciences Research Report, Volume 28. New York: Springer-Verlag, 1984.

The focus of this book is on polluting metals: their cycling and accumulation, metabolism and poisoning.

Nutt, A. R. <u>Toxic Hazards of Rubber Chemicals</u>. New York: Elsevier, 1984.

Epidemiological studies, bladder cancer, toxicity of many of the chemicals used in the rubber industry, and testing and monitoring methods are presented. An appendix lists sources of air monitoring equipment.

Obe, G., ed. <u>Mutations in Man</u>. New York: Springer-Verlag, 1984.

These articles discuss the relationship between environmental chemicals and radiation and mutagenesis in
humans. Among the topics covered are the biochemistry
of DNA damage and repair, the origin of chromosomal
aberrations, types of mutations, and the organization
of the human genome.

Occupational Exposure Limits for Airborne Toxic Substances:
    A Tabular Compilation of Values from Selected Coun-
    tries. 2nd ed. Occupational Safety and Health Series,
    No. 37. Geneva: International Labour Office, 1980.

This reference source presents in tabular form, the
guidelines used in a number of countries for about 1200
toxic substances.

Occupational Hazards from Non-ionizing Electromagnetic
    Radiation. Occupational Safety and Health Series, No.
    53. Geneva: International Labour Organization, 1985.

The types of non-ionizing radiation are described.
Interaction with biological tissues, and potential
health hazards are included.

Parmeggiani, Luigi, ed. Encyclopedia of Occupational Health
    and Safety 1983. 3rd ed. 2 volumes. Geneva: Inter-
    national Labour Office, 1983.

This comprehensive work contains entries for occupations which have hazards associated with them, as well
as substances, chemicals and concepts in occupational
health. It contains over 1150 articles and over 6000
bibliographic references.

Pearce, B. G., ed. Health Hazards of VDT's? New York:
    Wiley, 1984.

This book presents research to date on the subject of
VDT health hazards. It concludes that more research
needs to be done because the question posed in the
title is not conclusively answered. Subjects discussed
are facial rashes and cataracts, radiation emissions,
ergonomics, and occupational stress relating to VDT's.

Plaa, Gabriel L., ed. Toxicology of the Liver. New York:
    Raven, 1982.

Written for the toxicologist, this book describes the
effects of many types of organic and inorganic agents,
as well as synthetic agents, on the liver. Chapters on
subcellular events, and hepatocarcinogenesis are
included.

Preger, Leslie, ed. _Induced Disease: Drug, Irradiation,_
    Occupation. Orlando, FL: Grune and Stratton, 1980.

    This book was written to keep radiologists and physi-
    cians aware of the physiological effects of the disease-
    producing effects of hazardous materials. The parts on
    drug-induced and occupationally-induced disease are
    divided into chapters by type of disease. The part on
    irradiation-induced disease also covers management of
    radiation hazards in women of childbearing years.

Pucknat, A. W., ed. _Health Impacts of Polynuclear Aromatic_
    Hydrocarbons. Environmental Health Review, No. 5.
    Park Ridge, NJ: Noyes, 1981.

    PNA hydrocarbon levels in water and air must be moni-
    tored for public health risk. They are found in crude
    oil, petroleum, coal, and oil shale. Their fate in
    water, air, and soil, and health effects are covered in
    addition to current actions to protect human health,
    and methods for detection and analysis of PNA.

The Research Staff of F. & S. Press, with Michael H. Turk.
    Occupational Medicine: Surveillance, Diagnosis and
    Treatment. New York: F. & S. Press, 1982.

    The emphasis of this book is on describing screening
    methods presently used by industry to monitor workers
    who may be exposed to medical hazards. It is based on
    a survey done by Frost and Sullivan in which a question-
    naire was mailed to members of the American Occupational
    Medicine Association. Seventy tables list results of
    the study.

Rom, William and Victor Archer, eds. _Health Implications_
    of New Energy Technologies. Ann Arbor, MI: Ann Arbor
    Science, 1980.

    Subjects covered in this 785-page book include ionizing
    radiation, oil shale, coal combustion, and coal mining.

Sanders, Charles L. and Ronald L. Kathren. _Ionizing_
    Radiation: Tumorigenic and Tumoricidal Effects.
    Columbus, OH: Battelle Press, 1983.

    This book can be used as a textbook of radiation biolo-
    gy, and a reference book for personnel involved with
    the radiological aspects of cancer. The first half
    covers basic principles of radiation biology; the
    second half covers the radiation biology of specific
    organs, tissues, and systems.

Sax, N. Irving. <u>Cancer-causing Chemicals</u>. New York: Van
    Nostrand Reinhold, 1981.

    The emphasis of this book is on providing an exhaustive
    listing of known carcinogens. Two chapters deal with
    carcinogens in the workplace. Chapter 6 rates carcino-
    genicity of 2400 entries as "conclusive," "suggestive,"
    or "indeterminate." Each entry contains a NIOSH number
    and a CAS registry number. Chapter 2 has excellent
    tables on "Industrial Agents Associated with Cancer."
    Chapter 3 lists techniques for controlling exposure to
    carcinogens. In the chapter "Regulations Affecting Use
    of Carcinogens" a section describes regulations for
    the workplace, damages and compensation, and the issue
    of using developing countries as dumping and testing
    grounds is discussed.

Shaw, Susan. <u>Overexposure: Health Hazards in Photography</u>.
    Carmel, CA: The Friends of Photography, 1983.

    Potentially harmful chemicals and processes are identi-
    fied, as well as those involving little or no risk.
    The conclusion of the editor is that with precautions,
    black and white photographic processes have little
    risk, while color and many historical processes involve
    greater risk. With precautions and substitution on
    non-hazardous alternatives, photographers can minimize
    their risk. Charts list types of gloves to be worn
    with different photographic contaminants, processes
    which require exhaust ventilation, flammable and combus-
    tible liquids used, materials which should not be used
    and their suitable substitutes.

Shepard, Thomas H. <u>Catalog of Teratogenic Agents</u>. 4th ed.
    Baltimore: Johns Hopkins University Press, 1983.

    This book was written to summarize the findings on
    teratogenic agents and their effects. The listings
    include synonyms for the agent, and a summary of
    significant research findings.

Smith, Ivan C. and Bonnie L. Carson. <u>Trace Metals in the
    Environment--Cobalt</u>. Lancaster, PA: Technomic, 1981.

    A discussion of the chemical composition of cobalt, its
    uses and occurrence is followed by information on
    cobalt in the food chain, its physiological effects in
    humans, animals, and plants, and an assessment of its
    health hazards. The book resulted from research done
    by the National Institute of Environmental Health
    Sciences and the Midwest Research Institute.

Stellman, Jeanne and Mary Sue Henifin. Office Work Can Be
Dangerous to Your Health: a Handbook of Office Health
and Safety Hazards and What You Can Do About Them. New
York: Pantheon Books, 1983.

This is a comprehensive book on office health. In
addition to discussing VDT's, lighting, and indoor air
pollution, appendices list resource organizations,
model contract language, a bibliography on occupational
health and office safety hazards, a survey for office
health and safety, and "New Technology in the American
Workplace," which is the text of hearings by the Subcommittee on Education and Labor, U.S. House of Representatives Committee on Education and Labor, June 23,
1982. Numerous tables, charts, and statistics are
included.

Stich, Hans F., ed. Carcinogens and Mutagens in the
Environment. 5 volumes. Boca Raton, FL: CRC Press,
1982-1985.

This series covers assessment and harmful effects of
environmental carcinogens and mutagens. Individual
volumes discuss food products, naturally occurring
compounds, and the workplace.

Substances Used in Plastics Materials Coming into Contact
With Food. 2nd ed. Strasbourg: Council of Europe
Publications Section, 1982.

A study was done by the Council of Europe to identify
substances used in plastics materials which come into
contact with food, and to determine which are hazardous
and not hazardous to public health.

Suess, Michael J. Nonionizing Radiation Protection. WHO
Regional Publications. European Series, No. 10. Copenhagen: Regional Office for Europe, 1982.

The research for this material was done during the
1970's. The results are presented to provide information on biological effects of non-ionizing radiation
(ultraviolet, microwave, laser, infrared, ultrasound,
electric and magnetic fields). Standards for exposure
are recommended, and a glossary of internationally
standard terms are listed.

Susanne, C., ed. Genetic and Environmental Factors During
the Growth Period. NATO ASI Series, Series A, Life
Sciences, Volume 70. New York: Plenum, 1984.

A significant paper in this book is "Effects of Industrial Pollution on Somatic and Neuropsychological
Development," by Roland Hauspie, et al.

Thomas, John A., et al., eds. Endocrine Toxicology. New
York: Raven, 1985.

The endocrine system is especially susceptible to toxic
agents, and chemicals act on neuroendrocrine centers to
alter hormonal activity. The relationship between
chemicals and specific hormones is identified.

The U.N. Environmental Programme, World Health Organization,
and the International Radiation Protection Association.
Lasers and Optical Radiation. Environmental Health
Criteria, Volume 23. Geneva, World Health Organization,
1982.

This document was produced to provide criteria for
standards of protection measures against non-ionizing
radiation. Known adverse health effects of laser radiation are discussed in depth, especially photokeratitis,
retinal injury, erythema, and burns of the skin. Exposure limits, and eye and skin protection are recommended. There is an extensive bibliography and a glossary.

Wagner, Sheldon L. Clinical Toxicology of Agricultural
Chemicals. Park Ridge, NJ: Noyes, 1983.

In the first part of this book, the production and use
of agricultural chemicals, the background of the controversy about using these chemicals, and their toxicology and molecular biology are discussed. The second
part describes specific chemicals in depth.

Periodicals

American Industrial Hygiene Association Journal. American
Industrial Hygiene Association. Monthly, published
since 1958.

All of the articles in this journal are related to
health effects of hazardous substances in the industrial environment. Articles are peer-reviewed; book
reviews, meetings, courses, and conferences are listed.

American Journal of Industrial Medicine. Alan R. Liss, Inc.
Quarterly, published since 1980.

This journal contains reports of original research, case
reports, letters, and book reviews in the area of occupational medicine.

Annals of Occupational Hygiene. Pergamon. Quarterly, pub-
    lished since 1958.

    All of the articles in this publication of the British
    Occupational Hygiene Society are related to the health
    effects of hazards in the work environment. Book
    reviews and proceedings of the BOHS are included.

Archives of Environmental Health. Heldref Publications. Bi-
    monthly, published since 1950.

    International in scope, and including some animal stud-
    ies, the journal publishes articles on the relationship
    between the environment and human health. Effects such
    as carcinogenesis, mutagenesis, inflammation, infec-
    tion, or degenerative disease are considered.

Archives of Toxicology. Springer-Verlag. Two volumes per
    year, 3-4 issues per volume, published since 1930.

    Articles on all aspects of toxicology: pharmacodynamics,
    analysis, etc. are included. Animal studies and exper-
    imental studies are emphasized. This journal is the
    official organ of the European Society of Toxicology.

Art Hazard News. Center for Occupational Hazards. Published
    10 times per year, since 1978.

    This four-page newsletter contains brief articles about
    hazards to which artists and art students may be ex-
    posed and may be freely reprinted. It also contains
    articles on relevant legislation and regulations, and
    lawsuits.

At the Centre. Canadian Centre for Occupational Health and
    Safety. Quarterly, published since 1978.

    This newlsetter contains articles, news items, recent
    publications, and a listing of conferences, meetings
    and continuing education workshops.

British Journal of Industrial Medicine. TUC Centenary
    Institute of Occupational Health. Quarterly, published
    since 1944.

    The scope of the journal includes the medical treatment
    of occupational diseases, epidemiology, and toxicology
    of industrial and agricultural chemicals. Book reviews,
    brief reports, and case histories are included with
    original papers.

Bulletin of Environmental Contamination and Toxicology.
    Springer-Verlag. Monthly, published since 1966.

    The goal of this publication is to quickly publish
    short articles describing important discoveries in
    environmental contamination and toxicology. Most of
    the articles contain studies done with animals.

Canadian Occupational Health and Safety News. Corpus Infor-
    mation Services, Ltd. Weekly, published since 1978.

    Specific incidents relating to workers' health and
    safety are described, and government actions, govern-
    ment contracts awarded, and a legislative summary are
    included.

Canadian Occupational Safety.   Cash Crop Farming Publica-
    tions, Ltd. Bimonthly, published since 1963.

    Associated with the Canadian Society of Safety Engin-
    eers, this magazine contains articles on occupational
    safety and hazardous substances, letters, and a listing
    of meetings and conferences.

Computers in Safety and Health. Consulting Health Services.
    Published six times per year, since 1984.

    This newsletter publishes news items on computer appli-
    cations for industrial hygienists.

Drug and Chemical Toxicolgy. Dekker. Quarterly, published
    since 1978.

    Research papers, review articles, and short notes are
    included in this journal, which is published for the
    rapid communication of information on animal toxicolo-
    gy, teratology, mutagenesis, and carcinogenesis.

Ecotoxicology and Environmental Safety. Academic. Published
    six times per year, since 1977.

    Published under the auspices of the International
    Academy of Environmental Safety and the official jour-
    nal of the International Society of Ecotoxicology and
    Environmental Safety, this official organ contains
    articles discussing the effects of environmental health
    hazards on animals, plants, and microbes.

Environmental Health Letter. Environews. Bimonthly, pub-
    lished since 1961.

News items relating to legislation and government agencies are discussed in this newsletter.

Environmental Health Perspectives. Department of Health and Human Services, Public Health Service, National Institute of Health. DHHS Publication No. (NIH) 84-218. Published six times per year, since 1972.

Each issue of this journal covers a single topic in depth. The purpose is to communicate research findings on environmental health, and to inform researchers of possible health hazards. It publishes proceedings, reviews, summaries of toxicologic findings, and letters. Some topics that have been covered are pulmonary toxicology, asbestos, metallothionen and cadmium nephrotoxicity, tumor promotion, and health effects of toxic wastes.

Environmental Health Review. Canadian Institute of Public Health Inspectors. Quarterly, published since 1956.

This journal publishes short articles on public health topics with a focus on Canadian issues.

Environmental Health and Safety News. University of Washington, Department of Environmental Health. Monthly, published since 1951.

Each issue reports on a single topic in depth, giving research findings and case studies. References are listed for each issue. Some topics covered are "Indoor Air Problems - The Office Environment - A National Epidemic," and "Formaldehyde in Mobile and Conventional Homes."

Environmental Research. Academic. Bimonthly, published since 1967.

This journal focuses on international subjects relating to environmental medicine, and the effects of changes in the environment on humans.

Environmental Toxicology and Chemistry. Society of Environmental Toxicology and Chemistry. Quarterly, published since 1982.

The three sections in each issue of this journal are Environmental Chemistry, Environmental Toxicology, and Hazard Assessment. The short communications, letters to the editor, research papers, and review articles deal primarily with international studies on animals.

Hazards Review: an International Bulletin Reporting on Haz-
    ards Associated with the Production, Processing and Use
    of Rubber and Plastic Materials. Elsevier. Monthly,
    published since 1979.

    This newsletter provides brief articles on hazardous
    materials findings from selected countries.

Health and Safety in Industry and Commerce. Microinfo Ltd.
    Monthly, published since 1977.

    News items, listings of new publications including NTIS
    and WHO documents, upcoming meetings and workshops are
    features of this publication.

Health & Safety Information Bulletin. Eclipse Publications,
    Ltd. Monthly, published since 1976.

    Short articles on British health and safety matters are
    featured, emphasizing legislation, brief news items,
    results of relevant lawsuits, and reviews of new publi-
    cations.

IRCS Medical Science: Drug Metabolism and Toxicology. Inter-
    national Research Communications System. Bimonthly,
    published since 1973.

    Rapid publication and effective dissemination of ori-
    ginal articles are the goals of this refereed journal.
    Brief communications, letters to the editor, theoreti-
    cal and review articles, and meeting announcements are
    included. The journal is available in print format,
    microfiche, or online from BRS and DIMDI.

Industrial Hygiene Digest. Industrial Health Foundation.
    Monthly, published since 1937.

    Industrial health information from government agencies
    and journals is abstracted. Related upcoming meetings
    and book reviews are included.

Industrial Hygiene News. Rimbach Publishing, Inc.
    Bimonthly, published since 1978.

    This newspaper is provided free to the person responsi-
    ble for employee health in all industries and at all
    levels of government. It primarily contains reports on
    the newest industrial hygiene equipment, as well as a
    brief article and book review.

Industrial Hygiene News Report. Flournoy Publishers, Inc.
    Monthly, published since 1958.

This four-page report contains brief articles on
research involving the recognition, evaluation and control of workplace hazards. The newsletter format makes
it a source of very current information in this subject
area.

International Archives of Occupational and Environmental
    Health. Springer-Verlag. Published 8 times per year,
    since 1975.

Review articles, original papers, short communications,
and documents from international meetings and activities are included in this journal. The subject emphasis is on morbidity and mortality, clinical and epidemiological studies, environmental health effects in
humans and animals, and testing methodology.

International Journal of Clinical Pharmacology, Therapy and
    Toxicology. Dustri-Verlag. Monthly, published since
    1980.

This is the official publication of the International
Society of Chemotherapy, and deals with pharmacodyna-
mics, drug interactions, and metabolism.

JOM. Journal of Occupational Medicine. Fluornoy Publishers,
    Inc. Monthly, published since 1959.

This official publication of the American Occupational
Medicine Association and of the American Academy of
Occupational Medicine publishes original articles, book
reviews, letters to the editor, and a calendar of
meetings.

Job Safety Consultant. Business Research Publications.
    Monthly, published since 1973.

This newsletter has brief articles for managers on
occupational safety and health, along with a comment by
the editor, a question and answer column, and a brief
text of a talk for managers to deliver at safety meetings.

Job Safety and Health Report. Business Publishers, Inc.
    Biweekly, published since 1971.

Information on legislation, corporate policy changes,
conferences, seminars, and developments in occupational
health and safety is included in this newsletter.

Journal of the American College of Toxicology. Mary Ann
    Liebert, Inc. Quarterly, published since 1982.

This journal publishes reviews, major symposia, and articles in all areas of toxicology, such as risk assessment, carcinogenicity, teratogenicity, epidemiology, and clinical toxicology.

Journal of Analytical Toxicology. Preston Publications, Inc. Bimonthly, published since 1977.

Original papers and review articles on the isolation, identification, and quantitation of potentially toxic substances, monitoring environmental and industrial contaminants, clinical reports of poisonings, and forensic studies are included in this journal, as are book reviews, listings of meetings, and short communications.

Journal of Toxicological Sciences. Dokkukagaku-kai. Quarterly, published since 1976.

The Japanese Society of Toxicological Sciences publishes articles in Japanese and English and meeting announcements dealing mostly with the toxicology of various substances primarily in animals.

Journal of Toxicology and Environmental Health. Hemisphere. Monthly, published since 1975.

Human health in the workplace, epidemiological studies on workers, and environmental pollutants and their toxicology are the main topics covered in this journal. It includes articles, short communications, review articles, book reviews, symposia and conference proceedings.

Michigan's Occupational Health. Michigan State Department of Public Health, Division of Occupational Health. Quarterly, published since 1955.

This newsletter is free, and each issue deals with a specific subject relating to occupational health.

Microwave News: a monthly report on non-ionizing radiation. Microwave News. Monthly, published since 1980.

This is a newsletter which focuses on international safety standards, health effects, reports, and legislation related to non-ionizing radiation.

OSHA Up-to-Date Newsletter. National Safety Council. Monthly, published since 1982.

This four-page newsletter contains brief articles primarily about OSHA and other government agencies.

Occupational Hazards. Penton-IPC. Monthly, published since
1938.

Articles on laws, standards, regulations, new products,
and workers' compensation are included, as well as
letters to the editor, and a safety idea column.

Occupational Health. Bailliere Tindall. Monthly, published
since 1963.

The focus of this magazine is on British occupational
safety and health, although an "international" section
describes relevant events in other countries. There
is a listing of conferences and workshops. Each issue
focuses on a specific topic.

Occupational Health and Safety. Medical Publications, Inc.
Published 8 times per year, since 1976.

Short articles on various types of occupational health
hazards are published, along with a list of relevant
meetings.

Occupational Health & Safety Letter. Environews, Inc.
Bimonthly, published since 1971.

News on government regulations, legislation, and
actions of regulatory agencies is presented in brief
format.

Regulatory Toxicology and Pharmacology. Academic. Quarterly, published since 1981.

The scope of this journal encompasses the international, legal, and public health aspects of toxicological and pharmacological regulations in relation to
scientific knowledge, and the interpretation of regulatory decisions.

Risk Analysis: An International Journal. Plenum. Quarterly,
published since 1981.

The peer-reviewed articles include the analysis of risk
to health as well as mathematical and theoretical
aspects of risks, and social and psychological aspects
of risks.

Russian Pharmacology and Toxicology. Euromed Publications.
Bimonthly, published since 1967.

Selective articles from the following Russian publications are translated totally or in part: <u>Farmakologiya
i tolsikologiya</u>, <u>Khimiko-Farmatsevti-cheskii Zhurnal</u>,
and <u>Norye Lekarstvennye Preparaty</u>. Some of the articles
have been condensed or presented as abstracts. Reports
of the development of original single chemical entity
drugs are emphasized.

<u>SPAID News</u>. Society for the Prevention of Asbestosis and
Industrial Diseases. Published 3 times per year, since
1980.

A four-page newsletter of the Society serves to disseminate information on specific cases of asbestos-related
disease, government actions on such cases, and results
of conferences on the topic.

<u>Seminars in Occupational Medicine</u>. Thieme Inc. Quarterly,
published since 1986.

Each issue covers a specific problem faced by clinicians
in occupational medicine.

<u>Scandinavian Journal of Work, Environment and Health</u>.
Finnish Institute of Occupational Health. Quarterly,
published since 1975.

Symposium proceedings, book reviews, original papers,
and review articles are included in this journal.

<u>Toxicological and Environmental Chemistry</u>. Gordon & Breach.
Two volumes per year, four numbers per volume, published since 1981.

This international journal covers the analysis, metabolism, chemistry, and biochemistry of xenobiotic compounds and natural toxins in relation to the environment and human health. It also has book reviews.

<u>Toxicology</u>. Elsevier. Monthly, published since 1973.

This journal includes original scientific papers relating to biological effects of the administration of
chemical compounds on tissues. It includes chemical
contaminants, industrial chemicals, and pesticides.

<u>Toxicology and Applied Pharmacology</u>. Academic. Published
15 times per year, since 1959.

This official journal of the Society of Toxicology
publishes papers on alternations in tissue structure or

function due to the administration of chemicals and drugs in animals and humans.

Toxicology Letters. Elsevier. Monthly, published since 1977.

The goal of this journal is the rapid publication of articles in all areas of toxicology including environmental hazards to humans and animals, and experimental toxicity of industrial products and agricultural products.

VDT News: the VDT Health and Safety Report. Microwave News. Bimonthly, published since 1984.

The purpose of this newsletter is to inform the reader of current developments in international research, legislation, and standards involving health effects of VDT's. Meeting announcements are included.

Abstracts, Indexes and Guides to the Literature

Abstracts on Health Effects of Environmental Pollutants. BioSciences Information Service. Monthly, published since 1972.

Content summaries of research on the effects of environmental pollutants on health, especially human health, are contained in this abstracting journal. Topics covered include industrial medicine, and occupational health. There is a keyword subject index and an author index.

Bioresearch Today. Industrial Health and Toxicology. BioSciences Information Service. Monthly, published since 1972.

An abstracting journal which includes studies on physiological and pathological conditions affecting human workers, Industrial Health and Toxicology includes experimental, industrially-oriented studies of potentially harmful chemicals or agents and also industrial health studies, occupational allergies, physical, chemical and radiation work hazards, and control measures. Between 100 and 250 abstracts derived from Biological Abstracts are included in each issue.

CIS Abstracts. International Labor Office. Published 8 times per year, since 1974.

CIS Abstracts is produced by the International Occupational Safety and Health Information Centre. Each
issue contains 300 abstracts with a subject index. The
abstracts are in English, although the articles are
from journals and documents in all languages. The
final issue for the year is the annual cumulated subject and author index. CIS provides photocopies or
microfiche of the documents it abstracts. The documents are indexed using the CIS Thesaurus. CIS provides
a List of Periodicals Abstracted, Abstract Users'
Guide and CIS Bibliographies on up-to-date topics.
The scope of this abstracting journal covers international health and safety in the workplace.

Dworaczek, Marian, comp. Health and Safety Aspects of Visual
    Display Terminals: a Bibliography. 2nd ed. Public
    Administration Series, Bibliography P-1421. Monticello,
    IL: Vance Bibliographies, 1984.

    This is a comprehensive listing of publications in the
    subject area.

Excerpta Medica. Section 21. Developmental Biology and Tera-
    tology. Excerpta Medica. Published 10 times per year,
    since 1961.

    This abstracting periodical can be used to locate arti-
    cles on specific hazardous substances which cause birth
    defects.

Excerpta Medica. Section 30. Pharmacology and Toxicology.
    Excerpta Medica. Published 30 times per year, since
    1948.

    This section includes abstracts on industrial and
    domestic poisons.

Excerpta Medica. Section 35. Occupational Health and Indus-
    trial Medicine. Excerpta Medica. Published 10 times
    per year, since 1971.

    Especially applicable are sections on Physical Environ-
    ment, Chemical Environment, Environmental Sanitation,
    Skin, Respiratory Tract Infections and Toxins, Carcino-
    genesis, Mutagenesis, and Teratogenesis.

Index Medicus. National Library of Medicine. Monthly,
    published since 1960.

    By using the Medical Subject Headings (MESH) to iden-
    tify appropriate subject headings, this index can be
    searched for articles in the medical literature on the
    topic. The subheadings "adverse effects," "poisoning,"

and "toxicity" are especially useful when attached to particular substances, such as lead, cadmium, and arsenic.

Literature Search. National Library of Medicine. Irregular, published since 1967.

The NLM literature searches are reprints of data base searches on the MEDLARS system which have been frequently requested. They are available from the National Library of Medicine at no charge. The relevant recent bibliographies are:

NLM LS 81-1   Occupational Neurobehavioral Toxicology. Jan. 1975 through March 1981. 282 citations.

NLM LS 81-7   Lead Exposure in Children. Jan. 1977 through July 1981. 354 citations.

NLM LS 81-16   Male Reproductive Toxicology. Jan. 1977 through Aug. 1981. 230 citations.

NLM LS 81-22   Adverse Reactions to Radiographic Contrast Media. Jan. 1979 through Oct. 1981. 339 citations.

NLM LS 81-24   Chromium Toxicology. Jan. 1977 through Sept. 1981. 250 citations.

NLM LS 81-25   Arsenic Toxicology. Jan. 1977 through Oct. 1981. 354 citations.

NLM LS 82-4   Health Effects of Nuclear Radiation. Jan. 1979 through May 1982. 427 citations.

NLM LS 82-6   Asbestos Toxicology. Aug. 1977 through May 1982. 597 citations.

NLM LS 82-10   Zinc Toxicology. Jan. 1977 through July 1982. 204 citations.

NLM LS 82-11   Formaldehyde Toxicology. Jan. 1977 through June 1982. 213 citations.

NLM LS 83-19   Dioxin Toxicology (Including Agent Orange) Jan. 1981 through Sept. 1983. 128 citations.

NLM LS 84-28   Adverse Effects of Aluminum. Jan. 1977 through Aug. 1984. 427 citations.

Pantry, Sheila, comp. **Health and Safety: a Guide to Sources of Information**. CPI Information Reviews, No. 6. Edinburgh: Capital Planning Information, Ltd., 1983.

This guide is an excellent source for information on British publications. Chapters are divided by type of publication: periodicals, books, legislation, organisations, etc.

Pruett, J. G. and S. G. Winslow. **Health Effects of Environmental Chemicals on the Adult Reproductive System. A Selected Bibliography with Abstracts, 1963-1981**. Oak Ridge, TN: Toxicology Information Response Center, 1982.

The 305 references include citations on occupational or environmental exposure to chemicals.

"Recommended Library for Occupational Physicians." **JOM. Journal of Occupational Medicine** 26 (June 1984): 461-464.

This list of texts recommended by the American Occupational Medicine Association is divided into the categories "General Occupational Medicine," "Industrial Hygiene," "Toxicology, Epidemiology and Statistics," "Occupational Lung Disease," "Occupational Cancer," "Dermatology, Hearing Conservation," "Health Education," "General Preventive Medicine," and "Recommended Journals."

Teratology Lookout. Karoliska Institutet, Medical Information Center. Published 1969-1985.

This index has sections on environmental agents and perinatal effects, and includes citations from Biological Abstracts, Bioresearch Index, Chemical Abstracts, and the MEDLARS data bases. Each citation includes a reference to an abstract in Biological Abstracts or Chemical Abstracts, indexing terms, and the language of the original article.

Toxicology Abstracts. Cambridge Scientific Abstracts. Monthly, published since 1978.

5000 journals are screened for the 650 abstracts included monthly on effects of pharmaceuticals, radiation, metals, and nitrosamines on the animal kingdom.

Toxicology Research Projects Directory. Toxicology Information Subcommittee of the DHEW Committee to Coordinate Toxicology and Related Programs. DHEW Publication No. (OS) 76-50030. Quarterly, published since 1976.

Ongoing research projects in toxicology selected from Smithsonian Science Information Exchange are listed.

Proceedings

"3rd World Congress of Biological Psychiatry, Stockholm, Sweden, 1981." _Acta Psychiatrica Scandinavica_ 67 (1983).

Relevant topics of papers include occupational exposure to lead, jet fuel, and organic solvents.

"American Occupational Medicine Association 68th Annual Meeting." Washington, D.C., April 24-29, 1983. _JOM_ 25 (April 1983): 325-334.

The topics of this conference include the Merchant Marine Medical programs, community-based occupational health programs, diving medicine, epidemiology, toxicology, occupational health screening, wellness myths and methods, cardiovascular disease, occupational health program management, exercise programs, the hepatitis B vaccine, and clinical occupational medicine.

Bake, Bjorn, et al., eds.   "Chronic Bronchitis in Non-smokers." Gothenberg, September 24-25, 1981. _European Journal of Respiratory Diseases_ 63 (Supplement no. 118 1982).

Topics covered in these papers include epidemiology and environmental influence, including the role of infections; and the working environment: dust exposure, welding, construction work, cutting oil mist, machine-shop workers, and sulphur dioxide.

Berlin, A., et al.   _Assessment of Toxic Agents at the Workplace: Roles of Ambient and Biological Monitoring_. Boston: Martinus Nijhoff, 1984.

This book is based on papers presented at a seminar organized by the Commission of the European Communities, and the United States Occupational and Safety Health Administration and the National Institute for Occupational Safety and Health in Luxembourg in 1980. Monitoring of organic and inorganic hazards is covered; the role of the hygienist, nurse, physician, epidemiologist, engineer, chemist, economist, computer scientist, inspector, lawyer, and industrialist in prevention and health protection by monitoring; standardization and quality control; and education of the worker and manager are some of the topics covered.

Bitton, G., et al. <u>Sludge: Health Risks of Land Application</u>.
Ann Arbor, MI: Ann Arbor Science, 1980.

This book is the result of the Workshop on Evaluation
of Health Risks Associated with Animal Feeding and/or
Land Application of Municipal Sludge held in Tampa,
Florida from April 29-May 1, 1980. A relevant chapter
is "Occupational Hazards Associated with Sludge
Handling" by C. S. Clark, et al.

Bora, K. C., et al., eds. <u>Chemical Mutagenesis, Human
Population Monitoring, and Genetic Risk Assessment</u>.
Proceedings of the International Symposium held October
14-16, 1980, Ottowa, Canada. Progress in Mutation
Research, Volume 3. New York: Elsevier, 1982.

The emphasis is on human population monitoring.
Attempts to establish a link with mutagenicity testing.
The appendix is the "Report of the WHO Consultation on
Genetic Monitoring for Environmental Effects 17 October
1980."

British Association of Urological Surgeons. 37th Annual
Meeting. London, England, July, 1981. <u>British Journal
of Urology</u> 53 (June 1981).

"The Early Changes in the Development of Bladder Cancer
in Patients Exposed to Known Industrial Carcinogens" by
R. W. Glashan, et al., is a paper relevant to the topic
of this chapter.

Brown, Stanley S. and F. William Sunderman, Jr. <u>Nickel
Toxicology</u>. Proceedings of the 2nd International
Conference, September 3-5, 1980, Swansea, Wales.
Orlando, FL: Academic, 1980.

Organized by the Subcommittee on Environmental and
Occupational Toxicology of Nickel, Commission on
Toxicology, International Union of Pure and Applied
Chemistry, and the Association of Clinical Scientists.
A significant contribution of this conference was the
approval of a method for analyzing nickel in the serum
and urine by electrothermal atomic absorption
spectrometry. Most of the papers discuss nickel and
carcinogenesis in humans and animals. Metabolism,
pharmacology, and analysis are also covered.

Brown, Stanley S. and John Savory, eds. <u>Chemical Toxicology
and Clinical Chemistry of Metals</u>. Proceedings of the
2nd International Conference held in Montreal, Canada,
July 19-22, 1983. Orlando, FL: Academic, 1983.

This book contains about 50 of the 140 papers given at
this conference on metal toxicology. Articles are

divided into the following subject areas: analysis and
quality assurance, occupational and environmental exposures, speciation and interactions, and clinical and
experimental studies. Some specific topics include
iatrogenic aluminum poisoning, effect of manganese on
cerebral RNA polymerase and free ribosomal protein
synthesis, in utero and airborne lead exposure, sudden
infant death syndrome, and renal function and exposure
to mercury vapor.

The British Occupational Society. Annual Conference, 1981.
    Annals of Occupational Hygiene 24 (1981): 357-398.

    Some of the papers from the conference are included in
    this issue of the journal. The topics cover reproduc-
    tive hazards, immunological responses to inhaled dusts
    and chemicals, dust sampling and collecting, and using
    occupational health data in the coal industry.

Burchenal, J. H. and H. F. Oettgen. Cancer. Volume 1.
    Achievements, Challenges, and Prospects for the 1980's.
    New York: Grune & Stratton, 1981.

    This book is based on the 1980 International Symposium
    on Cancer held in New York City from September 14-18,
    1980. The relevant chapters include "Risk of Cancer
    Associated with Occupational Exposure in Radiologists
    and Other Radiation Workers" by G. M. Matanoski, "Indus-
    trial and Life-Style Carcinogens" by M. B. Shimkin,
    "Population-based Tumor Registries in the Identifica-
    tion of Occupational Carcinogens" by D. F. Austin, and
    "Three Decades of Environmental Concern" by D. P. Rahl.

Castellani, Amleto, ed. The Use of Human Cells for the
    Evaluation of Risk from Physical and Chemical Agents.
    NATO Advanced Science Institute Series, Series A, Life
    Sciences, Volumme 60. New York: Plenum, 1983.

    The emphasis of these papers is on the use of human
    cells because of the controversy surrounding animal
    experimentation and the possible false conclusions that
    can be made by relating animal studies to humans.

Chisolm, J. Julian, Jr. and David M. O'Hara, eds. Lead
    Absorption in Children: Management, Clinical, and Envi-
    ronmental Aspects. Baltimore: Urban and Schwarzenberg,
    1982.

    A conference on "Management of Increased Lead Absorp-
    tion in Children: Clinical, Social and Environmental
    Aspects" was the basis for this book, which is divided
    into sections on metabolism and toxicity of lead; envi-
    ronmental, nutritional, behavioral and social factors
    affecting lead toxicity in children; the relationship

between the health agency, analytical laboratory, environmental hygiene and clinical management; and research needs.

Coulston, Frederick and Francesco Pocchiari, eds. Accidental Exposure to Dioxins: Human Health Aspects. International Forum on Human Health Aspects of Accidental Chemical Exposure to Dioxins--Strategy for Environmental Reclamation and Community Protection, Bethesda, Maryland, 1981. Orlando, FL: Academic, 1983.

Taken from a conference sponsored by the International Academy of Environmental Safety and the International Society of Ecotoxicology and Environmental Safety, data is presented on populations which have been accidentally exposed to factory emmissions of dioxin. Decontamination procedures are described.

Cowser, K. E. and C. R. Richmond. Synthetic Fossil Fuel Technology: Potential Health and Environmental Effects. Ann Arbor, MI: Ann Arbor Science, 1980.

The relevant chapters from this first annual Oak Ridge National Laboratory Life Sciences Symposium held in Oak Ridge, Tennessee from September 25-28, 1978 include "Mortality Experience of 50 Workers with Occupational Exposure to the Products of Coal Hydrogenation Processes" by A. Palmer, and "Worker Protection in the Coal Carbonization Industry in the United Kingdom" by R. M. Archibald, et al.

Dunnom, D. D., ed. Health Effects of Synthetic Silica Particulates. A symposium sponsored by ASTM Committee E-34 on Occupational Health and Safety and the Industrial Health Foundation Benalmadena-Costa (Torremolinos), Spain, November 5-6, 1979. ASTM Special Technical Publication 732. Philadelphia: American Society For Testing and Materials, 1981.

Human health problems resulting from exposure to silica dust were studied by the ASTM committee, and are reported.

Englund, Anders, et al., eds. Occupational Health Hazards of Solvents. Advances in Modern Environmental Toxicology, Volume 2. Princeton, NJ: Princeton Scientific Books, 1982.

An international group of contributors presented epidemiological and toxicological information on health hazards of paints, solvents, and pigments at this International Symposium on Occupational Health Hazards Encountered in Surface Coating and Handling of Paints in the Construction Industry, which took place in Stockholm, Sweden in 1981.

Infante, Peter F. and Marvin S. Legator, eds. Proceedings of a Workshop on Methodology for Assessing Reproductive Hazards in the Workplace, April 19-22, 1978. DHHS (NIOSH) Publication, No. 81-100; HE20.7102 R29. Cincinnati: Department of Health and Human Services, Public Health Service, Center for Disease Control, National Institute for Occupational Safety and Health, Division of Surveillance, Hazard Evaluations and Field Studies, 1980.

Effects of exposure to lead, vinyl chloride, waste anesthetic gases, chloroprene, dibromochloropropane; birth defects and parental occupation; epidemiological studies for teratogen detection; semen assays for workplace monitoring, and spontaneous abortion studies are among the topics covered by papers in this book.

Jarvisalo, Jorma, et al., eds. Industrial Hazards of Plastics and Synthetic Elastomers. Proceedings of the International Symposium on Occupational Hazards Related to Plastics and Synthetic Elastomers, Espoo, Finland, November 22-27, 1982. Progress in Clinical and Biological Research, Volume 141. New York: Alan R. Liss, 1984.

The purpose of the symposium was to evaluate hazards in the plastics industry and define preventive measures. Synthetic polymers, polyvinyl, polyethylene, polypropylene, epoxy thermosets, styrene-polymers, and synthetic elastomers are discussed.

Kerr, J. W. and M. A. Ganderton. Proceedings of Invited Symposia: 11th International Congress of Allergology and Clinical Immunology. Barbican Center, London, England, October 17-22, 1982. New York: Macmillan Press, 1983.

The relevant papers from this congress include "Epidemiology of Occupational Lung Diseases-Asthma and Allergic Alveolitis" by H. Keskinen, "Occupational Asthma Induced by Phthalic-anhydride and Related Compounds" by I. L. Bernstein, "Occupational Asthma due to Colophony" by P. S. Burge, and "Bakers' Asthma-Epidemiological and Clinical Findings-Needs for Prospective Studies" by H. Thiel.

Kuratsune, Masanori and Raymond E. Shapiro, eds. PCB Poisoning in Japan and Taiwan: a Collection of Papers Dealing with the Effects of PCB's and Related Compounds. Progress in Clinical and Biological Research, Volume 137. New York: Alan R. Liss, Inc., 1984.

Dermatological and medical status of Yusho patients is
discussed along with causal agents of Yusho and PCB
poisoning in Taiwan: neurological studies, ocular manifestations, and "fasting cure." Clinical features of
the patients are emphasized. PCB's, PCDF's, PCQ's
retained in the blood and tissues, causes of the Yusho
and Taiwan incidents, and epidemiology of the Taiwan
incident are emphasized.

Leaverton, Paul E., ed. Environmental Epidemiology. New
York: Praeger, 1982.

Papers from the Yves Biraud Seminars held in France in
1979, 1980, and 1981 are included. Significant papers
include: "Cancer Mortality and Agricultural Activity--
an Association with Cotton Production and Large Farms"
by L. C. Clark, et al., "Epidemiological Surveillance
of the Health Effects of Environmental Hazards" by Z.
J. Brzezinski, and "Environmental and Occupational
Sentinels for Cardiovascular Diseases" by W. R. Harlau.

Li, A. P., ed. Toxicity Testing: New Approaches and Applications in Human Risk Assessment. New York: Raven
Press, 1985.

Based on a 1983 conference held at the Monsanto Company
in St. Louis, Missouri. Evaluation of genetic toxicology, immunologic and fetal toxicology are covered.

Mehlman, M. A. Carcinogenicity and Toxicity of Benzene.
Advances in Modern Environmental Toxicology, Volume 4.
Princeton, NJ: Princeton Scientific, 1983.

The Benzene Workshop, sponsored by the American College
of Toxicology in 1980, was the basis for this book.
The history of benzene carcinogenicity, developmental
toxicity of benzene, the relationship between benzene
toxicity and metabolism, its immunotoxicity and clinical hematotoxicity, occupational exposure and protection, and health effects in water are discussed.

Nicholson, William J. Management of Assessed Risk for
Carcinogens. Annals of the New York Academy of
Sciences, Volume 363. New York: New York Academy of
Sciences, 1981.

The papers in this volume resulted from a workshop held
at the New York Academy of Science in 1980 entitled
"Management of Assessed Risk for Carcinogens." One
relevant chapter is "The Elimination of Carcinogenic
Risks in Consumer Products" by P. W. Preuss. Others
deal with occupational carcinogen policy, risk management, and compensation for occupational illness.

Nicolini, Claudio, ed. <u>Chemical Carcinogenesis</u>. Inter-
national School of Pure and Applied Biostructure,
Ettore Majorana Center for Scientific Culture, 1981.
NATO Advanced Study Institute Series, Series A, Life
Sciences, Volume 52. New York: Plenum, 1982.

The purpose of this conference was to demonstrate to
scientists the current thinking on chemical carcinogenesis, presuming that most human cancer is related to
environmental exposure. Papers are included under the
following sections: "Chemicals as Carcinogens," "DNA
Adducts," "Chemicals as Promotors," and "Carcinogenesis
as Multistep Process." Each article has a list of
references. There is a transcription of a discussion
among the participants, and a list of participants and
their addresses.

Nordman, Henrik, ed. International Course on Occupational
Respiratory Diseases, Hanasaari Cultural Center, Espoo,
Finland, September 22-26, 1980. <u>European Journal of
Respiratory Diseases</u> 63 (Supplement No. 123, 1982).

This course was organized by the Institute of Occupational Health in Helsinki, and did not attempt to study
occupational lung diseases in depth, but focused on
basic mechanisms in occupational lung disease, occupational asthma, and extrinsic allergic alveolitis.

Porter, I. H. and E. B. Hook. <u>Human Embryonic and Fetal
Death</u>. Orlando, FL: Academic, 1980.

This book contains the proceedings from the 10th Annual
Birth Defects Institute Symposium held in Albany, New
York from October 29-30, 1979. One relevant chapter is
"Occupational Hazards and Fetal Deaths" by R. R. Monson.

Prasad, Ananda S., ed. <u>Clinical, Biochemical, and Nutri-
tional Aspects of Trace Elements</u>. Current Topics in
Nutrition and Disease, Volume 6. New York: Alan R.
Liss, Inc., 1982.

The papers are based on proceedings of a conference
sponsored by the International Union of Nutritional
Sciences, Wayne State University, and Harper-Grace
Hospitals in 1980. Effects of cadmium exposure are
among the topics discussed.

<u>Prevention of Occupational Cancer - International Symposium</u>.
Occupational Safety and Health Series, No. 46. Geneva:
International Labour Office, 1982.

This book contains the proceedings of the International
Symposium on the Prevention of Occupational Cancer held
in Helsinki from April 21-24, 1981. It was organized

by the Institute of Occupational Health of Helsinki,
Finland. The papers on occupational cancer cover epidemiology, methods of evaluation, prevention and control, and national prevention policies.

Radiological Protection--Advances in Theory and Practice,
    Vols. 1 and 2. Third International Symposium. London:
    Society for Radiological Protection, 1982.

    Many papers discuss exposure to radiation by workers:
    epidemiology, uranium mining, nuclear power plants, and
    nuclear medicine.

Sarkar, Bibudhendra, ed. Biological Aspects of Metals and
    Metal-related Diseases. New York: Raven, 1983.

    Compiled for researchers, this book contains the
    proceedings of a symposium, "Biological Aspects of
    Metals and Metal-related Diseases" which took place in
    Toronto in 1981. Metal metabolism in the body is
    discussed, as well as the role of metals in nutrition,
    Menkes' disease and Wilson's disease.

Seeberg, Erling and Kjell Kleppe, eds. Chromosome Damage
    and Repair. NATO Advanced Study Institutes Series,
    Series A, Life Sciences, Volume 40. New York: Plenum,
    1981.

    These proceedings were taken from a meeting held in
    Norway in 1980 by the NATO-EMBO advanced study insti-
    tute lecture course. Occupational exposure to carcino-
    gens is among the topics of the papers.

Smith, E. H. F. and A. G. Hillebrandt, eds. Health Evalua-
    tion of Heavy Metals in Infant Formula and Junior Food.
    New York: Springer-Verlag, 1983.

    This book consists of the proceedings of a symposium,
    and includes articles on the concentration of heavy
    metals in baby food; laws, suggestions for regulations;
    toxicology of heavy metals in children; and mineral
    metabolism in children.

Sorsa, M. and H. Vainio, eds. Mutagens in our Environment.
    12th Annual Meeting of the European Environmental
    Mutagen Society, Espoo, Finland, 1982. Progress in
    Clinical and Biological Research, Volume 109. New York:
    Alan R. Liss, Inc., 1982.

    These proceedings present the sources of mutagens in
    our environment, and discuss the relationship between
    carcinogenesis and genetic cell damage. Some of the
    topics of the papers are mutagenicity of typewriter

ribbons, mutagenicity in naval industry workers' urine, sister chromatid exchange and exposure to inhalation anesthetics, mutagens in food, and the relationship between motor vehicle exhaust and airborne particle mutagenicity.

Symposium on Occupational Immunologic Lung Disease, Washington, D.C., February 25, 1982. _Journal of Allergy and Clinical Immunology_ 70 (January 1982): 1-72.

The papers from this conference cover such topics as trimellitic anhydride disease, isocyanate-induced pulmonary disease, the relationship between western red cedar and asthma, silicosis and asbestosis.

Tordoir, W. F. and E. A. H. Van Heemstra-Lequin, eds. _Field Worker Exposure During Pesticide Application_. Proceedings of the 5th International Workshop of the Scientific Committee on Pesticides of the International Association on Occupational Health, The Hague, October 9-11, 1979. Studies in Environmental Science, Volume 7. New York: Elsevier, 1980.

The introductory papers of this conference discuss monitoring and assessment of exposure to pesticides. The technical papers are on the safe use of pesticides, worker exposure during pesticide application, and pesticide penetration through skin and clothing. Reports on international field studies are presented, and other papers document risk from combined exposure to different chemicals, and occupational medical aspects.

Tucker, Richard E., et al., eds. _Human and Environmental Risks of Chlorinated Dioxins and Related Compounds_. Proceedings of the International Symposium on Chlorinated Dioxins and Related Compounds held in 1981 at Arlington, Virginia. Environmental Science Research, Volume 26. New York: Plenum, 1983.

The purpose of the conference was to study the dioxin problem and make recommendations for future research. The sessions covered analytical chemistry, environmental chemistry, environmental toxicology, human observations, risk assessment, and laboratory safety and waste management.

Walton, W. H., ed. _Inhaled Particles V_. 1st ed. Proceedings of an International Symposium Organized by the British Occupational Hygiene Society, Cardiff, September 8-12, 1980. Elmsford, NY: Pergamon, 1982.

This large volume contains 69 papers presented at the conference, whose goal was to present research results on the entry, fate and effects of inhaled particles.

The sessions are divided into deposition and inhalation of particles; dust in human lungs; biological reactions to dust; carcinogenic and cytotoxic effects; airway response to aerosols; factors determining pneumoconiosis; and epidemiological studies. Also published as: Annals of Occupational Hygiene 26 (1-4) 1982.

Wolbarsht, Myron L. and David Sliney. Ocular Effects of Non-ionizing Radiation. Proceedings of the Society of Photo-optical Instrumentation Engineers, Volume 229. Bellingham, WA: Society of Photo-optical Instrumentation Engineers, 1980.

These papers discuss potential problems of high intensity radiation sources and regulation of these sources by NIOSH to inform designers and users of optical sources of the hazards with which they may be associated. Sessions dealt with industrial optical radiation, photochemical effects, and safety.

Ziegler, J. L. Carcinogenic and Mutagenic N-Substituted Aryl Compounds. NCI Monographs Volume 58. Bethesda, MD: Department of Health and Human Services, 1981.

Based on the International Conference on Carcinogenic and Mutagenic N-Substituted Aryl Compounds held in Rockville, Maryland from November 7-9, 1979, the relevant papers include "Occupational Exposure to Aromatic-amino-benzidine and Benzidine-based Dyes" by B. Walker and A. Gerber, and "Specific Aromatic-amines as Occupational Bladder Carcinogens" by D. B. Clayson.

Reports & Documents

Boeniger, Mark. The Carcinogenicity and Metabolism of Azo Dyes, Especially Those Derived from Benzidine. DHHS (NIOSH) Publication No. 80-119; HE20.7111.2 C17. Cincinnati: Department of Health and Human Services, Public Health Service, Center for Disease Control, National Institute for Occupational Safety and Health, Robert A. Taft Laboratories, 1980.

This publication reviews the literature on carcinogenicity and epidemiology of azo compounds and benzidine in humans and animals. It lists recommendations for future research. A bibliography contains 167 references. There is a table of 44 dyes exhibiting carcinogenic activity, their carcinogenic data and chemical structure. A consent form for participation in studies on exposure to benzindine-based dyes is included.

The Council on Scientific Affairs, Advisory Panel on Toxic
    Substances. The Health Effects of "Agent Orange" and
    Polychlorinated Dioxin Contaminants: Technical Report.
    Chicago: Department of Environmental, Public and Occu-
    pational Health, American Medical Association, 1981.

    This report describes the medical research involving
    toxic effects of Agent Orange and TCDD. Animal and
    human toxicity, mutagenicity, oncogenicity and terato-
    genicity are documented. Current and proposed studies
    by various agencies are described. There is an exten-
    sive list of references, and a table of occupational
    exposures to TCDD.

Criteria for a Recommended Standard ...Occupational Exposure
    to Styrene. DHHS (NIOSH) Publication No. 83-119;
    HE20.7110 St9. Cincinnati: Department of Health and
    Human Services, Public Health Service, Center for
    Disease Control, National Institute for Occupational
    Safety and Health, 1983.

    Specific recommendations are made regarding workplace
    air, clothing, equipment, sanitation, monitoring, and
    recordkeeping. Effects of exposure on humans and ani-
    mals are documented, as well as methods for determining
    exposure.

Evans, Linda S. and Stephanie A. Moyer, comps. Bibliography
    of Research Reports and Publications Issued by the Tox-
    ic Hazards Division, 1957-1982. Wright-Patterson Air
    Force Base, OH: Air Force Aerospace Medical Research
    Laboratory, 1983.

    This bibliography lists Air Force technical reports,
    U.S. government reports, journal articles and books
    written by members of the Toxic Hazards Division of the
    Air Force Aerospace Research Laboratory of which
    resulted from research contracts initiated by the
    Division. It is chronologically arranged.

Mushak, Paul, et al. Health Assessment Document for Inor-
    ganic Arsenic. Report #EPA-600/3-83-021A. PB83-232306.
    Research Triangle Park, NC: Environmental Protection
    Agency, Office of Research and Development, Environ-
    mental Criteria and Assessment Office, 1983.

    This report discusses properties of inorganic arsenic,
    and its metabolism and toxicology.

NIOSH Health Hazards Evaluation Summaries. Cincinnati:
    National Institute for Occupational Safety and Health,
    1980.

Synopses of NIOSH reports generated after the investigation of twenty-five workplaces across the country are presented. The history of the complaint, the possible health hazard, the methodology used to determine exposure, and the conclusion of the investigators are summarized. Full copies of the individual reports are available from NTIS.

Pedersen, David H., et al. A Model for the Identification of High Risk Occupational Groups using RTECS and NOHS Data. DHHS (NIOSH) Publication No. 83-117; HE20.7111.2 M72. Cincinnati: Department of Health and Human Services, Public Health Service, Center for Disease Control, National Institute for Occupational Safety and Health, Division of Surveillance, Hazard Evaluations and Field Studies, 1983.

Using the Registry of Toxic Effects of Chemical Substances and the National Occupational Hazard Survey, this book presents quantification of potential occupational health risks.

Recommended Health-Based Limits in Occupational Exposure to Heavy Metals. World Health Organization Technical Report Series, No. 647. Geneva: WHO, 1980.

This report of a WHO Study Group provides recommended exposure limits for lead, cadmium, manganese, and mercury.

Special Occupational Hazard Review: Alternatives to Di-2-Ethylhexyl Phtahalate ("DOP") in Respirator Quantitative Fit Testing. DHHS (NIOSH) Publication No. 83-109; HE20.7113 D54.2. Cincinnati: Department of Health and Human Services, Public Health Service, Center for Disease Control, National Institute for Occupational Safety and Health, Division of Standards Development and Technology Transfer, 1983.

Because of carcinogenicity in rodents exposed to DEHP, this report makes recommendations for reducing exposure to DEHP during respirator testing, and recommends alternatives to DEHP in the manufacture of respirators. Eighty references are listed in the bibliography.

Special Occupational Hazard Review for Benzidine-based Dyes. DHEW (NIOSH) Publication No. 80-109; HE20.7113 B44. Cincinnati, OH: Department of Health, Education, and Welfare, Public Health Service, Center for Disease Control, National Institute for Disease Control, National Institute for Occupational Safety and Health, 1980.

The Special Hazard Reviews are published to present
information on carcinogenicity, mutagenicity, or teratogenicity of a specific industrial chemical, process,
or physical agent. This Review presents evidence that
benzidine-based dyes are potential human carcinogens.
Characteristics and biological effects of the dyes,
characteristics of exposure to the dyes, monitoring
methods and recommendations for the workplace, and a
list of 79 references are included.

Data Bases

AGRICOLA
Producer: National Agricultural Library
Time span: 1970 to present
Coverage: International
Updated: Monthly
Vendor: DIALOG

The journal and monograph holdings of the NAL form the
contents of this data base.

AVLINE
Producer: National Library of Medicine
Time span: 1975 to present
Coverage: International
Updated: Weekly
Vendor: NLM
Print Counterpart: NLM Audio-visuals Catalog; Health
                   Sciences Audiovisuals

10,000 audiovisual programs in the health sciences can
be searched using the list of Medical Subject Headings.
Many formats are represented, and they include the
productions of 500 distributors.

BIOSIS PREVIEWS
Producer: BioSciences Information Service
Time span: 1969 to present
Coverage: International
Updated: Biweekly
Vendor: BRS, DIALOG
Print counterpart: Biological Abstracts and Biological
                   Abstracts/RRM

9000 primary journals and monographs, as well as symposia, reviews, reports, and other secondary sources are
included.

CATLINE
Producer: National Library of Medicine
Time span: 1901 to present
Coverage: International
Updated: Weekly
Vendor: NLM
Print counterpart: NLM Current Catalog

   Citations to 400,000 books can be searched using the
   Medical Subject Headings.

CHEMTOX
Producer: Van Nostrand Reinhold Information Services
Time Span: Unknown
Coverage: See below
Updated: Quarterly

   The CHEMTOX data base contains information on over 3200
   chemical substances that are hazardous and that are
   common to the environment and workplace.   Substances
   are indexed by Name, CAS number, RTECS number, DOT ID
   number, EPA (RCRA) ID number and STCC number.   Properties, toxicological data, and regulatory data are
   included.

   This data base is licensed to the purchaser by the
   producer for use with the customer's computer.   When
   used with portable equipment, the CHEMTOX data base may
   be taken on-site.

EMBASE
Producer: Excerpta Medica
Time span: 1974 to present
Coverage: International
Updated: Biweekly
Vendor: BRS, DIALOG
Print counterpart: Excerpta Medica (see "Indexes" section of
                   this chapter)

HSELINE
Producer: European Space Agency Information Retrieval
            Service
Time Span: Unknown
Coverage: International
Updated: Unknown
Vendor: ESA Information Retrieval Service
            Online Services Division
            ESRIN
            Via Galileo Galilei 00044
            Frascati, Rome, Italy

   HSELINE contains citations to international publications on occupational health and safety.   Journals,
   books, conference proceedings, and relevant legislation
   are included.

Life Sciences Collection
Producer: Cambridge Scientific Abstracts
Time span: 1978 to present
Coverage: International
Updated: Monthly
Vendor: DIALOG
Print counterpart: Life Sciences Collection of 17
                   Abstracting Journals

    Journals, books, conference proceedings and reports are
    covered.

MEDLINE
Producer: National Library of Medicine
Time span: 1966 to present
Coverage: International
Updated: Monthly
Vendor: BRS, DIALOG, NLM
Print counterpart: Index Medicus, International Nursing
                   Index, Index to Dental Literature

    The Medical Subject Headings can be used to find appro-
    priate terms to search for relevant articles from 3000
    biomedical and nursing journals.

Occupational Safety and Health
Producer: National Institute for Occupational
           Safety and Health
Time span: 1972 to present
Updated: Monthly
Vendor: DIALOG

    Citations to over 480 journal titles and 70,000 mono-
    graphs and technical reports are included.

RTECS
Producer: National Library of Medicine
Time span: 1979 to present
Updated: Quarterly
Vendor: NLM
Print counterpart: RTECS: Registry of Toxic Effects of
                   Chemical Substances (see "Books"
                   section of this chapter)

TSCA Plus, TSCA Initial Inventory
Producer: Office of Toxic Substances, Environmental
           Protection Agency
Time span: 1976 to present
Coverage: United States
Updated: Irregular, as needed
Vendor: SDC, DIALOG
Print counterpart: TSCA Initial Inventory, TSCA Cumulative
                   Supplement II, TSCA Initial Inventory,
                   Reporting Companies Section

This data base contains a list of 58 chemicals from the Toxic Substances Control Act inventory, and 3000 chemical companies. It lists chemical substances in use in the United States.

TOXLINE
Producer: National Library of Medicine
Time span: 1965 to present
Coverage: International
Updated: Monthly
Vendor: NLM
Print Counterpart: Chemical Abstracts Service: Chemical-Biological Activities, Biosciences Information Service: Abstracts on Health Effects of Environmental Pollutants, International Pharmaceutical Abstracts, NLM Toxicity Bibliography, Pesticides Abstracts, Environmental Mutagen Information Center file, Environmental Teratology Information Center file, Smithsonian Science Information Exchange: Toxicology/Epidemiology Research Projects, NTIS Toxicology Document and Data Depository, Hayes File on Pesticides, Toxic Materials Information Center file.

Publishes information on human and animal toxicity studies, effects of environmental chemicals and pollutants, adverse drug reactions and analytical methodology are included. Almost all of the citations have abstracts and CAS Registry numbers.

TOXNET
Producer: National Library of Medicine
Time Span: Unknown
Coverage: Unknown
Updated: Unknown
Vendor: NLM
Print counterpart: Variety of primary, secondary and tertiary sources

The TDB File consists of information on toxicology, chemistry, and hazardous waste information on 4100 chemicals; the HSDB File is a broader, more comprehensive file containing 144 data elements.

Audiovisual Materials

Coal Dust: Hazards and Controls. Washington, D.C.: Mine
    Safety and Health Administration, National Audiovisual
    Center, 1980. Color, 14 min.

    The hazards of coal mine dust to underground and sur-
    face miners are covered in this film.

Facts about Vinyl Chloride. Washington, D.C.: Occupational
    Safety and Health Administration, National Audiovisual
    Center, 1982. Videocassette, color, 35 min.

    This program shows the effects and potential hazards of
    working with vinyl chloride or polyvinyl chloride. It
    can be used to partially meet the OSHA training require-
    ment for industries which manufacture and use vinyl
    chloride and polyvinyl chloride.

Gray, Michael R., et al. The Plight of the Migrant.
    Tucson, AZ: Arizona Center for Occupational Safety and
    Health, 1980. Videocassette, color.

    Some health issues which migrant workers face, such as
    exposure to pesticides in the fields, are among the
    topics covered in this videotape, which is designed for
    nursing and medical students, and for continuing educa-
    tion of health professionals.

Greenberg, S. Donald. Asbestos-associated Pulmonary
    Diseases. New York: MEDCOM, 1981. 66 color slides, 1
    30-minute audiocassette, 1 guide.

    Designed for medical students and continuing education
    for physicians, three asbestos-associated diseases are
    discussed: asbestosis, lung cancer, and mesothelioma
    of the pleura.

Occupational Dermatoses: a Program for Physicians.
    Washington, D.C.: National Institute for Occupational
    Safety and Health, National Audiovisual Center, 1981.
    114 color slides, 38 min., audiocassette.

    Endorsed by the American Academy of Dermatology and the
    American Academy of Occupational Medicine, this slide
    tape discusses pathologic responses of the skin to the
    occupational environment.

Radio Frequency Radiation Hazards. Washington, D.C.: Air
    Force, National Audiovisual Center, 1982. Color, 18
    min.

Some hazards covered in this film are body damage,
ignition of fuel vapor, and detonation of devices.

Skin Diseases of Agricultural Workers. Washington, D.C.:
National Medical Audiovisual Center, National Audiovisual Center, 1982. 79 color slides, 28 min., audiocassette, user manual.

This slide tape covers five types of dermatoses,
sources of exposure, signs and symptoms, treatment and
prevention, and is designed for medical students and
physicians.

Over 50 audiovisual programs are offered for preview, rent
or sale by Industrial Training Systems Corp., 823 E. Gate
Drive, Mount Laurel, NJ 08054, (609) 234-2600. The subjects
include health hazard information, safety, and industrial
hygiene. A free quarterly newsletter describes new training
programs and contains articles by health and safety professionals. A catalog of existing training programs is
available.

Associations

American Association of Occupational Health Nurses
3500 Piedmont Road, NE
Atlanta, GA 30305   (404) 262-1162

Founded in 1942, this organization has more than 11,000
members who are occupational health nurses. Continuing
education is a priority of the association. It encourages research activities among its members, and is the
recognized source in matters of professional standards
in the field. Its annual meeting is held with the
American Occupational Medical Association and is called
the American Occupational Health Conference. Occupational Health Nursing is the official journal of the
organization. AAOHN News is a monthly newsletter
published for members.

American Conference of Governmental Industrial Hygienists
Building D-5
6500 Glenway Avenue
Cincinnati, OH 45211   (513) 661-7881

This group was organized in 1938 to provide a medium
for the exchange of ideas and experiences, and to
promote industrial health standards and techniques.
Members are professional personnel in government agencies or educational institutions engaged in occupational safety and health programs. Their publications in-

clude the ACGIH Annals Series, which contain the proceedings of conferences and symposia in the occupational health field.

American Occupational Medicine Association
2340 South Arlington Heights Road
Arlington Heights, IL 60005   (312) 782-2166

Physicians involved with workers' health promotion are
among the 4000 members of this association. Continuing
education programs are offered for members. It sponsors the American Occupational Health Conference annually with the American Association of Occupational
Health Nurses. Publications include JOM, the Journal
of Occupational Medicine, the AOMA Report (only available to members, and gives information about federal
regulatory developments), and a listing of reports and
reprints from JOM.

British Occupational Hygiene Society
c/o Esso Research Centre
Abingdon, Oxon OX13 6AE
England

The organization was founded in 1953 for the study of
environmental stresses which affect the health, com-
fort, and efficiency of workers and the community.
It publishes the Annals of Occupational Hygiene, a
Hygiene Standards Series, Hygiene Technology Guide
Series, and Technical Guide Series. It holds symposia
and annual conferences.

Canadian Centre for Occupational Health and Safety/Centre Canadien d'Hygiene et de Securite au Travail
250 Main Street East
Hamilton, Ontario L8N 1H6
Canada   (416) 523-2981

The Centre is an independent institute which exists to
promote occupational health and safety by gathering,
evaluating, and distributing information primarily
through its Information and Advisory Service. IAS
provides answers to telephone or mail inquiries through
staff scientists and in-house and acquired data bases.
It can provide project reports to boards and commis-
sions which are available to the public. At the Centre
is its quarterly newsletter. A list of bibliographies
and papers is available.

Center for Occupational Hazards
Five Beekman Street
New York, NY 10038   (212) 227-6220

The Center is a national clearinghouse for research and information on health hazards in the arts. The services provided are the Art Hazards Information Center, which gives telephone and written answers to questions on art hazards, and a list of relevant publications: the Art Hazards Newsletter (described in the "Journals" section of this chapter); a lecture program and course on art hazards, and a consultation program.

Human Ecology Action League
505 North Lake Shore Drive
Suite 6506
Chicago, IL 60611   (312) 836-0422

The members are interested in the effect of environmental, natural and synthetic substances on human health. It publishes a newsletter and a directory of physicians practicing clinical ecology.

Industrial Health Foundation
34 Penn Circle W
Pittsburgh, PA 15206   (412) 363-6600

The advancement of health in industry is the purpose of this research and service organization. Continuing education courses and information on relevant topics are provided. It publishes the Industrial Hygiene Digest, bulletins, technical papers, and symposia proceedings.

International Union of Electronic, Electrical, Technical, Salaried and Machine Workers
IUE Health and Safety Program
1126 16th Street, NW
Washington, D.C. 20036   (202) 296-1200

Some hazards for electrical workers are mercury, PCB's, trichloroethylene, radiation, and isocyanates. The IUE has developed a health and safety program which includes speakers, workshops, conferences, technical support and inspections, assistance with filing complaints to OSHA, and support in legal proceedings which might result from contested OSHA citations.

Midwest Center for Occupational Health and Safety
640 Jackson Street
St. Paul, MN 55101   (612) 221-3992

Informational sessions, short courses, and a two-week Graduate Occupational Health and Safety Institute are offered to all who are interested in occupational health and safety.

Ministry of Concern for Public Health
151 East Street
Buffalo, NY 14207   (716) 874-5955

The purpose of this organization is to eliminate disease caused by environmental pollution. It provides
information, consultation, and grants on occupational
health, and other effects of various substances on
human health.

Occupational Medical Administrator's Association
CN/VIA Medical and Health Services
P.O. Box 8100
Montreal, Quebec H3C 3N4
Canada   Phone #: not available

Full time administrators who participate in administrative policy decisions of occupational programs may
join this organization, which exists to stimulate
interest in and provide a forum for the discussion of
items related to occupational medical administration.
There are no membership dues, and a three-day meeting
is held each spring. They do not have any organizational publications.

Rocky Mountain Center for Occupational and Environmental
    Health
Building 512
Salt Lake City, UT 84112   (801) 581-5710

The Center was established in 1977 to coordinate and
develop training and research programs in occupational
safety and health. Occupational health professionals
are trained through an occupational medicine residency
program, master's degree programs, workshops and conferences. Consultants are available for on-site vis-
its, and a research facility and clinic is operated by
RMCOEH. It is based at the University of Utah, and has
received a five-year grant from the National Institute
for Occupational Safety and Health as one of 12 national NIOSH educational resource centers.

Society for Occupational and Environmental Health
2021 K Street, NW
Suite 305
Washington, D.C. 20006   (202) 737-5045

Members of this society are individuals who have an
interest in occupational or environmental health. They
include scientists and non-scientists, who try to improve the quality of our environment by holding open
forums for the exchange of information and ideas. Conferences and workshops are held, and their proceedings
are published.

The Society for the Prevention of Asbestosis and
    Industrial Diseases
38 Drapers Road
Enfield, Middx. EN2 8LU
England   01-366-1640

   This organization works actively to eliminate the use
   of asbestos, and helps victims of asbestosis and their
   families through letter-writing and phone campaigns,
   soliciting assistance from other organizations, fund
   raising, publicity on the effects of asbestos, and in-
   formal meetings.

System Safety Society
14252 Culver Drive, Suite A-261
Irvine, CA 92714   (714) 551-2463

   Members of this international society either have an
   interest in system safety, or are employed in this
   field. The journal Hazard Prevention and society-
   sponsored conferences and symposia are published, and
   the International System Safety Conferences are
   sponsored biennially.

Government Organizations

Alberta Workers' Health, Safety and Compensation
Occupational Health and Safety Division
Research and Education Services
10709 Jasper Avenue
Edmonton, Alberta, T5J 3N3
Canada   Phone #: not available

   Brochures, posters, and the O.H.S. Magazine can be
   requested from this agency. "Chemical Hazards and
   Custodial Workers" lists products and the hazardous
   chemicals they may contain, precautions and first aid
   measures. "Drycleaners Beware," "Code of Practice for
   Handling Sulphur," "Safe Operating Procedures for Seismic Drilling" and "Drilling Rig Health and Safety Committee Guidelines" are some of the relevant booklets
   available. The Occupational Health and Safety Heritage
   Grant Program supports research, training and education
   relating to accident prevention, occupational health,
   and health promotion for Alberta workers. Alberta
   residents or persons doing work that is significant to
   Alberta workers may apply. Educational programs, seminars, consultations, program development, and first aid
   training are offered.

Army Environmental Hygiene Agency
Edgewood Area
Aberdeen Proving Ground, MD 21010   (301) 671-4236

This agency provides consultations, support services,
investigations, and training to promote the health and
environmental programs of the Army. A staff of 400
medical officers, occupational health nurses, chemists,
engineers, and other scientific personnel provide services through five directorates: Environmental Quality,
Occupational and Environmental Health, Radiation and
Environmental Sciences, Laboratory Services, and Regional Activities. It publishes a number of technical
guides on health hazards. The library accesses NLM
data bases, and has a collection of texts on general
medicine, industrial and environmental health, and
allied preventive medicine. Industrial medicine and
occupational safety periodicals are held.

Consumer Product Safety Commission
1111 Eighteenth Street, NW
Washington, D.C. 20207
(202) 634-7740   TTY (800) 638-8270
(800) 638-2772   (in Maryland only) (800) 492-8104

The Commission was created in 1972 to help prevent
injuries and deaths associated with consumer products.
Its priority projects for 1984 included toxicity of
combustion products, hazards of gas-heating systems,
exposure to indoor air pollutants from fuel-fired
appliances and from pressed wood products, and exposure
to chlorocarbons.

Environmental Teratology Information Center
Oak Ridge National Laboratory
P.O. Box Y, Building 9224
Oak Ridge, TN 37831   (615) 574-7871

In 1975, the Center was organized to collect, organize
and disseminate information on the evaluation of chemical, biological, and physical agents for teratogenic
activity. The ETIC maintains a data base which is part
of the National Library of Medicine's TOXLINE system
and the Department of Energy's RECON system.

International Labor Office
Washington Branch
1750 New York Avenue, NW
Washington, D.C 20006   (202) 376-2315

One purpose of the ILO is to help improve working
conditions. It sponsors an annual conference and the

International Institute for Labour Studies in Geneva, and publishes books in the area of labor, occupational safety and health.

National Center for Toxicological Research
Jefferson, AZ 72079   (501) 541-4000

Some areas in which research programs are conducted are biological effects on humans, human risk assessment, and preventive health. A book describing specific research areas, capabilities of the Center, methods used, types of study for specific chemicals is available, as well as a list of final reports which are kept in the Center's archive.

National Institute for Occupational Safety and Health
5600 Fishers Lane
Rockville, MD 20857   (301) 443-2140

The Institute works to prevent work-related illness and injury through health protection and promotion, and preventive health services. There are 10 regional offices in the country. The NIOSH Educational Resource Centers are located within universities in the 10 HHS regions of the U.S. They offer continuing education courses.

Occupational Health and Safety Council
P.O. Box 1900
42 Great George Street
Charlottetown, Prince Edward Island C1A 7N5
Canada   (902) 892-0941

This newly formed organization was formed to draft an Occupational Health and Safety act for Prince Edward Island. They do not yet have a list of publications. The Council is an advisory body to the Minister of Labour. Its library of journals and reference books is open to the public.

Occupational Safety and Health Administration
Department of Labor
Washington, D.C. 20213   (202) 634-7960

OSHA was created in 1970 to encourage employers and employees to reduce workplace hazards and to implement safety and health programs, to maintain a reporting and recordkeeping system to monitor occupational illness and injury, to develop occupational safety and health standards, and to monitor state occupational safety and health programs. Information on standards can be found in the Federal Register. Employers are responsible for informing employees about safety and health matters. Unscheduled on-site inspections may be performed by

OSHA personnel. Free on-site consultations are provided to help employers determine hazardous conditions
and implement corrective measures. They will provide a
list of publications and audiovisuals on OSHA regulations, employee information, manager information, occupational health and safety, reports of investigations,
and training programs.

Toxicology Information Response Center
Oak Ridge National Laboratory
P.O. Box X
Oak Ridge, TN 37831   (615) 576-1743

The Center is a focal point for toxicology information
and uses online data bases and its library collection
to provide reference services on topics such as food
additives, pharmaceuticals, industrial chemicals,
environmental pollutants, heavy metals, and pesticides.
It was founded in 1971.

Libraries, Information Centers

Alberta Worker's Health Safety and Compensation
Occupational Health and Safety Library
5th Floor
10709 Jasper Avenue
Edmonton, Alberta T5J 3N3
Canada   (403) 427-4671

Special collections in this library include U.S.
National Institute for Occupational Safety and Health
recommended standards, and U.S. Bureau of Mines Reports
of Investigations. The library is open to the public
with restrictions.

American Cyanamid Company
Environmental Health Library
One Cyanamid Plaza
Wayne, NJ 07470   (201) 831-4379

The library holds 2000 books, 2000 pamphlets, 12 vertical file drawers of reprints, and 132 serial subscriptions on toxicology, occupational medicine, industrial
hygiene and safety.

Canadian Asbestos Information Centre
1130 Sherbrooke Street W.
Suite 410
Montreal, Quebec H3A 2M8
Canada   (514) 844-3956

The Center was founded in 1982 to disseminate information and promote the safe use of asbestos throughout
Canada and the world. It encourages governments and
international bodies to institute adequate standards
and control measures for asbestos use.

Commonweal-Research Institute Library
Box 316
Bolinas, CA 94924   (415) 868-0970

This library holds 1200 books, 16 vertical file drawers, 300 government publications, and 153 serial subscriptions on environmental and occupational health,
and environmental toxicology.

Department of Labor
OSHA Technical Data Center
200 Constitution Avenue, NW
Room N-2439 Rear
Washington, D.C. 20210   (202) 523-9700

This library was established in 1972 to provide technical information to the OSHA national office and field
staffs. It can be used by the public during official
working hours for reference purposes. Other libraries
and government agencies may borrow materials through
interlibrary loan. Their holdings include 8000 books
and bound periodical volumes, 105,000 fiche, 2000 technical documents, 3000 standards and codes, and 200
serial subscriptions in the area of occupational safety, industrial hygiene, toxicology, hazardous materials
and carcinogens.

DuPont de Nemours (E.I.) and Co., Inc.
Haskell Laboratory for Toxicology and Industrial Medicine
Library
P.O. Box 50, Elkton Road
Newark, DE 19711   (302) 366-5225

The Haskell Laboratory functions to determine toxicity
of chemical compounds, and the safety precautions which
should be taken when using them. The library's holdings are in the area of industrial medicine and toxicology. It is open for use by qualified persons.

Eastman Kodak Company
Health, Safety, and Human Factors Laboratory
Library
Kodak Park Division Building 320
Rochester, NY 14650   (716) 458-1000

Toxicology, occupational medicine and environmental
hygiene are the subject areas on which this library

concentrates, with 8000 books and government publications, and 365 serial subscriptions.

Exxon Corporation
Medicine and Environmental Health Department
Research and Environmental Health Division Library
Mettlers Road
Box 235
East Millstone, NJ 08873   (201) 474-2506

Toxicology, industrial hygiene, and industrial medicine are the subject emphasis of this library, which holds 1000 books and 303 serial subscriptions.

Exxon Corporation
Medical Library
1251 Avenue of the Americas
New York, NY 10020   (212) 398-2504

Nine hundred books, 1900 pamphlets, and 80 serial titles are held in this library in the area of industrial medicine and hygiene.

Geomet Technologies, Inc.
Information Center
1801 Research Boulevard, 6th Floor
Rockville, MD 20850   (301) 424-9133

This library has holdings on toxicology, environmental health, and industrial and occupational medicine. They include 500 books, 15,000 reprints, 3000 translations, 50 serial subscriptions, and 10 newspapers. The library is open to the public by appointment.

Manville Service Corporation
Health, Safety and Environment Library and
    Information Center
Box 5108
Denver, CO 88217   (303) 978-2580

In addition to a special collection on asbestos, silica, and man-made vitreous fibers, this library has materials on occupational health, toxicology, industrial hygiene and safety, and carcinogens. There are 5000 books, and 193 serial subscriptions.

Massachusetts Institute of Technology
Environmental Medical Service
Library
77 Massachusetts Avenue, Room 20B-23B
Cambridge, MA 02139   (617) 253-7983

The library has 9 verticle file drawers of clippings
and reprints on industrial hygiene and toxicology,
radiation protection, and clinical toxicology. It is
open to the public with permission.

Massachusetts State Dept. of Labor and Industries
Division of Occupational Hygiene
Special Technical Library
39 Boylston Street
Boston, MA 02116   (617) 727-3982

The subject areas in this collection include radiation
and occupational medicine. There are 500 books, 180
pamphlets, and 28 serial subscriptions.

Ontario Ministry of Labour
Library
400 University Avenue
10th Floor
Toronto, Ontario M7A 1T7
Canada   Phone #: not available

Over 70,000 books and reports, 1500 journal titles, 500
Statistics Canada reports, Canadian, American and
international standards, annual reports and other serials, government documents, and vertical file materials
are held by this library, in the area of employment,
labor relations, working conditions, equal employment,
human rights, women, occupational health and safety,
environmental health, and radiation protection. Over
13,000 pre-1900 reports and reprints on occupational
health are held in an archival collection. The general
public may use the library's reference services. A
Library Bulletin is published.

Toxicology Information Response Center
    see listing under "Government Organizations"

Research Centers and Industrial Laboratories

B.C. Cancer Research Centre
Environmental Carcinogenesis Unit
601 West 10th Avenue
Vancouver, B.C. V5Z IL3
Canada   (604) 873-8401

Environmental carcinogenesis is studied, and weekly
seminars are held.

Bureau of Radiological Health
5600 Fishers Lane
Rockville, MD 20857   (301) 443-4690

The Bureau plans, conducts, and supports research on
health effects of radiation exposure.   The Bureau of
Radiological Health Technical Report Series is published irregularly. A semiannual symposium series is
held.

Centers for Disease Control
1600 Clifton Road, NE
Atlanta, GA 30333   (404) 329-3291

Infectious diseases and occupational health are the
main areas of research.

Colorado State University
Institute for Rural Environmental Health
College of Veterinary Medicine and Biomedical Sciences
Fort Collins, CO 80523   (303) 491-6228

Rural environmental health and safety of residents of
Colorado and the Rocky Mountain region engaged in agriculture is emphasized.

Drexel University
Institute for Environmental Studies
32nd and Chestnut Streets
Philadelphia, PA 19104   (215) 895-2265

Occupational health, environmental toxicology, and
environmental health are researched at the institute.

Health Effects Research Laboratory
Environmental Protection Agency
26 W. St. Clair Street
Cincinnati, OH 45268   (513) 684-7406

Harmful effects of pollutants are studied, emphasizing
airborne pollutants.

Health Effects Research Laboratory
Environmental Protection Agency
Research Triangle Park, NC 27711   (919) 541-2281

The biological effects of various pollutants and environmental health problems are studied.

Health Research, Inc.
1315 Empire State Plaza Tower Building
Albany, NY 12237   (518) 474-1689

Environmental studies to determine long-range health
hazards of pesticides, industrial discharges, and
radioactive and sewage wastes are conducted.

Inhalation Toxicology Research Institute
P.O. Box 5890
Albuquerque, NM 87185   (505) 264-6835

Inhalation toxicology of airborne particulates, fibers,
vapors, and gases is studied. Seminars, professional
meetings, and conferences are held; a postdoctoral
training program is offered.

Laboratory of Radiobiology and Environmental Health
University of California, San Francisco
San Francisco, CA 94143   (415) 666-1636

This laboratory is owned by the Department of Energy
and conducts research in radiobiology, cytogenetics,
biophysics, and developmental biology.

Michigan State University
Center for Environmental Toxicology
C231 Holden Hall
East Lansing, MI 48824   (517) 353-6469

Behavioral toxicology, biochemical toxicology, carcinogenesis, ecotoxicology, genetic toxicology, and terrestrial toxicology are researched. Doctoral and postdoctoral training in environmental toxicology is
offered. Workshops, symposia, seminars, and an annual
conference are sponsored.

National Institute of Environmental Health Sciences
P.O. Box 12233
Research Triangle Park, NC 27709   (919) 541-3201

Behavioral and toxicological effects of environmental
exposure to toxic substances, development of systems
for monitoring the human population, studies on environmental mutagens and the accumulation of chemicals in
living organisms are the principal areas of research.
Environmental Health Perspectives is published by the
Institute.

New York University
Institute of Environmental Medicine
550 First Avenue
New York, NY 10016   (212) 340-5280

The areas of research include toxicology, carcinogenesis, respiratory disease, epidemiology, and environ-

mental health hazards.   Courses are offered for professionals in medicine, engineering, and public health.

North Carolina State University
Interdepartmental Toxicology Program
1541 Gardner Hall
Raleigh, NC 27607   (919) 737-2276

   Environmental health sciences are emphasized.

Ohio State University
Environmental Health Laboratory
Sisson Hall, Room 239
1900 Coffey Road
Columbus, OH 43210   (614) 422-1206

   The Laboratory evaluates heavy metal exposure in human
   and animal populations in Ohio.

Oregon State University
Environmental Health Sciences Center
317 Weniger Hall
Corvalis, OR 97331   (503) 754-3608

   The effect of environmental chemicals on human health
   is studied.   Conferences, symposia, and meetings for
   student training and communication with the public are
   held.

Purdue University
Department of Medicinal Chemistry
Bionucleonics Division
Pharmacy Building
West Lafayette, IN 47907   (317) 494-1435/6

   Biological effects of radiation, evaluation of environmental radiation hazards, radiation risk assessment,
   environmental toxicology, inhalation toxicology, occupational health, and industrial hygiene are the activities researched by this department.

University of California, Davis
Laboratory for Energy-related Health Research
Davis, CA 95616   (916) 752-1341

   Twenty researchers study the biomedical effects of
   long-term low level exposure to nuclear and non-nuclear
   related effluents from energy production.

University of California, Davis
Occupational and Environmental Health Unit
TB 136
Davis, CA 95616   (916) 752-3317

The epidemiology of occupational and environmental
health problems is studied, and instruction in occupational medicine is given.

University of Cincinnati
Kettering Laboratory
3223 Eden Avenue
Cincinnati, OH 45267   (513) 872-5701

Specific research areas include effects of trace metals,
behavioral toxicology, environmental carcinogenesis,
teratogenesis, and mutagenesis. Professional seminars
and postgraduate courses are held.

University of Colorado-Denver
Center for Environmental Sciences
Campus Box 136
1100 14th Street
Denver, CO 80202   (303) 629-3460

The effects of mineral and energy production on health
and the environment are studied.

University of Michigan
Institute of Environmental and Industrial Health
School of Public Health
Ann Arbor, MI 48109   (313) 764-3188

Industrial and environmental health is studied; services to industry and patients with occupational health
problems is offered. The annual Selby Discussional on
mutual problems and new procedures in occupational
health is offered by invitation for industrial hygienists.

University of Texas
Institute of Environmental Health
P.O. Box 20186
Houston, TX 77025   (713) 792-4425

An ecological orientation to epidemiological studies of
disease distributions is emphasized in the research
performed at the Institute.

Vanderbilt University
Environmental Health Science Center
21st Avenue South and Garland
Nashville, TN 37208   (615) 322-2262

Biochemical and chemical toxicology, pesticides, toxic metals, mutagenic, carcinogenic, and teratogenic agents are studied. A monthly Toxicology Center Seminar is held.

Windsor Occupational Safety and Health Council
824 Tecumseh Road East
Windsor, Ontario N8X 2S3
Canada   (519) 254-4192

Occupational and environmental health research is published in pamphlets, books and audiovisuals. The Council sponsors the Women's Occupational Safety and Health Conference and the Occupational Cancer Conference, and maintains a library of 2000 volumes.

8.

Disposal of Non-Radioactive Hazardous Wastes

Patricia Ann Coty

Review: Terminology and Scope

The United States Environmental Protection Agency has
estimated that between 28 and 54 million metric tons of
federally-regulated hazardous waste were generated in the
United States in 1980. Some states are even stricter than
the federal government in their definitions of "hazardous"
waste. If wastes further defined as hazardous by the
various states are added, the amount jumps to between 255
and 275 million metric tons of regulated hazardous waste
generated per year. Most of these wastes - 70 to 85 per
cent - are managed on the site where they are generated, the
great majority being land-disposed.

Waste type is an important factor in choosing a technology
for waste disposal. For wastes considered hazardous because
of their reactivity, corrosiveness or ignitability, physical-
chemical treatments are common. For toxic materials, whether
organic, inorganic or metallic, treatment and disposal is
more complicated. Toxicity is considered the capacity of a
substance to produce injury, while hazard is a measure of
potential risk which incorporates information on intended
use, anticipated route of entry, and frequency and duration
of exposure. Texts on hazardous wastes usually include
coverage of toxic wastes, as well as flammable, explosive,
biological, and radioactive wastes. This chapter will focus
on information sources in the disposal of non-radioactive
hazardous wastes, as regulated within Section C of the
Resource Conservation and Recovery Act. For the most part,
industrial wastes, not municipal or federal, will be covered.

The Resource Conservation and Recovery Act (RCRA) defines
hazardous waste as a subset of "solid waste", solid waste
being garbage, refuse or sludge from waste treatment, water
supply treatment, or air pollution control facilities.
These can be solid, liquid, semisolid, or gaseous, even
though RCRA uses the generic term "solid waste". Those
solid wastes which fall into the category of "hazardous" are
wastes which can cause or significantly contribute to

mortality or illness, or those which create a present or potential hazard to human health or the environment. A few exceptions, regulated by agencies other than the Environmental Protection Agency (EPA), include solid or dissolved materials in domestic sewage or in irrigation flows, industrial discharges that are point sources, or special nuclear, source, or by-product materials of nuclear industries.

An estimated 7,785 hazardous waste management facilities were operating in the United States in 1981. These facilities fell under nine major technology types: injection wells, landfills, land treatment, surface impoundments, waste piles, incinerators, storage containers, storage tanks, and treatment tanks.

"Pretreatment processes" partially alter, accumulate, or reduce the volume of hazardous wastes, in preparation for further treatment, storage or disposal. Pretreatment processes include flocculation-sedimentation, filtration, precipitation, ammonia stripping, evaporation, centrifugation, carbon sorption, solidification/fixation, solvent extraction, neutralization, dialysis, reverse osmosis, ion exchange, and calcination. "Source segregation" is the separation of hazardous constituents from large volumes of waste, resulting in a small volume of concentrated hazardous waste. "Recovery" is the separation of a substance from a mixture, and "recycling" is the use of recovered materials.

Technologies that reduce the hazards in waste, rather than the volume, can be grouped into two general categories: treatment and disposal. Treatment technologies generally decompose hazardous wastes into non-hazardous components, and include solidification, incineration, pyrolysis, molten salt reactors, plasma arc, wet oxidation, and biological treatments. Disposal technologies include landfills, surface impoundments, land treatment, deep-well injection sites, land burial, ocean dumping, and ocean incineration. Many of the technologies do not actually alter the hazardous characteristics of the wastes, but attempt to hold the hazardous substances in long-term isolation from the surrounding environment. Some technologies rely on eventual biodegradation of the wastes, rendering them free from hazard.

The source of much of the above data, and an excellent introduction for the researcher with little or no familiarity with hazardous waste management and disposal, is

Technologies and Management Strategies for Hazardous Waste
        Control. Washington, D.C.: Congress of the U.S.,
        Office of Technology Assessment, 1983. Report
        #OTA-M-196. (SuDoc Y3.T 22/2:2 T 22/8).[1]

---

[1]Where possible throughout this chapter, Superintendent of Documents Classification numbers will be provided for federal publications, for those readers having access to Federal Depository libraries.

The purpose of this text is an assessment, at the request of Congress, of the technological options for managing hazardous wastes at operating facilities; the technical means to address the problem of uncontrolled and abandoned hazardous waste sites; and the technical adequacy of the Federal regulatory program. The report presents five policy options for future hazardous waste programs. This book is valuable for the beginner because it reviews the current scope of hazardous waste treatment and disposal in the United States. A summary edition (55 pages) is also available (Report #OTA-M-197). Both can be purchased from the Government Printing Office.

Other texts which offer a solid starting point for the serious student unfamiliar with the terminology and technology of hazardous waste disposal include:

Epstein, Samuel S., et al. <u>Hazardous Waste in America</u>. San Francisco: Sierra Club Books, 1983.

A look at hazardous waste disposal from the environmentalists' point of view. Cites numerous instances of improper waste management; examines the policies of the EPA; explores alternatives to landfilling. A helpful appendix lists sources and composition on non-radioactive wastes, compares state programs, lists bibliographic sources in hazardous waste exchange technology, and provides other data. Includes bibliographical references and index.

<u>Everybody's Problem: Hazardous Waste</u>. Washington, D.C.: EPA, Office of Solid Waste, 1980. SW-826. (SuDoc EP1.17:826).

A booklet featuring striking color photographs of various problem sites in the nation, with non-technical explanations of the various hazardous waste management options.

<u>Hazardous Waste Management</u> and <u>Chemical Risk: A Primer</u>, from the Information Pamphlet Series of the American Chemical Society. Single copies free from the Office of Federal Regulatory Programs, ACS Department of Government Relations and Science Policy, 1155 16th Street, NW, Washington, D.C. 20036.

These pamphlets discuss science-related regulatory issues in a clear manner, for a general audience. <u>Hazardous Waste Management</u> describes this complex issue - from waste generation through disposal - and discusses the federal regulations affecting it.

Chemical Risk: A Primer deals with the scientific
issues involved in determining the health risks arising
from exposure to chemicals.

Siting Hazardous Waste Management Facilities: a Handbook.
Prepared by the members of the Hazardous Waste Dialogue
Group; under the sponsorshop of the Program for Environ-
mental Dispute Resolution, The Conservation Foundation.
Washington, D.C.: Conservation Foundation, 1983.

An educational handbook designed to promote public
participation in evaluating the suitability of proposed
hazardous waste management facilities. Written by a
group including representatives of environmental and
public interest organizations, trade associations,
industries, and state governments. An excellent introduction for the beginner. Includes a glossary and
bibliographic references.

Additional texts of a general nature, but more technically
advanced, include:

Allegri, Theodore H. Handling and Management of Hazardous
Wastes. London: Chapman & Hall; Distributed in U.S. by
Methuen, Inc., 1985.

Describes the manner in which most hazardous materials
must be handled throughout their life cycles, and
discusses the legal and ethical aspects of their use.
Explains how certain chemicals must be identified,
labeled, transported, and stored, and the documentation
that must accompany them.

Cashman, John R. Management of Hazardous Waste Treatment/
Storage/Disposal Facilities. Lancaster, PA: Technomic,
1986.

The author, surveying over 20 major American facilities, presents a detailed examination of the methods in
use today in waste treatment, storage and disposal.
Actual operational procedures and management are
addressed. Included is a chapter on "Research Sources
and Resources."

Dawson, Gaynor W. and Basil W. Mercer. Hazardous Waste
Management. New York: Wiley, 1986.

Provides guidance for the technological and policy
considerations of facilities design and location, waste
site reclamation, treatment processes, incineration
alternatives, landfills, salt dome disposal, and survey
techniques. Offers state-of-the-art solutions to
hazardous waste problems.

Edwards, B. H., et al. <u>Emerging Technologies for the Control of Hazardous Wastes</u>. Park Ridge, NJ: Noyes, 1983.

A review and assessment of emerging technologies for
controlling hazardous wastes, with an emphasis on
organic substances. Three technologies - molten salt
combustion, fluidized bed incineration, and UV/ozone
destruction - are covered in detail, while others such
as catalyzed wet oxidation, and catalytic hydrogena-
tion-dechlorination of PCBs, are surveyed. Includes
bibliographic references and index.

Francis, Chester W. and Stanley I. Auerbach, eds. <u>Environ-
ment and Solid Wastes: Characterization, Treatment and
Disposal</u>. Boston: Butterworth, 1983.

Proceedings of the Fourth Life Sciences Symposium,
Environment and Solid Wastes, Gatlinsburg, Tennessee,
October 4-8, 1981; sponsored by the Oak Ridge National
Laboratory, the Department of Energy, the EPA, and the
Electric Power Research Institute. Representatives
from major industries, state and federal agencies,
leading universities and community action groups evaluate the impact of solid waste management practices of
the utility, chemical, mining and municipal waste
generators on human health and the environment.
Bibliographies included.

Greenberg, Michael R. and Richard F. Anderson. <u>Hazardous
Waste Sites: The Credibility Gap</u>. New Brunswick, NJ:
Center for Urban Policy Research, 1984.

Intended for planners, environmental scientists, and
public health officials, this text covers methods used
to identify wastes, the federal laws that control
wastes, and possible dangers to health and the environment from hazardous wastes. An extensive bibliography
is included.

Highland, Joseph H., ed. <u>Hazardous Waste Disposal</u>. Ann
Arbor, MI: Ann Arbor Science, 1982.

Contains a broad overview of the hazardous waste disposal problem and the impact that past disposal practices have had on environmental health. Includes
bibliographic references and index.

Kiang, Yen-Hsiung and Amir A. Metry. <u>Hazardous Waste Processing Technology</u>. Ann Arbor, MI: Ann Arbor Science,
1982.

The state-of-the-art of hazardous waste processing
technology is presented. Covers thermal incineration
systems and chemical, physical and biological treatment

technologies. Discusses safety, process control, site
selection, equipment descriptions, operating characteristics, and applications of the various methods.
Includes bibliographic references and index.

Lehman, John P., ed. <u>Hazardous Waste Disposal</u>. NATO/CCMS
Symposium on Hazardous Waste Disposal. NATO Challenges
of Modern Society, Volume 4. New York: Plenum, 1983.

Proceedings of the NATO/CCMS (North Atlantic Treaty
Organization/Committee on the Challenges of Modern
Society) Symposium on Hazardous Waste Disposal held in
1981. An excellent, comprehensive survey of the international status of hazardous waste disposal, with useful information ranging from general reviews to advanced primary research, written by expert contributors
from nine countries. References included with some
papers. Indexed.

Pojasek, Robert B., ed. <u>Toxic and Hazardous Waste Disposal</u>.
4 volumes. Ann Arbor, MI: Ann Arbor Science, 1979-1980.

Volume 1: <u>Processes for Stabilization/Solidification</u>.
1979.

Volume 2: <u>Options for Stabilization/Solidification</u>.
1979.

Volume 3: <u>Impact of Legislation and Implementation of
Disposal Management Practices</u>. 1980.

Volume 4: <u>New and Promising Ultimate Disposal Options</u>.
1980.

Volumes 1 and 2 examine stabilization/solidification in
depth. Volume 3 compares RCRA and other hazardous
waste legislation with legislation worldwide. Volume 4
reviews ultimate disposal options such as secure landfills, incineration, ocean incineration and composting.
All have bibliographic references and indexes.

Sweeny, Thomas L., et al, eds. <u>Hazardous Waste Management
for the 80's</u>. Ann Arbor, MI: Ann Arbor Science, 1982.

An outgrowth of the Second Ohio Environmental Engineering Conference held in Columbus, Ohio in March 1982.
The book's intent is to provide practical and useful
information on hazardous waste management to engineers,
scientists, lawyers, government officials, and managers
in waste-generating organizations. 34 papers deal with
varied aspects of hazardous waste disposal. Some papers
list references. Indexed.

An overview of the directions in which hazardous waste
research will move in the next few years is provided in
"Hazardous Waste Research" by Julian Josephson (in __Environ-
mental Science and Technology__ vol. 18, no. 7, 1984).

Regulatory/Legislative Background

In 1976, the Resource Conservation and Recovery Act (Public
Law 94-580) established a national hazardous waste regulatory program. It amended the Solid Waste Disposal Act of
1965 and was the first comprehensive national regulation of
solid waste. The principal concern of the Resource Conservation and Recovery Act (RCRA), covered in Section C, is
hazardous waste disposal, although the promotion of resource
recovery and management of non-hazardous wastes are also
covered. The Environmental Protection Agency (EPA) was
directed by this law to develop and administer criteria to
determine which wastes are hazardous, and to establish standards for siting, design and operation of disposal facilities. States are encouraged to use EPA guidelines to
develop their own regulatory programs.

RCRA is specifically about the disposal of hazardous wastes,
whereas previous legislation had indirectly impacted hazardous waste disposal by regulating its effects on air and
water quality. Preceeding laws included the Federal Water
Pollution Control Act as amended in 1972 (PL 92-500), the
Clean Water Act of 1977 (PL 95-217), the Safe Drinking Water
Act of 1974 (PL 93-523), the Toxic Substances Control Act of
1976 (PL 94-469), and the Occupational Safety and Health Act
of 1970 (PL 91-596).

In 1980, the Comprehensive Environmental Response, Compensation and Liability Act (PL 96-510), also known as CERCLA or
"Superfund", established the collection of funds from waste
generators, transporters, site owners and site operators, to
finance the cleanup of sites which pose immediate dangers to
the public. A national priority list has been established
to identify such sites.

Currently, RCRA requires that generators of hazardous
wastes, excluding households, agricultural concerns using
wastes as fertilizers, and small-quantity generators (less
than 1,000 kg. per month), must provide a written notification - a "manifest" - describing the contents of wastes
hauled from a facility. Wastes are determined to be hazardous by consulting EPA's extensive listings or by testing
the waste for ignitable, corrosive, reactive, or toxic characteristics. EPA's listing of hazardous wastes can be
obtained from any of the ten regional EPA offices.

State regulations for hazardous waste disposal often differ
from the federal regulations, with more stringent definitions of "hazardous waste" and greater limitations on the

types of disposal allowed. State regulatory agencies should be consulted for further information.

Published guides to the laws regulating hazardous waste disposal abound, and the practitioner of hazardous waste disposal will want to have at least one of them within reach. Generally, they review the laws, often reorganizing the various regulations in a practical manner.

The following is a listing of some of the guides currently available:

Environment Regulation Handbook. New York: Special Studies Division of the Environment Information Center, 1973- .

> A four-binder, comprehensive guide of about 3,000 pages. Updated monthly, it organizes all significant environmental statutes.

Groziak, Michael, ed. Pollution Control Guide. Chicago: Commerce Clearing House, 1973- .

> Available in looseleaf format, with updating service. Covers the environmental laws related to pollution, including hazardous waste disposal.

Handbook of Hazardous Waste Regulations. Volume 1: Introduction to RCRA Compliance. Madison, CT: Bureau of Law and Business, Inc.

> Complete guide to regulations, with annotated laws.

Hazardous Waste Reference Service. Silver Spring, MD: Business Publishers, Inc.

> A looseleaf text which is updated as regulations are added or modified. Contains the full texts of hazardous waste and Superfund statutes and regulations and important court decisions.

Keller, J. J. and Associates. Hazardous Waste Management Guide: Identification, Monitoring, Treatment and Disposal. Neenah, WI: J. J. Keller and Associates, Inc., 1980- .

> A subscription service, divided into three sections: management and compliance guidelines; regulations, proposed regulations, and a synopsis of state regulations; and a reference section, with information on waste exchanges and waste spills. Includes a reprint of RCRA.

McCoy and Associates. _RCRA and HSWA: 1985 Index to Hazardous Waste Regulations_. 2nd ed. Lakewood, CO: McCoy and Associates, 1985.

Covers all RCRA regulations that were in effect as of June 30, 1985; includes a keyword index.

Mallow, Alex. _Hazardous Waste Regulations: An Interpretive Guide_. New York: Van Nostrand Reinhold, 1981.

Covers RCRA Section C, then adds five appendices: a glossary of words and phrases that appear in the regu- lations; a complete copy of RCRA; all recently promul- gated regulations relating to hazardous waste manage- ment; proposed hazardous waste regulations; and tables of contents of transportation regulations. Indexed. Although somewhat dated, it presents a clear synopsis of the regulations.

1983/84 Hazardous Materials, Substances and Wastes Compli- ance Guide: References EPA/DOT, CFR 40--Parts 117- 260/261, CFR 40--Parts 171/172, CFR 40--Proposed Part 302. Kutztown, PA: Transportation Skills Program, Hazardous Materials and Waste Management Association, 1982.

Provides copies of the _Code of Federal Regulations_ parts listed in the title, with additional information appended.

Watson, Tom, et al. _Hazardous Wastes Handbook_. 5th ed., partial revision, 1985. Loose-leaf. Rockville, MD: Government Institutes, Inc., 1985.

Analyzes developments under RCRA and Superfund through 1984. Provides practical suggestions for compliance. Appendices include the RCRA and CERCLA (Superfund) statutes, _CFR (Code of Federal Regulations)_ reprints, and a list of _Federal Registers_ published under RCRA. Includes bibliographic references.

Weinberg, David B., et al. _Hazardous Waste Regulation Hand- book, a Practical Guide to RCRA and Superfund_. New York: Executive Enterprises Publications Company, Inc., 1982.

Reviews RCRA and Superfund; reviews the identification of hazardous wastes; lists the obligations of genera- tors, transporters, and the owners and operators of treatment, storage and disposal facilities. RCRA per- mitting requirements are explained. Enforcement of the laws, including penalties for non-compliance, are listed. Additional appendices include summary charts.

Questions about hazardous waste management activities under
RCRA and the Superfund can be submitted directly to the
EPA via their toll-free telephone hotline service, at
(800) 424-9346. Washington, D.C. callers should dial (202)
382-3000.

The major instruments of promulgation of EPA rules and
regulations are the <u>Federal Register</u> and the <u>Code of Federal
Regulations</u>, and these publications should be readily
available in most libraries. Both have periodic subject
indexes which guide the researcher to applicable sections.
Some companies reprint selected topical <u>Federal Register</u>
announcements, such as the reprint service for <u>Federal
Register</u> coverage of hazardous materials, chemicals and
substances published by the Hazardous Materials Publishing
Company. The online computer service <u>CSI Federal Register
Data Base</u> (FEDREG), available through DIALOG and SDC, provides access to regulatory activity as published in the
<u>Federal Register</u>.

Two texts have been published recently, which offer helpful
suggestions for compliance with federal regulations:

Keller, J. J. and Associates. <u>Hazardous Waste Audit Pro-
     gram: A Regulatory and Safety Compliance System</u>. Loose-
     leaf. Neenah, WI: J. J. Keller and Associates, Inc.,
     1982.

     This guide is designed to assist generators and owners
     and operators of treatment, storage and disposal facil-
     ities in determining their own compliance with EPA
     regulations. It includes evaluation guidelines, moni-
     toring procedures, checklists and forms for evaluating
     compliance, and also includes an employee training
     program.

New England Legal Foundation. <u>Managing Hazardous Wastes:
     Proceedings on the Practical Aspects of Hazardous Waste
     Regulation</u>. Hartford, CT: New England Legal Founda-
     tion, 1982.

     Proceedings of a day-long seminar in Hartford, Connec-
     ticut, on November 20, 1981, on the practical aspects
     of compliance with federal and state hazardous waste
     regulations. Gives a review of the overall status of
     regulatory issues, and concrete suggestions for legal
     compliance. Appendices include a glossary of hazardous
     waste terminology; summary of federal regulations; and
     a summary of state laws for Maine, Connecticut, Massa-
     chusetts, New Hampshire, Rhode Island, and Vermont.

The following three publications provide a survey of various
state regulations enacted under RCRA:

Keller, J. J. and Associates. Hazardous Waste Regulatory
    Guide: State Waste Management Programs. Neenah, WI:
    J. J. Keller and Associates, Inc., 1980- .

    Gives information on identifying, monitoring, treating
    and disposing hazardous wastes as defined by each state
    agency responsible for hazardous waste administration.
    Also indicates differences between state and federal
    regulations. Includes an updating service.

National Conference of State Legislatures' Solid and Hazar-
    dous Waste Project. Hazardous Waste Management: A
    Survey of State Legislation 1982. Denver, CO: National
    Conference of State Legislatures, 1982.

    Contains abstracts of laws enacted in the fifty states.

Speer, R. D. and Gerard A. Bulanowski, eds. Speer's Digest
    of Toxic Substances State Law 1983-84: Trends, Summar-
    ies and Forecasts. Boulder, CO: Strategic Assessments,
    Inc., 1983. (Published annually).

    Reviews state trends, summarizes new legislation, and
    forecasts probable developments.

The Environmental Protection Agency: its Services and Publications

The EPA is the largest single publisher in the field of
hazardous and toxic waste disposal. EPA literature ranges
from general items for students and the public to complex
manuals for the practitioner. The EPA is generally an
inexpensive source of information, and many of its services
and publications are free. Since the EPA funds much of
the current research in the field, their publications often
offer the most recent available data. In fact, many books
and articles published in the private sector rely on EPA
documents for their data. Refer to the introductory chapter
for information on accessing EPA publications.

The National Technical Information Service, described in the
introductory chapter, offers 28 weekly abstract newsletters
covering specific subject areas, which give summaries of
documents from EPA and other sources. Additionally, the
NTIS Environmental Pollution Control, An Abstract Newsletter, includes waste disposal in its coverage. Most
documents cited in the NTIS publications can be purchased in
microfiche or paper from NTIS.

NTIS also produces bibliographies via computer searches of
various data bases. These bibliographies are listed and
indexed by subject in the Published Searches Master Catalog

available from NTIS. The following bibliographies are currently offered for sale:

Hazardous Materials Waste Disposal 1980-January 1983. 260
     abstracts from the NTIS data base. Excludes radio-
     active. NTIS #PB83-802660/CAC.

Ocean Waste Disposal August 1980 - September 1983. 139 ab-
     stracts from the NTIS data base.   Excludes thermal
     effluents. NTIS #PB84-800135/CAC.

Ocean Waste Disposal 1977-January 1983. 138 abstracts from
     the Selected Water Resources Abstracts data base. NTIS
     #PB83-859041/CAC.

Government Regulations for Industrial Emissions January
     1970-June 1981. 114 abstracts from the Pollution
     Abstracts data base. NTIS #PB81-867830/CAC.

The Office of Solid Waste of the EPA offers a free list
of its publications, including reprints of Federal Register
notices. This list can be obtained by writing to RCRA
Docket, Office of Solid Waste (WH-562), 401 M Street SW,
EPA, Washington, D.C. 20460.

The Center for Environmental Research Information offers a
list of documents produced by the Office of Research and
Development of the EPA, the ORD Publications Announcement.
Published three or four times a year, this list is free, and
cited documents can also be requested at no charge.   Order
the listing from ORD Publications, P.O. Box 14249B,
Cincinnati, OH 45214.

There are hundreds of EPA publications, many of them
focusing specifically on hazardous waste disposal, but space
permits only a representative few to be listed here.   The
researcher should keep in mind, however, that any thorough
search of the literature on hazardous waste disposal must
include a survey of EPA publications, since the EPA is the
primary source of regulation and research in this field.   A
sampling of EPA documents follows:

Bonner, T. et al.   Engineering Handbook for Hazardous Waste
     Incineration. 1981. NTIS #PB81-248163
     (SuDoc EP1.18:En3/3).

Damages and Threats Caused by Hazardous Material Sites.
     1980. EPA/430/9-80/004 (SuDoc EP1.2:H33/5).

     A partial compilation of damages and threats from hazar-
     dous waste sites in the United States. More than 350

site descriptions which include ground water contamination, drinking water well closures, fish kills, property damage from fires and explosions, and kidney disorders, cancer and death.

<u>Disposal of Hazardous Waste</u>. Proceedings of the Sixth
  Annual Research Symposium, Chicago, Illinois, March 17-
  20, 1980. EPA-60019-80-010 (PB80-175086). Cincinnati,
  OH: Environmental Protection Agency, 1980.

Hall, Charles V. et al. <u>Safe Disposal Methods for
  Agricultural Pesticide Wastes</u>. 1981. EPA-600/2-81-074;
  NTIS #PB81-197584 (SuDoc EP1.23/2:600/2-81-074).

<u>Hazardous Waste Generation and Commercial Hazardous Waste
  Management Capacity: An Assessment</u>. 1980. SW-894
  (SuDoc EP1.17:894).

Organized into three parts: 1) estimates of amounts of
wastes likely to be processed by waste generators them-
selves, and amounts likely to be treated or disposed by
the hazardous waste management industry at off-site
facilities; 2) existing and planned capacity of off-
site facilities to manage hazardous waste received; and
3) comparison of part 1, waste generation estimates,
with part 2, off-site capacity.

<u>Hazardous Waste Management: A Guide to the Regulations</u>.
  Developed by Angela S. Wilkes and compiled by Irene
  Kiefer. 1980. SW-876 (SuDoc EP1.17:876).

Somewhat mistitled, this guide actually consists of a
series of questions, and answers, concerning RCRA, with
references to appropriate locations in the actual law.

<u>Hazardous Waste Sites: Descriptions of Sites on Current
  National Priorities List, October 1984</u>. 1984.
  (SuDoc EP1.2:H33/11).

<u>Inventory of Open Dumps</u>. 1983. SW-964 (SuDoc EP1.17:964).

<u>Land Disposal of Hazardous Waste</u>. Proceedings of the Annual
  Research Symposium. Cincinnati, OH: EPA.

  7th--Philadelphia, PA, March 16-18, 1981
    EPA-600/9-81-0026 (PB81-173882)
    (SuDoc EP1.23/6:600/9-81-0026)

  8th--Fort Mitchell, KY, March 8-10, 1982
    EPA-600/9-82-002 (PB82-173022)
    (SuDoc EP1.23/6:600/9-82-002)

9th--Fort Mitchell, KY, May 2-4, 1983
  EPA-600/9-83-018 (PB84-118777)
  (SuDoc EP1.23/6:600/9-83-018)

10th--Fort Mitchell, KY, April 3-5, 1984
  EPA-600/9-84-007 (PB84-177799)
  (SuDoc EP1.23/6:600/9-84-007)

11th--Cincinnati, OH, April 29-May 1, 1985
  EPA-600/9-85-013 (PB85-196376)
  (SuDoc EP1.23/6:600/9-85-013)

12th--Cincinnati, OH, April 21-23, 1986
  (Report and order numbers not available
  at time of publication)

The above documents are Proceedings of the Annual
Research Symposiums sponsored by the EPA, Office of
Research and Development, Municipal Environmental
Research Laboratory, Solid and Hazardous Waste Research
Division, which has been holding annual symposiums
since 1975. These are important publications for those
in the field. Separate annotations for each symposium
are included in Chapter 4.

Research Outlook. 1976- . Published annually.
  (SuDoc EP1.81/2:[date]).

Annual reports to Congress which describe the direction
of EPA's environmental research program with a five
year outlook.

Resource Conservation and Recovery: Current Reports. 1980.
  SW-536 (SuDoc EP1.17:536).

A bibliography which lists published and unpublished
resource conservation and recovery information.

"Technical resource documents". A series of documents that
  represent the best engineering judgement for the
  design, operation and closure of hazardous waste facilities. A compilation of the research efforts to date,
  developed to assist in implementing 40 CFR, parts 264,
  265 and 267, concerning hazardous waste disposal facilities (landfills, surface impoundments and land treatment). Technically, not policy, oriented. Not regulatory in design but intended to provide guidance in
  decision-making. Those currently available are:

  TRD 1 Evaluting Cover Systems for Solid and Hazardous
         Waste. Revised, 1982. SW-867.

  TRD 2 Hydrologic Simulation on Solid Waste Disposal
         Sites. Revised, 1982. SW-868.

TRD 3   Landfill and Surface Impoundment Performance
        Evaluation. Revised, 1983. SW-869.

TRD 4   Lining of Waste Impoundment and Disposal Facil-
        ities. Revised, 1983. SW-870.

TRD 5   Management of Hazardous Waste Leachate. Revised,
        1982. SW-871.

TRD 6   Guide to the Disposal of Chemically Stabilized
        and Solidified Wastes. Revised, 1982. SW-872.

TRD 7   Closure of Hazardous Waste Surface Impoundments.
        Revised, 1982. SW-873.

TRD 8   Hazardous Waste Land Treatment.   Revised, 1983.
        SW-874.

Treatability Manual. 1980; Revised 1981, 1982, 1983. EPA-
    600-8-80-042[a-e] (SuDoc EPl.18:T71/v.1-5). 5 volumes.

An exhaustive manual for evaluating the treatment of
specific pollutants, and the potential effectiveness
and costs of proposed effluent treatment systems.

Volume 1, Treatability Data, is a compendium of treatability data, industrial occurrence data, and pure
species descriptions for specific pollutants, including
metals, ethers, phthalates, nitrogen compounds, PCB's,
and pesticides. Emphasized in this volume are the 129
priority pollutants, prevalent in industrial wastewaters, that do not readily degrade in the environment.

Volume 2, Industrial Descriptions, contains industrial
wastewater discharge information compiled by categories
of industry; guidelines and standards for treatment
technology; and tabulated information on individual
plants.

Volume 3, Technologies for Control/Removal of Pollutants, summarizes performance data for existing wastewater treatment technologies.

Volume 4, Cost Estimating, gives information on capital
and operating costs of the treatments described in
Volume 3.

Volume 5, Summary, reviews the first four volumes and
outlines their applicability. Additionally, it includes a master bibliography and a glossary of abbreviations and terms used in the Manual.

The Treatability Manual is designed for use by National
Pollutant Discharge Elimination System (NPDES) permit
writers, in developing their wastewater pollution control and monitoring requirements.

Waste Treatment Manuals

The Treatability Manual of the EPA, cited above, is the authoritative guide to treatment options for certain wastes. Another publication that offers data on chemical properties, handling, storage, and disposal of hazardous and toxic wastes is the Toxic and Hazardous Industrial Chemicals Safety Manual for Handling and Disposal with Toxicity and Hazard Data (Tokyo, Japan: The International Technical Information Institute, 1982). Dangerous Properties of Industrial Materials by N. Irving Sax (6th ed. New York: Van Nostrand Reinhold, 1984) and the Handbook of Environmental Data on Organic Chemicals by Karel Verschueren (2nd ed. New York: Van Nostrand Reinhold, 1983), address the environmental fate of wastes and other chemicals. Additional sources are cited in the introductory chapter.

The Handbook of Hazardous Waste Management by Amir A. Metry (Westport, CT: Technomic Publishing Company, 1980), offers broad coverage on the definition, regulation, treatment, transportation, disposal, and overall management of hazardous industrial wastes. Bibliographies are included as additional sources of information.

The Handbook of Industrial Residues by Jon C. Dyer and Nicholas A. Mignone (Park Ridge, NJ: Noyes, 1983), provides data on the 34 categorical industries (i.e. coal mining, foundries, textile mills) listed by the EPA as regulated under current standards. Part 1 lists industries and management options for the 34 industries, and part 2 covers pretreatment technologies. Bibliographic references are included.

Additional manuals which address the waste disposal options within particular industries are published by EPA, and can be identified in the indexes cited in Chapter 1 (EPA Publications Bibliography Quarterly, Government Reports Announcements and Index, and the Monthly Catalog of United States Government Publications).

The American Chemical Society has published a handbook discussing laboratory chemical management as a means to achieve waste reduction, Less is Better. Single copies are available free from the Office of Federal Regulatory Programs, ACS Department of Government Regulations and Science Policy, American Chemical Society, 1155 Sixteenth Street NW, Washington, D.C. 20036.

The Water Pollution Control Federation publishes a number of manuals and directories on wastewater, which while not specifically addressed to hazardous and toxic waste disposal, do include coverage of these types of wastes. A few of special interest include Pretreatment of Industrial Wastes (1981), Glossary: Water and Wastewater Control

Engineering (Revised and updated, 1981), Units of Expression
for Wastewater Management (1982), and Water Pollution
Control Market Facts (1981).

Directories

A practical directory focusing on hazardous waste disposal
services available in the United States is Industrial and
Hazardous Waste Management Firms (Minneapolis: Environmental
Information Ltd., 1985). This work lists transportation
firms, recycling facilities, and treatment and disposal
facilities. Included are tables which help locate facilities that accept various types of wastes, and maps that show
distances to waste facilities, to help in estimating transportation costs. A listing of firms offering alternatives
to conventional land disposal, such as solvent recovery, PCB
decontamination, and mobile treatment, is included.

The Hazardous Waste Management Directory, 4th ed. (Philadelphia: Pennsylvania Environmental Council, Inc., 1985), is a
national listing of resources for the treatment, storage and
disposal of hazardous waste. This comprehensive text identifies special waste handling experts, methods of management
at each facility listed, specific wastes processed at each
facility, emergency spill services, transport services,
pollutant testing laboratories, and industrial facility and
process designers.

The Hazardous Waste Services Directory (Neenah, WI: J. J.
Keller and Associates, 1984) is a looseleaf service which
provides listings for over a thousand firms in services
related to the management of hazardous materials, including
haulers, processors, disposal sites, operators, laboratories, and consultants. Updating is done on a quarterly
basis.

A "Directory of Commercial Hazardous Waste Facilities"
appears in the March/April 1985 issue of the Hazardous Waste
Consultant (Lakewood, CO: McCoy and Associates). This directory includes a state-by-state assessment of commercial
treatment, storage and disposal facilities; a listing of
facilities; an alphabetical index by company name; and a
master map of facilities.

"Hazardous Waste Sites in the United States", edited by H.
Pishdadazar and A. Alan Moghissi, appeared in 1981 as a
special issue of the journal, Nuclear and Chemical Waste
Management (vol. 1 no. 3-4). This list, arranged alphabetically by state, cites hazardous waste disposal site name,
owner's name, site address, information on disposed materials, quantity of disposed material, and disposal technique
used. The information in this list was synthesized from
surveys produced by Congress and the EPA.

Other surveys of hazardous waste facilities in the United
States include <u>Notification to EPA of Hazardous Waste
Activities</u> (ten volumes, one for each of the ten EPA
regions. 1980. SW897.1 - SW897.10) and <u>Inventory of Open
Dumps</u> (1983. SW-964), both EPA publications.

<u>World Wastes</u>, a monthly trade publication, also produces
periodic directories, such as the "World Wastes Equipment
Catalog" and the annual "Buyer's Guide".

Two useful guides to European activities in waste disposal
are available. These are the <u>Directory of Pollution Control
Equipment Companies in Western Europe</u> (4th ed. London: European Directories, Inter Company Comparisons Limited, 1982)
and the <u>Directory of Solid Waste and Chemical Waste</u> by the
Environmental Research Projects in the European Communities
(Hitchin Herts, England: Peter Peregrinus, 1980. 2 volumes).
The first title lists firms in 17 European countries, and
indexes them by type of service offered; sections on consultants, advisory bodies and institutes of pollution are
especially helpful. The second title is of narrower scope,
focusing on research projects in solid and chemical waste.

A timely review of some of Europe's facilities is given in
"A Grand Tour of Europe's Hazardous-Waste Facilities" by
Bruce Piasecki and Gary A. Davis (in <u>Technology Review</u> vol.
87, no. 5, July 1984, pp. 20-29). Described are the Kommu-
nekemi of Denmark; the Gesellschaft zur Beseitingung von
Sondermull in Bayern, West Germany; and Sweden's national
waste-managment firm, SAKAB. Also useful is <u>Overview of Six
European Waste Management Facilities</u> by M. J. Mudar (Albany,
NY: State Environmental Facilities Corporation, 1982).
Reprints are available from the Northeast Industrial Waste
Exchange (90 Presidential Plaza, Suite 122, Syracuse, NY
13202).

The May/June 1984 issue of <u>Water and Pollution Control</u> (vol.
122 no. 3) includes a "Government Reference Manual" for
Canada. Listed are an "Environmental Who's Who", a selected
list of environmental publications from federal and provin-
cial government sources, a guide to Canadian environmental
information sources which includes information on the computerized data bases WATDOC and MUNDAT, and a list of booksel-
lers who stock Canadian government publications.

Indexes, Abstracts, and Data Bases

The universe of literature on hazardous and toxic waste
disposal consists of books, technical documents, government
publications, audiovisual media, and articles published in
periodicals. An important step in any research undertaking
is the searching of relevant indexes and data bases to
identify the existing literature on the topic of research.
There is no one single index which lists all of the litera-

ture of hazardous and toxic waste disposal, but there are a
number of indexes and data bases which offer wide coverage
of the field. The serious researcher will want to take
advantage of all of the indexes and data bases to which he
or she has access.

A specialized index designed for researchers in waste
management is Waste Management Resources Recovery Information Data Base (generally, coverage from 1971 to present.
Eagan, MN: International Research and Evaluation, available
online from IRE). This file contains citations, with
abstracts, to the world-wide literature covering solid,
liquid, hazardous and nuclear waste management; water quality; toxic substances; land reclamation; and resource recovery. An "Information Database", consisting of the microfiche
text of over 5,000 of the publications indexed, is available
to supplement the data base.

From the same publisher, a one-volume print index is available, A Reference Guide to Environmental Management, Engineering, and Pollution Control, by Randall L. Voight (Eagan,
MN: International Research and Evaluation Press, 1983).
Reprints of all documents listed in this index are available
for purchase from IRE.

Other indexes to the literature are broader in scope, but
include publications in hazardous and toxic waste disposal.
Two cited above, the Government Reports Announcements and
Index and the Monthly Catalog of United States Government
Publications, as well as the EPA Publications Bibliography
Quarterly, are important starting points since they index
EPA documents and other federal publications.

The EPA Clearinghouse Technical Reports Database (CLEAR),
available online through CIS, contains references and
descriptions of over 17,000 technical reports, 370 data
bases and information files, and 170 computer models that
have been produced or funded by EPA. Searchable by individual substance names, name fragments, or Chemical Abstracts Service Registry Numbers, it is a valuable source of
information for the researcher interested in specific hazardous or toxic chemicals.

Other indexes to the literature which include coverage of
hazardous and toxic waste disposal are cited in the introductory chapter.

The Toxic Substances Sourcebook (Annual. New York: Environment Information Center, Inc.), reviews legislation, books,
films, conferences, NIOSH literature, periodicals, and data
bases related to toxic substances. There are subject,
author, and geographical indexes. Philip Wexler's Information Resources in Toxicology (New York: Elsevier/North
Holland, 1982) is a concise guide to the literature of
toxicology. Although not exclusively addressing toxic waste
disposal, these two sources offer useful general guidance in
the field of toxicity and toxicology.

Two of the Current Contents publications, Current Contents: Engineering, Technology, and Applied Sciences and Current Contents: Physics, Chemistry and Earth Sciences, scan the contents of journals publishing articles on hazardous and toxic waste disposal. These reproductions of tables of contents are useful for current awareness of the immediate publishing that has not yet been indexed through other sources.

Some computerized data bases available today are of a different nature than those cited above; rather than offering bibliographic citations to the literature, they actually provide raw data or text, and may give additional bibliographic citations for further research. Data bases of this type with particular relevance to hazardous and toxic waste disposal are:

Compliance Alert: Federal Register Digest. 1981- . Madison, CT: Bureau of Law and Business, Inc. (available in print as a twice-monthly digest and index, or online through CompuServe and NewsNet).

Summarizes federal regulations pertaining to the environment, waste management, energy, and Occupational Safety and Health Administration (OSHA). Covers relevant sections of the Federal Register and the Code of Federal Regulations.

Environmental Fate Data Bases. 1979- . Syracuse, NY: Syracuse Research Corporation. (Available online only, through Syracuse Research Corporation and CIS).

Contains three files of information on the environmental release and fate (transport and degradation) of chemicals in the environment. The first two files contain actual data, with the third file being bibliographic. File 1 indexes the type of fate data pertinent to a particular chemical, with over 33,000 chemicals listed; file 2 contains the actual fate data; and file 3 provides bibliographic references to files 1 and 2.

Hazardline. New York: Occupational Health Services, Inc. (available online through Occupational Health Services, CompuServe and BRS).

Provides data on safety and regulatory information for over 3,000 hazardous substances, gathered from regulations issued by states and U.S. government agencies (OSHA, NIOSH, EPA, FDA, etc.), current court decisions, books, and journal articles. Comprehensive data for disposal of substances is included. Bibliographic references are provided for locating additional information.

OHM-TADS (Oil and Hazardous Materials - Technical Assistance
     Data System). Falls Church, VA: Computer Sciences Corp.
     (available online through CIS, DIALOG, MEDLARS, and
     SDC).

     This file, produced by the EPA, Emergency Response
     Division, includes data on over 1,000 hazardous sub-
     stances. Up to 126 data fields, some textual and some
     numeric, may be present for each substance. Data
     fields can include information on identification, phy-
     sical properties, uses, toxicity, handling precedures,
     and recommended disposal methods.

The NIH/EPA Chemical Information System (CIS), which pro-
vides access to the OHM-TADS file, also offers over a dozen
other data bases, including the Registry of Toxic Effects of
Chemical Substances (RTECS), the Federal Register Search
System (FRSS), and the Environment, Retrieval and Estimation
(SPHERE) data base.

Periodicals

Much of the literature on hazardous and toxic waste disposal
is found in the journals of chemistry, engineering, technology, and industry. The best route of access to these articles is through the indexes cited above.

A number of periodicals limit their scope to hazardous and
toxic waste management and disposal. The following list
includes most such journals, newsletters and trade magazines
published currently in the United States.

From the State Capitals... Waste Disposal and Pollution Con-
     trol. Wakeman/Walworth. Monthly, published since
     1982.

     A newsletter that analyzes state and municipal legisla-
     tive and regulatory trends of nationwide significance
     in the management of hazardous wastes, air and water
     pollution controls, pesticide regulation, insurance
     protection, and resource recovery projects.

Hazardous Materials and Waste Management. Hazardous Mater-
     ials and Waste Management Association. Bimonthly, pub-
     lished since 1983.

     A newsy publication addressed to generators of hazar-
     dous waste. Includes feature articles and regular
     departments such as "Washington Newsletter", "State
     Regulatory Trends", "New Products, Literature and Ser-
     vices". Free to qualified personnel.

Hazardous Materials Bulletin for Supervisors. Bureau of Law
and Business, Inc. Monthly, published since 1982.

Practical information for supervisors involved with
staff training required by RCRA.

Hazardous Materials Intelligence Report. World Information
Systems. Weekly, published since 1980.

A newsletter covering developments in state and federal
regulations, recent court decisions, new products and
services for hazardous waste treatment and disposal,
case histories of waste-site clean-ups and spill
responses, bid opportunities, technical reports, and
announcements of conferences and seminars.

Hazardous Waste. Mary Ann Liebert, Inc. Quarterly, published since 1984.

A new research journal proposing to serve as a central
source of information for the purpose of advancing
technology and providing methodologies for the regulation and management of hazardous waste. Topics will
include technology, health effects, environmental
effects, and policy and decision-making. An editorial
board of 37 experts with academic, industrial and
governmental backgrounds review papers for inclusion.

Hazardous Waste Consultant. McCoy and Associates. Bimonthly, published since 1983.

An excellent source of information for the practitioner
and researcher. Each issue includes Technology/ Economics, Regulatory Issues, Legal Issues, Status of Projects, Bibliography of Recent Hazardous Waste References, and Coming Events, plus Executive Summary and
"Ask the Consultant".

Hazardous Waste News. Business Publishers, Inc. Weekly,
published since 1979. (Available electronically via
NewsNet).

A newsletter aimed at the hazardous waste industry.
News items, ranging in length from a paragraph to a
page, deal with pending legislation, statements and
recommendations of EPA officials, Congressional
activities, industrial trends and innovations, court
cases, federal and state funding, and business and
industry news.

Hazardous Waste Report. Aspen Systems Corp. Biweekly,
published since 1979.

A timely newsletter which surveys developments in RCRA
and Superfund (CERCLA): proposals, guidelines, industry
developments, state policies and legislation, enforcement and litigation. "Special issues" periodically
review specific topics in depth, and "Alerts" are published periodically for important developments whose
publication cannot be delayed.

Hazardous Waste Training Bulletin for Supervisors.
John F. Brady, publisher. Monthly, published since
1982.

A newsletter addressed to practitioners who supervise
hazardous waste management. Included are news briefs
on current legislative developments, notices of training institutes and seminars, and practical suggestions
for compliance with current regulations. Written in
non-technical terms, and meant to assist those not
familiar with the complexities of the law.

Management of World Wastes. (formerly, Solid Wastes Management). Communication Channels. Monthly with a 13th
issue in December, published since 1983.

Deals with all types of waste, hazardous and non-
hazardous. In addition to feature articles, regular
departments such as "Calendar of Events", "New Products", "News Briefs", "People in the News", and clas-
sified ads make this publication of interest to practitioners. The July issue includes a "World Wastes Buy-
ing Guide", and the "Sanitation Industry Yearbook" is
also published yearly as a supplement to the journal.

Nuclear and Chemical Waste Management. Pergamon. Quarterly,
published since 1980.

Subtitled "An International Journal of Hazardous Waste
Technology", this is a research journal which encompasses the entire scope of hazardous waste management,
including new technologies of disposal.

Pollution Control Guide. Commerce Clearing House. Weekly,
published since 1973.

Covers federal and state laws relating to water and air
pollution, solid waste, noise pollution, pesticides,
radiation, and toxic substances.

State Regulation Report. Business Publishers Inc. Biweekly,
published since 1981. (Available electronically via
NewsNet).

A newsletter reporting current issues in the states' regulation of toxic substances and hazardous wastes. Directed toward hazardous waste and toxic substances generators, transporters, and disposers, who need to be aware of state regulations.

Wasteline. Hazardous Waste Services Association. Biweekly, published since 1980.

A membership newsletter of the HWSA, reviewing legislative, regulatory, legal, technical, and industry developments.

In addition to the titles listed above, a number of journals and newsletters with broader scope offer periodic coverage of hazardous and toxic waste disposal. Some of these are cited below. See Chapter 1 for additional titles.

Air Pollution Control. Bureau of National Affairs, Inc. Published since 1980.

Air Pollution Control Technology. Casey Publishing Co. Published since 1980.

Air/Water/Pollution Report. Business Publishers, Inc. Published since 1963. (Available electronically via News-Net).

Archives of Environmental Contamination and Toxicology. Springer-Verlag. Published since 1972.

ASTM Standardization News. American Society for Testing and Materials. Published since 1961. (See special issue on hazardous wastes, March 1984).

Bulletin of Environmental Contamination and Toxicology. Springer-Verlag. Published since 1966.

ChemEcology. (formerly Currents). Chemical Manufacturers Association. Published since 1983. (Free).

Consulting Engineer. Technical Publishing. Published since 1952. (See special issue, "Disposing of Hazardous Wastes", March 1984).

Effluent and Water Treatment Journal. Thunderbird Enterprises Ltd. Published since 1961.

*Environmental Analyst: Federal, State and Regional Issues
for Industry*. Executive Enterprises Publications Co.
Published since 1980.

Environmental Compliance Update. Ridge, NY: High Tech
Publishing Co. Published since 1985.

Environmental Manager's Compliance Advisor. Bureau of Law
and Business, Inc. Published since 1985.

Environmental Pollution, Series A: Ecological and Biological
and *Environmental Pollution, Series B: Chemical and
Physical*. Elsevier. Both published since 1970.

Environmental Pollution Management. London: Polcon Publishing Ltd. Published since 1971.

Environmental Progress. American Institute of Chemical
Engineers. Published since 1982.

HMCRI Forum. Hazardous Materials Control Research Institute.
Published since 1979.

Haz-Mat Technology. Medford, NJ: Teaberry Associates.
Published since 1986.

Industrial Wastes. Scranton-Gillette Communications, Inc.
Published since 1971.

Industrial Water Engineering. Wakeman/Walworth. Published
since 1964.

Journal of Environmental Engineering. American Society of
Civil Engineers. Published since 1983. Issued under
various titles since 1956.

*Journal of Environmental Science and Health: Part A,
Environmental Science and Engineering*. Marcel Dekker
Journals. Published since 1976.

Plant/Operations Progess. American Institute of Chemical
Engineers. Published since 1982.

Pollution Equipment News. Rimbach Publishing, Inc. Published since 1968. (Free to qualified personnel).

Sludge Newsletter. Business Publishers, Inc. Published
    since 1978. (Available electronically via NewsNet).

Solid Waste Report. Business Publishers, Inc. Published
    since 1970. (Available electronically via NewsNet).

Toxic Materials News. Business Publishers, Inc. Published
    since 1974. (Available electronically via NewsNet).

Toxic Substances Journal. Executive Enterprises Publica-
    tions. Published since 1979.

Waste Age. National Solid Wastes Management Association.
    Published since 1970.

Waste Management and Research. Academic. Published since
    1983. (Journal of the International Solid Waste and
    Public Cleansing Association, ISWA, of London).

Waste-to Energy Report. McGraw-Hill. Published since 1985.

Water and Pollution Control. Southam Business Publications.
    Published since 1966.

Water and Wastes Digest. Scranton Gillette Communications.
    Published since 1961. (Free to qualified personnel).

Water Pollution Research Journal of Canada. Pergamon.
    Published since 1980.

Books and Proceedings

Book publishing in the field of hazardous and toxic waste
disposal is flourishing, with interest in this area spurred
by regulatory and legislative developments of the last
decade. Publishers who are especially active in this subject include Ann Arbor Science Publishers and Noyes Data
Corporation. Free catalogs of available publications can be
requested from these publishers (addresses are listed at the
back of this book).

The following representative list of monographs attempts to
present an overview of the literature published in monographic form since 1980, with particular emphasis on general
works in hazardous and toxic waste disposal, and only

partial coverage of books that deal with specific wastes,
specific tehcnologies, or specific industries. Titles that
have been cited previously in this paper are not repeated.
Space does not permit abstracting of all titles, but some of
particular interest have been highlighted.

Ackerman, D. G., et al. Destruction and Disposal of PCBs by
    Thermal and Non-thermal Methods. Park Ridge, NJ:
    Noyes, 1983.

    Two EPA reports were combined to make this book: Guide-
    lines for the Disposal of PCBs and PCB Items by Thermal
    Destruction (EPA report 600/2-81-022) and Interim
    Guidelines for the Disposal/Destruction of PCB and PCB
    Items by Non-Thermal Methods (EPA Report 600/2-82-069).

Alternatives to the Land Disposal of Hazardous Wastes: An
    Assessment for California. Sacramento, CA: Toxic Waste
    Assessment Group, Governor's Office of Appropriate
    Technology, 1981. Principal investigator S. Kent
    Stoddard.

Bhatt, Harasiddhiprasad G., et al. Management of Toxic and
    Hazardous Wastes. Chelsea, MI: Lewis Publishers, 1985.

    Updated presentation of papers presented at the Third
    Ohio Environmental Engineering Conference (ASCE) of
    1983. Deals with techniques, monitoring systems, in-
    process controls, process substitutions, cleanup
    priorities, site plans, waste exchanges, liability,
    etc.

Bonner, T., et al. Hazardous Waste Incineration Engineering.
    Park Ridge, NJ: Noyes, 1981.

    Based on EPA's 1980 draft of Engineering Handbook for
    Hazardous Waste Incineration (final edition published
    in 1981, NTIS #PB81-248163).

Borchardt, Jack A., et al. Sludge and its Ultimate Disposal.
    Ann Arbor, MI: Ann Arbor Science, 1981.

Brown, Kirk W., et al., eds. Hazardous Waste Land Treatment.
    Boston: Butterworth, 1983.

Bruce, A. M., et al. Disinfection of Sewage Sludge:
    Technical, Economic and Microbial Aspects. Reading,
    MA: D. Reidel, 1983.

Brunner, Calvin. Incineration Systems. Washington, D.C.:
    Government Institutes, 1983.

    Evaluates the state-of-the-art of incineration systems,
    their costs, environmental considerations, energy con-
    siderations, and their relative merits compared to
    alternative disposal methods.

Castaldini, C., et al. Disposal of Hazardous Wastes In
    Industrial Boilers and Furnaces. Park Ridge, NJ:
    Noyes, 1986.

Champ, Michael A. and P. Kilho Park. Global Marine Pollution
    Bibliography: Ocean Dumping of Municipal and Industrial
    Wastes. New York: Plenum, 1982.

Cherry, Kenneth F. Plating Waste Treatment. Ann Arbor,
    MI: Ann Arbor Science, 1982.

Chivers, G. E. The Disposal of Hazardous Wastes. Northwood,
    Middlesex, England: Science Reviews Limited, 1983.

Cleaning Up Hazardous Wastes: An Overview of Superfund
    Reauthorization Issues: Report to the Congress.
    Washington, D.C.: General Accounting Office, 1985.
    (SuDoc GA1.13:RCED-85-69).

Conway, R. A. and W. P. Gulledge, eds. Hazardous and
    Industrial Solid Waste Testing: Second Symposium. A
    Symposium Sponsored by ASTM Committee D-34 on Waste
    Disposal. Philadelphia: American Society for Testing
    and Materials, 1983.

    At the second annual symposium held by this committee
    in 1982, improved methods of evaluating hazardous and
    industrial solid wastes in terms of hazard degree and
    disposal options are discussed.

Cope, C. B., et al. The Scientific Management of Hazardous
    Wastes. Cambridge, England: Cambridge University Press,
    1983.

    Surveys the science and administration of waste dispo-
    sal in the United Kingdom, and includes useful biblio-
    graphic references.

Cushnie, George C. Jr., ed. Removal of Metals from Waste-
    water, Neutralization and Precipitation. Park Ridge,
    NJ: Noyes, 1984.

A manual of design and operating precedures for the
removal of metals from wastewater by neutralization and
precipitation. Also covered are methods for handling
and disposal of residues from the treatment process.

DeRenzo, D. J., ed. Biodegradation Techniques for Indus-
    trial Organic Wastes. Park Ridge, NJ: Noyes, 1980.

Dillon, A. P., ed. Pesticide Disposal and Detoxification,
    Processes and Techniques. Park Ridge, NJ: Noyes, 1981.

Duedall, Iver W., et al. Wastes in the Ocean. New York:
    Wiley, 1983- .

    Volumes include: Volume 1, Industrial and Sewage Wastes
    in the Ocean; Volume 2, Dredged-Material Disposal in
    the Ocean; Volume 3, Radioactive Wastes in the Ocean;
    Volume 4, Energy Wastes in the Ocean; Volume 5, Deep-
    Sea Waste Disposal; and Volume 6, Near-Shore Waste
    Disposal.

Dyer, Jon D., et al. Handbook of Industrial Wastes Pre-
    treatment. New York: Garland STPM Press, 1981.

    Examines the General Pretreatment Regulation (40 CFR
    Part 403). Provides information on the development of
    local pretreatment programs. Includes case histories.

Ehrenfeld, John and Jeffrey Bass. Evaluation of Remedial
    Action Unit Operations at Hazardous Waste Disposal
    Sites. Park Ridge, NJ: Noyes, 1984.

Everett, L. G., et al. Vadose Zone Monitoring for Hazardous
    Waste Sites. Park Ridge, NJ: Noyes, 1984.

Exner, Jurgen H., ed. Detoxication of Hazardous Waste. Ann
    Arbor, MI: Ann Arbor Science, 1982.

    Text which evolved from a symposium of the Division of
    Environmental Chemistry of the American Chemical
    Society. Focuses on detoxication of hazardous waste as
    an alternative and/or a complement to storage and
    destruction techniques.

Fawcett, Howard H. Hazardous and Toxic Materials: Safe
    Handling and Disposal. New York: Wiley, 1984.

Freeman, Harry. Innovative Thermal Hazardous Organic Waste
    Treatment Processes. Park Ridge, NJ: Noyes, 1985.

Discusses 21 thermal processes identified by the EPA as innovative processes for treating or destroying hazardous organic wastes.

Fung, R., ed. _Protective Barriers for Containment of Toxic Materials_. Park Ridge, NJ: Noyes, 1980.

Haxo, Henry E. Jr., et al. _Liner Materials for Hazardous and Toxic Wastes and Municipal Solid Waste Leachate_. Park Ridge, NJ: Noyes, 1985.

Presents studies which assess the relative effectiveness and durability of a wide vaiety of liner materials.

Hazardous Solid Waste Testing: First Conference: A Symposium Sponsored by ASTM Committee D-34 on Waste Disposal. Philadelphia: American Society for Testing and Materials, 1981.

At the first annual symposium of this newly formed committee, topics covered included laboratory testing of hazardous and industrial solid wastes, evaluation of land disposal sites and materials, and risk assessment approaches.

Hazardous Waste Incineration: Selected Papers from an APCA Annual Meeting. Pittsburgh: Air Pollution Control Association, 1984.

Contains seven technical papers which were presented at the APCA's annual meeting in 1984. All deal with the incineration of hazardous wastes.

Hazardous Waste Management: Recent Changes and Policy Alternatives. Washington, D.C.: Congress of the U.S., Congressional Budget Office, 1985. (SuDoc Y10.2:H33).

Hazardous Waste "Problem" Sites: Report of an Expert Seminar. Paris: Organisation for Economic Co-operation and Development, distributed by OECD Publication and Information Center, Washington, D.C., 1983.

Hollander, H., ed. _Thesaurus on Resource Recovery_. Philadelphia: American Society for Testing and Materials, 1983.

A compendium of terms which, while not all specifically in the realm of "hazardous" or "toxic", are of general use in the field of recovery of solid waste resources.

Holmes, John R., ed. Practical Waste Management. New York: Wiley, 1983.

A comprehensive summary of current developments in all phases of waste management; legislation, new codes and standards of practice, and technological advances are covered. Includes four chapters on hazardous waste and its disposal.

Hooper, G. V. Offshore Ship and Platform Incineration of Hazardous Wastes. Park Ridge, NJ: Noyes, 1981.

Johnston, J. B. and S. G. Robinson. Genetic Engineering and New Pollution Control Technologies. Park Ridge, NJ: Noyes, 1984.

An EPA-sponsored study documenting new pollution controls using genetic technology, particularly in phosphorous removal, ammonia oxidation, flocculation, and in situ treatment.

Ketchum, Bostwick H., et al. Ocean Dumping of Industrial Wastes. New York: Plenum, 1981.

Lester, James P. and Ann Bowman. The Politics of Hazardous Waste Management. Durham, NC: Duke University Press, 1983.

All major aspects of the hazardous waste situation are discussed. Comprehensive and current.

Lindgren, Gary F. Guide to Managing Industrial Hazardous Wastes. Boston: Butterworth, 1983.

Written for industrial personnel, but in non-technical language. Outlines recommended procedures and actions for compliance with hazardous waste regulations.

Long, F. A. and Glenn E. Schweitzer, ed. Risk Assessment at Hazardous Waste Sites: Based on a Symposium Sponsored by the ACS Committee on Environmental Improvement... Washington, D.C.: American Chemical Society, 1982.

McCormick, R. J. et al. Costs for Hazardous Waste Incineration: Capital, Operation and Maintenance, Retrofit. Park Ridge, NJ: Noyes, 1985.

The book covers the relationships between capital and operation and maintenance costs for hazardous waste

incineration, and the various waste-specific, design-
specific, and operational factors that affect these
costs.

McGlashan, J. E., ed. Modern Trends in Sludge Management.
New York: Pergamon, 1983.

Reprints from Water Science Technology journal. Covers
experiences in sludge disinfection in different countries and developments in sludge conditioning and de-
watering. Sea and land disposal practices currently
used in various parts of the world are described,
including relevant legislation.

Medine, Allen and Michael Anderson, eds. 1983 National Conference on Environmental Engineering: Proceedings of
the ASCE Specialty Conference... New York: American
Society of Civil Engineers, 1983.

Sessions specifically addressing hazardous waste disposal include "Innovations in Hazardous Wastes", with
five papers; "Hazardous Waste Transport in Groundwater", five papers; "Air Pollution Aspects of Hazardous Waste Land Treatment and Disposal Facilities",
five papers; and "Hazardous Waste Case Studies", three
papers.

Mercer, James W., et al., eds. Role of the Unsaturated
Zone in Radioactive and Hazardous Waste Disposal. Ann
Arbor, MI: Ann Arbor Science, 1983.

Explores the problems caused by hazardous and low-level
radioactive waste disposal in the unsaturated zone of
soil.

Morrell, David and Christopher Magorian. Siting Hazardous
Waste Facilities: Local Opposition and the Myth of
Preemption. Cambridge, MA: Ballinger Publishing Co.,
1982.

National Conference on Hazardous and Toxic Wastes Management: Proceedings of the Conference at the New Jersey
Institute of Technology... Newark, NJ: Institute for
Toxic Waste Management, New Jersey Institute of Technology, 1980.

Proceedings of a 1980 Conference which explored waste
enforcement programs, waste generation, waste characteristics, minimizing of wastes, disposal/control alternatives, risk assessment, civil and criminal liability,
and site selection of treatment facilities.

National Conference on Management of Uncontrolled Hazardous
     Waste Sites. (1980-1985). Silver Spring, MD: Hazar-
     dous Materials Control Research Institute, 1981-1985.

     For the past six years, the EPA and the H.M.C.R.I. have
     co-sponsored, in affiliation with the American Society
     of Chemical Engineers and other groups, this important
     conference. The series of proceedings presents the
     most current information available for assessing and
     correcting dangerous waste sites. Contents include
     site investigation; remedial response activities;
     subsurface migration, sampling and interception;
     personnel safety; geohydrology; treatment; ultimate
     disposal; research and development; case histories;
     risk decision and analysis; and legal and regulatory
     actions. Bibliographic references are included.

New Business Opportunities in Hazardous Waste Management: A
     Literature Survey. By the Chemcontrol Group. Copen-
     hagen, Denmark: Chemcontrol, 1984.

     An international survey of available sources for infor-
     mation on alternative and new hazardous waste manage-
     ment technologies, arranged by country and covering
     North America, Scandanavia, Europe, Asia, and Africa.

North Atlantic Treaty Organization/Committee on the
     Challenges of Modern Society (NATO/CCMS) pilot study on
     hazardous waste disposal, published reports:

     (Phase 1)
     #52   Chromium Recycling (NTIS #PB-279152)
     #55   Manual on Hazardous Substances in Special Wastes
           (NTIS #PB-270591)
     #62   Recommended Procedures for Hazardous Waste
           Management (NTIS #PB-276555).
     #63   Organization (NTIS #PB-276559).
     #64   Landfill (NTIS #PB-276811).
     #68   Transportation (NTIS #PB-279682).
     #77   Pilot Study Final Report (NTIS #PB-286050/AS).

     (Phase 2)
     #118 Thermal Treatment (NTIS #PB82-114521).
     #119 Chemical, Physical, and Biological Treatment
           (NTIS #PB82-114539).
     #120 Landfill (NTIS #PB82-114547).
     #121 Metal Finishing Wastes (NTIS #PB82-114554).
     #122 Final Report: NATO/CCMS Pilot Study on Disposal of
           Hazardous Wastes (Phases 1 and 2).
           (NTIS #PB82-114562).

Overcash, Michael R. Techniques for Industrial Pollution
     Prevention: A Compendium for Hazardous and Nonhazardous
     Waste Reduction. Chelsea, MI: Lewis Publishers, 1986.

Parr, J. F., et al., eds.   Land Treatment of Hazardous
    Wastes.   Park Ridge, NJ: Noyes, 1983.

Peirce, J. Jeffrey and P. Aarne Vesilind.   Hazardous Waste
    Management.   Ann Arbor, MI: Ann Arbor Science, 1981.

    Examines problems in hazardous waste management, tech-
    nological factors in the management process, and hazar-
    dous waste management case studies.

Petitions to Delist Hazardous Wastes: A Guidance Manual.
    Cambridge, MA: Industrial Economics, Inc. (Prepared for
    the Environmental Protection Agency), 1985. Available
    from National Technical Information Service as PB 85-
    194488.

    This guidance manual describes all necessary data which
    must be included in a delisting petition submitted
    under 40 CFR 260.22.   It presents a step-by-step
    approach to compiling a delisting petition.

Porteus, Andrew, ed.   Hazardous Waste Management Handbook.
    Boston: Butterworth, 1985.

Proceedings of the Mid-Atlantic Industrial Waste Conference.
    Ann Arbor, MI: Ann Arbor Science, annual.

    Proceedings of annual conferences designed to provide a
    forum for the exchange of technical and non-technical
    ideas to pose solutions for pollution control problems.
    The 17th Conference Proceedings, Toxic and Hazardous
    Wastes (Edited by Irwin J. Kugelman; available from
    Technomic Publ. Co. 1985), includes 47 reports on
    siting, biological treatment, pretreatment, physical
    chemical treatment, ground water, land applications and
    liners, resource recovery, hazardous waste, metals and
    industrial waste, sludge management, and safety and
    right-to-know.   The Proceedings of the 16th Conference,
    Toxic and Hazardous Waste (1984) (Edited by Michael D.
    LaGrega and David A. Long; available from Technomic
    Publ. Co. 1984), covers similar topics, with sections
    on international developments and hazardous waste site
    cleanup.   The 15th Conference Proceedings, titled Toxic
    and Hazardous Waste (edited by Michael D. LaGrega and
    Linda K. Hendrian, 1983), presents the latest technol-
    ogy for hazardous waste management in industry, and
    includes a five-year index to the Proceedings.   The
    Proceedings of the 14th Conference, titled Industrial
    Waste (edited by James E. Alleman and Joseph T.
    Cavanaugh, 1982), covers industrial residuals manage-
    ment, the handling of toxic and hazardous materials,
    and innovative waste treatments.

Proceedings of the National Conference on Risk and Decision
    Analysis for Hazardous Waste Disposal. Silver Spring:
    MD: Hazardous Materials Control Research Institute,
    1981.

    Conference held in 1981 in Baltimore, Maryland, in-
    cludes information on fate, monitoring, land applica-
    tion, disposal, and movement of hazardous materials.

Proceedings of the National Hazardous Waste Siting Confer-
    ence, March 1983. Denver, CO: National Conference of
    State Legislatures, 1983.

Prudent Practices for Disposal of Chemicals from Labora-
    tories. Washington, D.C.: National Academy Press,
    1983.

Purdue Industrial Waste Conference Proceedings. Ann Arbor,
    MI: Ann Arbor Science, 1977- . (Meetings prior to 1977
    were published as a subseries of the Purdue University,
    Lafayette, Indiana: Engineering Extension Department
    Series, "Industrial Waste Conference, Purdue Univer-
    sity, Lafayette, Indiana, Proceedings").

    A comprehensive series covering the annual Purdue
    Industrial Waste Conference, which offers the state-of-
    the-art in industrial waste impoundment, handling and
    disposal. Proceedings of the last few years consist of
    close to a thousand pages each, with just under a hun-
    dred papers grouped by types of waste (i.e. chemical
    wastes, metal wastes, explosive wastes). The Proceed-
    ings of the 37th Conference, published in 1983, include
    a ten-year index to Conference Proceedings.

Research Needs Associated with Toxic Substances in Waste-
    water Treatment Systems. Washington, D.C.: Water
    Pollution Control Federation, 1982.

Rishel, H. L., et al. Costs of Remedial Response Actions at
    Uncontrolled Hazardous Waste Sites. Park Ridge, NJ:
    Noyes, 1984.

    Based on an EPA report by the authors, originally
    published in 1982.

Rogoshewski, P., et al. Remedial Action Technology for Waste
    Disposal Sites. Park Ridge, NJ: Noyes, 1983.

    Also published as EPA Report #EPA-625/6-82-006.
    Describes methods for controlling, containing, treating
    or removing contaminants from uncontrolled hazardous
    waste sites.

Steeler, Jonathan H. A Legislator's Guide to Hazardous
     Waste Management. Denver, CO: National Conference on
     State Legislatures, 1980.

Subramanian, S. K. Symposium on Disposal and Recycling of
     Industrial Hazardous Wastes. Tokyo, Japan: Asian
     Productivity Organization, 1983.

     Treatment and disposal of hazardous wastes from
     industry: some experiences. Report of the Symposium on
     Disposal and Recycling of Industrial Hazardous Wates,
     Tokyo, Japan, 1982.

Supervisor's Guide to Hazardous Waste and Materials Manage-
     ment. Madison, CT: Bureau of Law and Business, Inc.,
     n.d.

     Self-paced training on hazardous waste materials, emer-
     gencies, spill response, protective equipment; with
     checklists and case studies.

Theodore, Louis and Anthony J. Buonicore, eds. Air Pollution
     Control Equipment: Selection, Design, Operation and
     Maintenance. Englewood Cliffs, NJ: Prentice-Hall,
     1982.

Vance, Mary A. Industrial Waste Disposal: A Bibliography.
     Monticello, IL: Vance Bibliographies, 1982.

Warner, Don L. and Jay H. Lehr. Subsurface Wastewater
     Injection: The Technology of Injecting Wastewater into
     Deep Wells for Disposal. Berkeley, CA: Premier Press,
     1981.

     Photographic reproduction of An Introduction to the
     Technology of Subsurface Wastewater Injection issued in
     1977 by the EPA (#EPA-600/2-77-240). For use by all
     involved in planning, design, construction, operation
     and abandonment of injection wells.

Waste Discharge into the Marine Environment: Principles and
     Guidelines for the Mediterranean Action Plan. Prepared
     in collaboration with the Institute of Sanitary Engin-
     eering, Polytechnic of Milan, Italy. Elmsford, NY:
     Pergamon, 1982.

Wilson, David C. Waste Management: Planning, Evaluation,
     Technologies. New York: Oxford University Press, 1982.

Young, Richard A. and Frank Cross, eds. <u>Specifying Air Pollution Control Equipment</u>. New York: Dekker, 1982.

Audiovisual Materials

Few audiovisual titles on the topic of hazardous and toxic
waste disposal are available, but this is an area where we
should expect to see new materials being produced, as disposal policies and processes gain attention. Audiovisual
materials are especially useful in training or educational
situations, where visual impact can improve the effectiveness of the presentation.

Here are listed some audiovisual materials which examine
hazardous waste disposal. Materials that have more general
scope, covering, for instance, pollution and its effects,
are not included in this list.

<u>The Burial Ground</u>. Boston: Thomas McCann and Associates;
SCA Services, 1982. 16mm film, sound, color, 30 min.
Also available as video cassette.

A fictionalized account of an investigation and prose-
cution of a violation of state and federal hazardous
waste disposal laws.

<u>Complete Hazardous Waste Training Course</u>. LaPlata, MD:
Haztrain, n.d. Training workbooks with a separate set
of 106 35mm color slides.

Includes actual samples of state and EPA forms.

<u>Hazardous Waste: The Search for Solutions</u>. Los Angeles:
Direct Cinema, Ltd, 1984. 16mm film; color, sound, 35
min. Also available as videocassette.

A documentary about concerned citizens organizing at
the local level to clean up some of the toxic chemical
dumps.

<u>Hazardous Waste Options</u>. Falls Church, VA: Stuart Finley,
Inc., 1981. 16mm film, color, sound, 22 min.

Describes how recycling, waste treatment and conversion, secure landfills, high temperature incineration,
deep well injection and land treatment can constitute
the appropriate methods of handling hazardous wastes.

The New R.C.R.A. Washington, D.C.: The Office of Solid
Waste, Environmental Protection Agency, 1984. Distributed by the National Audiovisual Center. Videocassette, color, sound, 180 min. (also issued in 56
min. version).

Explains the major provisions of the R.C.R.A. of 1984.
A training seminar for employees on how to comply with
the new Hazardous and Solid Waste Amendments.

RCRA/Hazardous Waste Training Series. Madison, CT: Bureau
of Law and Business, Inc., n.d. Available as 16mm film
or videocassette.

Module 1, Employee Introduction to RCRA, 25 minutes.
Module 2, Keeping Track of Hazardous Wastes, 22 minutes. Module 3, Safety Training, Protective Clothing
and Equipment, 25 minutes. Module 4, Contingency Planning and Spill Response, 22 minutes.

Associations and Special-Interest Groups

There are a number of associations and special-interest
groups in the United States for individuals, agencies and
corporations who have a serious and continuing interest in
the disposal of hazardous and toxic wastes. Membership in
these groups often provides a forum for the exchange of
ideas among members, through newsletters, seminars, and
other group publications and activities. Below are listed a
few of the groups that provide such services.

The American Society for Testing and Materials (1916 Race
Street, Philadelphia, PA 19103) has recently created a Committee on Waste Disposal (Committee D-34) to improve testing
methods to support waste management technology such as recycling, land disposal, incineration, and waste treatment.
The Committee has since 1981 been sponsoring symposiums in
this area (see citations listed in "Books and Proceedings"
section of this chapter, i.e. Hazardous and Industrial Solid
Waste Testing: Second Symposium.)

The Association of State and Territorial Solid Waste
Management Officials (44 North Capitol Street, Room 343,
Washington, D.C. 20001), partially funded by the EPA,
surveys state needs in waste management programs and legislation, and promotes intrastate and interstate cooperation
in the handling of hazardous waste materials.

The Environmental Hazards Management Institute (P.O. Box
283, 45 Pleasant Street, Portsmouth, NH 03801) evolved from
the Northeast Conferences on Hazardous Waste held in 1979
and 1980. Incorporated in 1981, the EHMI provides media
production, topic research, press and political relations,

management counseling, and the organization of conferences and workshops. The Institute assists such diverse groups as communities with contaminated water; industries with waste disposal problems; and workers in need of training in the handling of hazardous materials and waste disposal. They sponsor a 3 1/2 month hazardous materials/waste training program, conducted at the University of New Hampshire. Members of EHMI's Northeast Hazardous Waste Advisory Council include representatives of industry, government, the academic community and the public. Mail and telephone inquiries are welcome.

The Hazardous Materials and Waste Management Association (Box 308, Kutztown, PA 19530), is a new group, formed in 1984. The goal of the Association is to offer a legislative, educational and communication network for generators and transporters of hazardous waste, and operators of treatment, storage and disposal facilities. In addition to their journal, Hazardous Materials and Waste Management, they publish reprints of the Federal Register and Code of Federal Regulations changes pertinent to hazardous wastes, and sponsor regional trade shows and seminars.

The Hazardous Materials Control Research Institute (9300 Columbia Boulevard, Silver Spring, MD 20910) offers membership to individuals and corporations that have a vested interest in hazardous material management and control. The Institute offers conferences, seminars, equipment exhibitions, and publications, including the HMCRI Forum newsletter. The HMCRI is a non-profit organization that seeks to "promote the establishment and maintenance of a reasonable balance between expanding industrial productivity and an acceptable environment".

The Hazardous Waste Services Association (1333 New Hampshire Avenue, NW, Suite 1100, Washington, D.C. 20036), is a nonprofit trade association with over 100 member companies involved in hazardous waste management. This group provides legal assistance to its members, and publishes the biweekly newsletter, Wasteline. They also produce a national membership directory, and organize meetings and seminars.

The National Solid Waste Management Association (Tenth Floor, 1730 Rhode Island Avenue, NW, Washington, D.C. 20036), represents many of the major commercial hazardous waste disposal firms. They publish the monthly Waste Age and sponsor the annual WasteExpo conference, the Hazwaste Forum, and other meetings. Special-interest sections of NSWMA include the Institute of Chemical Waste Management, the Waste Equipment Manufacturers Institute, the Liquid Industrial Control Association, the Waste Haulers Council, and the Ontario (Canada) Waste Management Association.

A number of clearinghouses for the exchange of wastes with reuse value are in operation in the United States. These clearinghouses publish listings of "available" and "wanted" wastes, and attempt to put waste users in contact with waste producers. Two such agencies are the Northeast Industrial

Waste Exchange (90 Presidential Plaza, Suite 122, Syracuse, NY 13202), co-sponsored by the New York State Environmental Facilities Corporation and the Ohio Environmental Protection Agency, and the Illinois Environmental Protection Agency's Industrial Material Exchange Service (Division of Land Pollution Control, 2200 Churchill Road, Springfield, IL 62706). To locate similar activities in other states, contact individual state environmental regulatory agencies or the EPA regional offices, or refer to "Directory of North American Waste Exchanges", in the May/June 1984 issue of The Hazardous Waste Consultant (vol. 2, issue 2, pp. 4-11 to 4-16).

A "National Conference on Waste Exchange" has been sponsored annually since 1984 by the Western Waste Exchange at Arizona State University, the Illinois Environmental Protection Agency's Industrial Material Exchange Service, and the Southern Waste Information Exchange at Florida State University. Proceedings of these conferences are available from the National Conference on Waste Exchange, P.O. Box 6487, Tallahassee, FL 32313.

9.
Radioactive Materials

Erich J. Mayer

Introduction

Ionizing radiation and radioactive waste management are
sensitive subjects in today's society. When one hears the
terms "radiation" or "radioactivity", a particular set of
images comes to mind. These images are either positive
(nuclear power generation, radiation therapy) or negative
(thermonuclear war) depending upon your viewpoint. They are
almost never neutral.

Within the past decade, radioactive waste management has
become a topic of national and international concern.
Legislators are now being forced to deal with both socio-
economic and scientific issues and are expected to make
decisions that will affect many future generations. To do
this properly, legislators and anyone who deals with radioactive wastes need information.

From the moment that the first nuclear reactor was built in
1942, promises have been made concerning the safe management
of radioactive wastes. Today, scientists and engineers in
France, Germany, Great Britain, Japan and the United States
are making good on those promises.

However, not everyone believes that radioactive wastes can
be managed in a safe, cost effective manner. These people
need information, too.

Each side of the radioactive waste management issue requires
and deals with the same sets of information. They often
come to opposite conclusions based on this data. This does
not make the librarian's role of answering reference
questions on radioactive waste management, in an inpartial
manner, an easy one. You have to be able to rise above the
issue and your own views on the subject and provide
unrestricted access to both positive and negative viewpoints.

This chapter on radioactive materials includes sources from both sides of the radioactive waste management issue. Although my workplace gives me a positive bias, I have included every source I could find on radioactive waste management, both positive and negative.

The larger portion of this chapter deals with radioactive waste management from a positive viewpoint. This is an unavoidable bias. There is more information being published by nuclear engineers and scientists than by antinuclear ecologists and sociologists.

Most of the sources in this chapter are intended for practicing engineers and scientists and are technologically oriented. Most of the negative sources tend to have an ecological or sociological viewpoint. More statistics and opinions, less hard data of pratical, every day use.

Two major sets of publications have purposefully been omitted from this chapter. They are the Canadian and United States federal governments and their nongovernment contractors (i.e. Atomic Energy of Canada, Ltd., Electric Power Research Insitute, Office of Nuclear Waste Isolation). They have been listed as sources of information but, because of the number of documents they publish each year, I decided to exclude them rather than try to pick just a few.

The Abstracts and Indexes and Data Bases section includes sources (i.e. DOE/ENERGY and Energy Research Abstracts) that list documents published by the federal government and their contractors.

I have included as much foreign literature as possible. The Council of European Communities, the International Atomic Energy Agency and the Nuclear Energy Agency all publish numerous books and technical reports each year on radioactive waste management.

For the most current, up-to-date data, look at the Proceedings section. Although some series date back to the 1970's, the meetings are held each year and thus are constantly being updated by engineers and scientists in the field.

Only one data base, DOE/ENERGY, has been included as an information source. At present, the Department of Energy's RECON data base system is scheduled to be shut down by January 1987. As yet, there are no specific plans on how or when the RECON data bases will be available to the public via commerical vendors.

New information on radioactive waste management is being published daily. It is an international problem that is being solved on an international scale. This chapter is meant to be a guide to data published since 1980. It does not include everything that has been published but it is a good beginning for anyone searching for information on radioactive waste management.

Books

Bartlett, Donald L. and James B. Steele. Forevermore: Nuclear
    Waste in America. New York: W. W. Norton & Co., 1985.

This book is an outgrowth of a series of articles that
appeared in the Philadelphia Enquirer in November 1983.
Information from the original series has been updated
and expanded to include data on defense wastes, fuel
reprocessing outside the U.S. and uranium mill tailings. Illustrated with a subject index.

Brill, A. Bertrand, ed. Low-Level Radiation Effects: A Fact
    Book. New York: The Society of Nuclear Medicine, 1982.

The book provides a summary of radiobiologic data compiled from scientific and technical journals as well as
national and international radiological and scientific
organizations. Includes many figures and tables, a
question and answer section, and a list of references.

Brodsky, Allen. CRC Series in Radiation Measurement and
    Protection. Boca Raton, FL: CRC Press, Inc., 1982.

The purpose of this handbook series is to provide data
and methods for dealing with the design and evaluation
gradiation measurement instruments; monitoring and
survey methods; selection of protective facilities;
equipment and procedures for handling radioactive
materials; radioactive waste disposal; transportation
of radioactive materials; and many other related subjects. Each volume contains a list of cited references
and a subject index.

Brookins, Douglas G. Geochemical Aspects of Radioactive
    Waste Disposal. New York: Springer-Verlag, 1984.

The focus of this book is on geochemistry with an emphasis on the emplacement of radioactive wastes. It
includes background material on radioactive wastes and
geotoxicity as well as geology and geochemistry. This
book is an attempt to summarize the literature on
radioactive waste management that has been published
from 1970 to 1980.

Campara, Robert J. and Sidney Langer. Nuclear Power and the
    Environment, Book 1. Radiation: Questions and Answers.
    La Grange Park, IL: American Nuclear Society, 1982.

Presents information on the overall U.S. energy outlook
and, in particular, on electricity generated from
nuclear energy. Includes facts and opinions, actual

experiences, estimates of current and future energy
needs and projections about the future. If the topic
is controversial, the book includes all viewpoints. A
bibliography follows each chapter citing additional
information sources.

Carley-Macawly, K. W., et al, eds. Radioactive Waste:
Advanced Management Methods for Medium Active Liquid
Waste. Radioactive Waste Management, Volume 1. New
York: Harwood, 1981.

This study is a bibliographic survey and assessment of
the principle methods of purifying liquid waste of
medium activity, taking into consideration methods
recently introduced into industrial operation, or newly
developed in the laboratory or on the pilot plant
scale. These include: chemical precipitation, ion
exchange adsorption, membrane processes, electrical
processes, foam separation, metabolization and magnetic
separation. Includes references.

Colglagier, William E., ed. The Politics of Nuclear Waste.
New York: Pergamon, 1982.

This book is an outgrowth of an Aspen Institute conference on radioactive waste management. The chapters and
appendix include some of the papers presented at that
conference as well as a sampling of the diverse views
involved in waste management as given by all sides of
the issue. Illustrated with a subject index.

Committee on the Biological Effects of Ionizing Radiations.
The Effects on Populations of Exposure to Low Levels of
Ionizing Radiation: 1980. Washington, D.C.: National
Academy Press, 1980.

This is the third Biological Effects of Ionizing
Radiations (BEIR) Committee report dealing with the
scientific basis of effects of low-dose radiation. It
includes a review and evaluation of scientific knowledge since the first BEIR report concerning radiation
exposure of human populations was published in 1972.
This report attempts to present all sides in a balanced
fashion. Includes references and a glossary of terms
used.

Commission of the European Communities. Research and Development on Radioactive Waste Management and Storage: Progress Reports of the European Community Programme 1980-
1982. Radioactive Waste Management, Volumes 4, 8, and
12. New York: Harwood, 1982/83.

These progress reports cover a program that deals with
the joint development and improvement of a radioactive

waste management system produced by the nuclear industry which, at its various stages, ensures the safety
and protection of both man and his environment.

D'Alessandro, Marco and Arnold Bonnes. Radioactive Waste
Disposal into a Plastic Clay Formation. Radioactive
Waste Management, Volume 2. New York: Harwood, 1981.

This book deals with the geological containment of
radioactive waste from the biosphere. The problem
studied is: how effective is geological containment?
How can the barrier fail? What is the probability of
failure? Towards this problem, Fault Tree Analysis of
a specific site is utilized. Includes references and a
subject index.

Diamant, R. M. E. Atomic Energy. Lancaster, PA: Technomic,
1982.

This is a basic text providing a review of the science
and technology of atomic energy. Includes fuels,
reactors, health, safety and environmental aspects, and
applications. Illustrated with references.

Duffy, Joan Irene. Treatment, Recovery and Disposal
Processes for Radioactive Wastes: Recent Advances.
Park Ridge, NJ: Noyes, 1983.

The information in this book is based on U.S. patents
issued from January 1975 through May 1982 that relate
to radioactive waste treatment, recovery, and disposal
processes. Thus it provides detailed technical information as well as serving as a guide to U.S. patent
literature. Illustrated, with inventor and patent
number indexes.

The Economics of the Nuclear Fuel Cycle. Washington, D.C.:
Nuclear Energy Agency, Organisation for Economic Co-
Operation and Development. 1985.

This book examines, in detail, the costs of the various
stages of the nuclear fuel cycle and the methodology
employed for comparative calculations of disaggregated
and total costs provided by seventeen OECD countries
and four international organizations. Illustrated,
with a list of references.

Edelson, Edward. The Journalist's Guide to Nuclear Energy.
Bethesda, MD: Atomic Industrial Forum, 1985.

This book is meant to serve as a ready reference primer
on the civilian nuclear energy program for reporters
covering the industry. The facts are presented in a

concise, easy to read format and could be used by anyone who was looking for an overview of the nuclear
industry today and for the near future. Illustrated,
with a glossary of terms used.

Gollnick, Daniel A.   Basic Radiation Protection Technology.
Temple City, CA: Pacific Radiation Press, 1983.

A textbook designed to aid radiation protection technologists studying for the National Registry of Radiation Protection Technologists examination. Can also
serve as a broad, indepth introduction to radiation
protection as well as a reference text for experienced
personnel. Includes a set of questions at the end of
each chapter as well as a list of organizations and a
subject index.

Hebel, W. and G. Cottone, eds.   Methods of Krypton-85
Management. Radioactive Waste Management, Volume 10.
New York: Harwood, 1984.

Krypton-85 is a noble gas that occurs during nuclear
fuel reprocessing. The present practice is to discharge this gas into the atmosphere after dilution.
The purpose of this study is to make a comparative
evaluation of the various strategies applied to the
management of Krypton-85 as regards both its retention
and its storage and disposal. The study considers both
the technical and economic issues involved and the
safety of the different methods of management.
Illustrated.

Jedruch, Jacek.   Nuclear Engineering Data Bases, Standards,
and Numerical Analysis. New York: Van Nostrand
Reinhold, 1985.

This book focuses on the sources and repositories of
information, data computing methods, regulations and
standards of the nuclear industry and the methods of
information transfer between them. It is intended as a
guide for practicing nuclear engineers, personnel of
public utilities, state and federal regulatory bodies
and students of nuclear engineering. Illustrated with
a bibliography and a subject index.

Jensen, B. Skytte.   Migration Phenomena of Radionuclides
into the Geo-sphere. Radioactive Waste Management,
Volume 5. New York: Harwood, 1982.

This book is divided into 4 sections. Part A deals
with the chemical state of the radioelements in solution in groundwater. Part B deals with the composition
of groundwaters in light of newer theories taking
chemical equilibria with rock minerals into account.

In Part C, quantitative aspects of sorption processes
are discussed and in Part D, the interplay of all
effects are discussed in setting up the differential
equation describing the migration of solutes with
groundwater. Includes references.

Jordan, Julie M. and L. Gail Melson. A Legislator's Guide
    to Low-Level Radioactive Waste Management. Denver, CO:
    National Conference of State Legislators, 1981.

    This book is meant to familiarize state legislators and
    legislative staff with the legal, institutional, finan-
    cial, environmental and technological issues associated
    with low-level waste management. Includes a list of
    state laws, the 1980 Low-Level Radioactive Waste Policy
    Act, a glossary of terms used and a list of references.

Kaku, Michio and Jennifer Trainer, eds. Nuclear Power -
    Both Sides. W. W. Norton & Company, Inc., 1982.

    This book is meant for persons who are undecided about
    the relative benefits/risks of nuclear power. To help
    them to decide, the editors have broken down the con-
    troversy into six main parts with an introduction that
    provides information and raises the issues. The essays
    then attempt to answer those issues. Includes referen-
    ces and a subject index.

Kasperson, Roger E. Equity Issues in Radioactive Waste
    Management. Cambridge, MA: Velgeschlager, Gunn & Hain,
    Publishers, 1983.

    This book deals with the issue of fairness in radio-
    active waste management. Is it fair that one state
    will hold all of the others' high-level wastes? The
    issue of the nuclear fuel cycle: should the U.S. repro-
    cess spent fuel? Why are there inequal exposure limits
    to ionizing radiation for workers vs. the public?
    Illustrated, with a subject index.

Kathren, Ronald L. Radioactivity in the Environment:
    Sources, Distribution, and Surveillance. New York:
    Harwood, 1984.

    This book is intended to be a reference book for
    scientists, engineers and regulators, both within and
    outside the nuclear field, who may be involved with the
    application and/or control of nuclear power or other
    sources of radioactivity. Illustrated with references
    and a subject index.

Knief, Ronald Allen. <u>Nuclear Criticality Safety: Theory and Practice</u>. La Grange Park, IL: American Nuclear Society, 1985.

This book is intended to provide a comprehensive description of the nuclear criticality safety discipline. It deals not only with the theoretical basis for criticality safety, but also details practical applications in use today at facilities across the United States. Illustrated with references and a subject index.

Krissher, W. and R. A. Simon, eds. <u>Testing, Evaluation and Shallow Land Burial of Low and Medium Radioactive Waste Forms</u>. Radioactive Waste Management, Volume 13. New York: Harwood, 1984.

This book presents the preliminary results of the Commission of the European Communities program on the charaterization of low and medium active waste forms and the safe management and disposal of alpha bearing wastes and other long-lived radionuclides. Illustrated.

Lapp, Ralph E. <u>The Radiation Controversy</u>. Greenwich, CT: Reddy Communications, Inc., 1979.

This book concerns itself with radiation health physics: Radiation risks, how to prevent/avoid them and the publics' perception of radiation. Based on first hand experience in handling radioactivity, it is both a factual account of radiation measurement and protection as well as an attack on the media and public for misrepresenting and not understanding the relative risks of radiation.

Lipschutz, Ronnu D. <u>Radioactive Waste: Politics, Technology and Risk</u>. Cambridge, MA: Ballinger Publishing Co., 1980.

This book by the Union of Concerned Scientists provides information about both the technical and non-technical issues required for meaningful public involvement in the issue of radioactive waste management. It includes information on history, catagories and production of radioactive wastes, and the federal program set up to manage radioactive wastes. Illustrated, with a bibliography, a glossary of terms used and a subject index.

<u>Managing the Nation's Commercial High-Level Radioactive Waste</u>. Office of Technology Assessment. New York: UNIPUB, 1985.

This report presents the findings and conclusions of
the Office of Technology Assessment's analysis of
federal policy for the management of commercial high-
level radioactive waste. It updates and expands upon
an earlier 1982 document published during the debate
leading to the passage of the Nuclear Waste Policy Act
of 1982. Illustrated, with references.

McKinlay, A. F. Thermolunminesence Dosimetry. Bristol,
England: Adam Hedger Ltd., 1981.

The purpose of this book is to provide an introduction
to the use of thermoluminescent dosimeters (TLDs) in
ionizing radiation dose measurement with emphasis on
clinical dosimetry. Written for graduate scientists,
it would be useful to anyone routinely dealing with
TLDs. Includes references and a subject index.

Meehan, Richard L. The Atom and the Fault. Cambridge, MA:
The MIT Press, 1984.

This book focuses on the long-running controversy over
suspected earthquake faults near the United States'
first commercially owned nuclear test reactor at
Vallecitos, California. In the process, it attempts to
probe the nature of scientific truth and its relationship to the determination of public safety. Illustrated with an annotated bibliography.

Miller, William H., et al. Nuclear Power and the Environment, Book 4. Energy Altervatives: Questions and
Answers. La Grange Park, IL: American Nuclear Society,
1981.

Provides information on the many energy alternatives
that may become viable in the future, i.e. solar and
wind power and fusion and breeder reactors, etc. It
then puts these alternatives into perspective with
respect to the current U.S. energy situation, the
projected energy demands of the future and the total
impact the alternatives will have on these energy
needs. Includes references intended to cite as well as
guide the reader to additional information.

Murdock, Steve H., et al., eds. Nuclear Waste: Socioeconomic
Dimensions of Long-Term Storage. Boulder, CO: Westview
Press, 1983.

This book examines the socio-economic implications of
radioactive waste management and repository siting
primarily for rural areas in the United States. It
attempts to propose solutions to the problems of packaging and disposing of high-level radioactive wastes

from a socio-economic rather than technical point of
view. Includes a bibliography and a subject index.

Murray, Raymond L. <u>Nuclear Energy</u>. Oxford, NY: Pergamon,
1980.

A textbook that provides a factual description of basic
nuclear phenomena, devices and processes that involve
nuclear reactions and social and economic problems and
opportunities in the nuclear age. Includes references,
problem sets and answers to the problems.

Murray, Raymond L. <u>Understanding Radioactive Waste</u>. Columbus, OH: Battelle Press, 1982.

Provides a brief but concise overview of the past,
present and future of radioactivity and radiation,
nuclear power, radiation applications and radioactive
waste management. This book seeks to answer the most
commonly asked questions on radiation and radioactive
waste management. Includes a glossary of terms used
and a chapter specific bibliography.

National Research Council, Panel on Social and Economic
Aspects of Radioactive Waste Management, et al.
<u>Social and Economic Aspects of Radioactive Waste
Disposal</u>. Washington, D.C.: National Academy Press,
1984.

This book is the report of a panel established under
the aegis of the Board on Radioactive Waste Management
to study the non-technical criteria for siting geologic
repositories for high-level nuclear wastes. Includes
many charts and tables and a list of references at the
end of each chapter.

National Research Council, Panel on Social and Economic
Aspects of Radioactive Waste Management. <u>Social and
Economic Aspects of Radioactive Waste Disposal: Consid-
erations for Institutional Management</u>. Washington,
D.C.: National Academy Press, 1984.

This study identifies: the major socio-economic considerations in the location, construction, and operation
of a generic radioactive waste repository, an assessment of what is known about these considerations, and
how to incorporate socio-economic considerations into
the repository selection process. Illustrated, with
references.

Park, P. Kilho, et al., eds. <u>Wastes in the Ocean, v. 3.
Radioactive Wastes and the Ocean</u>. New York: Wiley,
1983.

The objectives of this book are to present a comprehensive overview of the state of man's knowledge concerning the disposal of radioactive wastes in the ocean and to present new and original contributions to the evaluation of the impact of the disposal of radioactive wastes on human, as well as marine life. Illustrated with an author and subject index.

Platt, A. M., et al., eds. <u>Nuclear Fact Book</u>. 2nd ed. New York: Harwood, 1985.

This book may be used as a handy reference for highlights and summaries of facts in: energy production, consumption and costs, nuclear energy production, nuclear fuel cycle, and nuclear wastes. Illustrated with a list of references and a glossary of terms used.

Resinkoff, Marvin. <u>The Next Nuclear Gamble: Transportation and Storage of Nuclear Waste</u>. New York: Council on Economic Priorities, 1983.

This book deals with short-term management of radioactive waste storage from an anti-nuclear viewpoint. It attempts to examine the real risks, costs and alternatives for handling irradiated nuclear fuel. Includes many illustrations and is extensively referenced.

Roy, Rustum. <u>Radioactive Waste Disposal. Volume 1 : The Waste Package</u>. Elmsford, NY: Pergamon, 1982.

This book deals with one component of the technological subsystem for radioactive waste disposal, the waste package. Emphasis is placed on the waste form, the container for the waste, the overpack and waste/rock interactions. Illustrated, with a list of references at the end of each chapter.

Russ, George D. <u>Nuclear Waste Disposal: Closing the Future</u>. Bethesda, MD: Atomic Industrial Forum, 1984.

The book briefly surveys the origin of high-level nuclear waste in power generating reactors as well as its treatment, packaging and transportation for eventual disposal. Emphasis is placed on the techniques and technologies employed in high-level waste management. Includes a list of groups that may be contacted for further information.

Schleien, Bernard and Michael S. Terpilak, eds. and comps. <u>The Health Physics and Radiological Health Handbook</u>. Ohey, MD: Nuclear Lectern Associates, 1984.

Intended for use by health physics practitioners, tech-
nicians, and students as an easy to use, comprehensive
handbook on health physics and radiological health
data. Includes many charts, graphs, and tables as well
as a glossary of terms used and a subject index.

Schrayer, Fred, comp. Radioactive Waste: Issues and Answers.
    Arvada, CO: American Institute of Professional Geolo-
    gists, 1985.

The disposal of radioactive waste is currently a focus
of public interest. Prudent public policy concerning
disposal of radioactive waste requires a good understanding of the scientific, technical and social issues
involved. The purpose of this booklet is to provide
policy-makers, legislators and the general public with
information to better understand the issues, particularly geological considerations. Illustrated, with a
list of references and a glossary of terms used.

Seaborg, Glenn T. and Walter Loveland, eds. Nuclear Chem-
    istry. Stroudsburg, PA: Hutchinson Ross Publishing
    Co., 1982.

This book contains a collection of reprints of bench-
mark papers in the field of nuclear chemistry from the
19th century to the present with the original references. The editors have added annotations, comments, and
an author and subject as well as an index of cited
authors.

Shapiro, Fred C. Radwaste. New York: Random House, Inc.,
    1981.

One reporter's search for questions and answers to the
problems of nuclear waste management. Includes a
subject index but no references.

Stewart, Donald C. Data for Radioactive Waste Management
    and Nuclear Applications. New York: Wiley, 1985.

This book brings together, in a single reference
source, useful scientific data on radioactive waste
management as well as other nuclear applications.
Includes a subject index and a list of references.

Walker, Charles A., et al. Too Hot to Handle? New Haven,
    CT: Yale University Press, 1983.

This book provides basic information on radioactive
waste management. It covers the history of waste management, the processes that produce wastes, methods
proposed for management and the political institutions

and processes that govern radioactive waste management.
Illustrated with a subject index.

Weinberg, Alvin, et al., eds. The Nuclear Connection. New
    York: Paragon House Publishers, 1985.

    This book deals with nuclear power's potential linkage
    to weapons production and the subsequent proliferation
    in the number of nations having nuclear weapons. The
    main theme is "international collaboration in nuclear
    power commerce". Includes a subject index and referen-
    ces.

Whicker, F. Ward and Vincent Schultz. Radioecology: Nuclear
    Energy and the Environment. Boca Raton, FL: CRC Press,
    Inc., 1982.

    This book is intended as a broad survey of the field of
    radioecology. Each chapter deals with one part of the
    overall topic: natural and man made, radioactivity, the
    movement of such materials on the environment and the
    effects of ionizing radiation populations and biotic
    communities. Contains a list of references and a sub-
    ject index.

Periodicals

Bartlett, Donald L. and James B. Steele. "Forevermore:
    Nuclear Waste in America."   The Philadelphia Enquirer
    (Nov. 13-20, 1983).

    "Forevermore..." is a series of investigative reporting
    articles written for the Philadelphia Enquirer exam-
    ining the current state of radioactive waste management
    and the nuclear fuel cycle in the United States. Al-
    though the information presented is accurate, the writ-
    ing style is inflamatory and tends to negate the arti-
    cles' impact.

Fairhurst, Charles and Donald Gillis, eds.   "Managing
    Nuclear Waste: The Underground Perspective."   Under-
    ground Space 6, nos. 4-5 (1982).

    This is an entire issue of Underground Space dedicated
    to raising and answering questions on radioactive waste
    management from a governmental, technological and
    public policy oriented point of view. The articles are
    written by advocates and opponents of waste management,
    thus providing a balanced viewpoint to the issues
    raised.

Inside N.R.C. McGraw-Hill. Alternate Mondays, published
    since 1979.

    Contains information on administrative, judicial,
    legislative and contract activities and decisions
    within the Nuclear Regulatory Commission.

International Atomic Energy Agency. Bulletin. International
    Atomic Energy Agency. Quarterly, published since 1959.

    Contains articles, conference reports and news briefs
    on the international nuclear industry. Also lists
    conference announcements, forthcoming reports from the
    IAEA and job vacancies within the IAEA.

Meetings on Atomic Energy. Vienna: International Atomic
    Energy Agency. Quarterly, published since 1969.

    Each issue presents a list of future conferences,
    exhibitions and training courses that will deal with
    subjects directly or indirectly related to nuclear
    energy and other peaceful uses of the atom.

NEA Newsletter. Nuclear Energy Agency, Organisation for
    Economic Co-Operation and Development. Semiannually,
    published since 1979.

    Published simultaneously in French and English, it is
    meant to provide information about the major issues
    involved in the peaceful development of nuclear power,
    the NEA program and the direction of the agency's
    future activities.

Nuclear and Chemical Waste Mangagement. Pergamon. Quarter-
    ly, published since 1979.

    Covers the entire field of hazardous chemical and nu-
    clear low, high and transuranic waste management from
    generation to storage and disposal. Articles and pa-
    pers on laws and regulations as well as standards and
    operating experiences are also included.

Nuclear Engineering International. I.P.C. Electrical-
    Electronic Press Ltd. Monthly, published since 1956.

    Includes a review of newsworthy events, editorial com-
    ments and announcements of meetings of all types as
    well as articles on the design and instruction, opera-
    tion and maintenance, and research and development of
    nuclear power plants, reprocessing facilites and stor-
    age and disposal areas.

Nuclear Fuel Cycle. NTIS. Bimonthly, published since 1982.

Announces the current worldwide information available
on all aspects of the nuclear fuel cycle except reactor
properties and the performance of fuels and the management of radioactive waste.

Nuclear Law Bulletin. Nuclear Energy Agency, Organisation
for Economic Co-Operation and Development. Semiannually, published since 1968.

International in scope, NLB contains information on
legislative and regulatory activities, case law and
international organisations and agreements (i.e. IAEA,
NEA, Euratom, etc.) as well as the actual texts of laws
recently amended or effected in any OECD country.

Nuclear News. American Nuclear Society. Monthly, published
since 1959.

Contains short communications and editorials and comments on the nuclear industry in the United States with
some coverage of international news. Also includes a
calander of events and announcements of new products
and services and new standards and technical reports.

Nuclear Plant Safety. Equipment Qualification Engineering
Services Ltd. Monthly, published since 1983.

Articles and comments on safety and quality assurance
for the entire nuclear fuel cycle. Each issue emphasizes a particular problem or topic. Also includes
news briefs, new product announcements and new, safety
related standards, catalogs and technical reports.

Nuclear Reactor Safety. NTIS. Bimonthly, published since
1982.

Announces the current worldwide information available
on all safety-related aspects of reactors, including:
accident analysis, safety systems, radiation protection, decommissioning and dismantling and security
measures.

Nuclear Safety. Department of Energy. Bimonthly, published
since 1959.

Contains papers and review articles relevant to the
analysis and control of hazards associated with the
generation of nuclear energy. Primary emphasis is
placed on safety in reactor design, construction and
operation. Articles and papers on safety in the entire
nuclear fuel cycle are also included.

Nuclear Science and Engineering. American Nuclear Society. Monthly, published since 1956.

Contains papers and short communications on unreported research in the nuclear sciences and nuclear engineering. Also includes descriptions of new computer programs available through computer code centers. International in scope, providing practical and theoretical information.

Nuclear Technology. American Nuclear Society. Monthly, published since 1965.

Contains review papers, short communications, book reviews and unreported work on all phases of applications of fundamental research to nuclear technology, International in scope with emphasis placed on the United States and Europe.

Nuclear Waste News. Business Publishers, Inc. Weekly, published since 1981.

A six to eight page newsletter covering local, state and federal activities affecting the generation, packaging, transportation, processing and disposal of all types of nuclear waste.

Radiation Protection Dosimetry. Nuclear Technology Publishing. Irregular, published since 1981.

Includes papers and short communications dealing with all aspects of radiation monitoring and the use of dosemeters to record exposures. Also includes announcements of meetings dealing with radiation monitoring. Emphasis is on practical, work-related information.

Radiation Protection Management. The Techrite Company. Quarterly, published since 1983.

RPM contains articles of practical, work-related information of interest to applied radiation protection professionals. The journal covers the full spectrum of applied health physics topics with special emphasis on technical and managerial subjects of interest to anyone working in a radiation protection program.

The Radioactive Exchange. Exchange Publications. Irregular, published since 1981.

Contains news briefs, editorials and comments, interveiws, lists of new technical reports and a calander of events all relating to nuclear waste management. Empha-

sis is placed on state and federal legislation affecting low-level waste management.

Radioactive Waste Management. NTIS. Bimonthly, published
    since 1981.

Announces the current worldwide information available
on the critical topics of spent-fuel transportation and
storage, radioactive effluents from nuclear facilities,
techniques of processing radioactive wastes and their
ultimate disposal.

Radioactive Waste Management and the Nuclear Fuel Cycle.
    Harwood. Quarterly, published since 1980.

Includes original papers and reports dealing with all
aspects of nuclear waste management: inventory, collection, treatment, transportation, R&D, interim storage,
ultimate storage, etc. Major emphasis is placed on
waste from nuclear energy production. Waste disposal
problems from the use of radioisotopes in industrial,
scientific and medical fields are also included.

Radwaste News. Radwaste News. 20 issues per year, pub-
    lished since 1979.

Contains news and views on anything related to radioactive waste management: legislation and litigation, pro
and anti-nuclear group activities, new publications and
new research. International in scope, main emphasis is
on the United States.

USCEA Update. Committee for Energy Awareness. Bimonthly,
    published since 1984.

A newsletter published to promote the use and generation of electricity especially by nuclear power plants.
Includes articles on electricity demands, international
use of nuclear power and a column containing excerpts
from editorials in favor of nuclear energy.

The Waste Paper. Sierra Club Radioactive Waste Campaign.
    Quarterly, published since 1979.

The WP is a newspaper covering any issue dealing with
radioactive waste. Includes editorials, guest
features, book reviews, editorial cartoons and news,
"what's happening" and citizen's protests about radioactive waste management. Particular emphasis is given
to transportation of radioactive wastes.

Reports

Acceptance Criteria for Disposal of Radioactive Wastes in
    Shallow Ground and Rock Cavities. Vienna: Inter-
    national Atomic Energy Agency, 1985.

This report provides an overview of basic information
related to waste acceptance criteria for disposal in
shallow ground and rock cavity repositories, consisting
of a discussion of the regulatory aspects, a brief
description of available disposal options, and a
discussion of acceptable waste types. Includes a list
of references.

Air Cleaning in Accident Situations. Washington, D.C.:
    Nuclear Energy Agency, Organisation for Economic Co-
    Operation and Development, 1984.

Although safe operation of nuclear facilities requires
containment behind suitable barriers of the radionuclides and/or other chemical toxins involved, it is
sometimes necessary to breach the containment to provide ventilation. This report reviews the performance
of off-gas cleaning systems in accident situations, and
outlines outstanding problems and their safety significance. Illustrated with a list of references.

Basic Safety Standards for Radiation Protection-1982
    Edition. Vienna: International Atomic Energy Agency,
    1982.

This report includes standards that provide guidance
for the protection of man from undue risks of the
harmful effects of ionizing radiation, while still
allowing beneficial practices involving exposure to
radiation. Includes a subject index.

Concepts and Examples of Safety Analyses for Radioactive
    Waste Repositories in Continental Geological
    Formations. Vienna: International Atomic Energy
    Agency, 1983.

The safe management and disposal of radioactive wastes
that arise from the nuclear fuel cycle are important
aspects of nuclear power development. This report is
intended for authorities and specialists responsible
for or involved in planning, performing and/or
reviewing safety assessments of underground radioactive
waste repositories. Illustrated, with references.

Conditioning of Low- and Intermediate-Level Radioactive
    Wastes. Vienna: International Atomic Energy Agency,
    1983.

    This technical report primarily describes the technolo-
    gies available for the conditioning (i.e. immobiliza-
    tion and packaging) of low- and intermediate-level
    radioactive wastes for transport, storage and disposal.
    Includes references and a glossary of terms used.

Control of Radioactive Waste Disposal into the Marine
    Environment. Vienna: International Atomic Energy
    Agency, 1983.

    The development of nuclear programs in many countries
    throughout the world requires that measures be taken to
    effectively manage the radioactive wastes that arise
    from such programs. This report provides information
    on the background, technical procedures and control of
    the marine disposal of radioactive wastes and on rela-
    tive national laws and international recommendations.
    Includes a list of references and a glossary of terms
    used.

Control of Semivolatile Radionuclides in Gaseous Effluents
    at Nuclear Facilities. Vienna: International Atomic
    Energy Agency, 1982.

    This report, based on the results of a Technical Com-
    mittee Meeting on retention of Semivolatile Radionu-
    clides at Nuclear Facilities combines the results of
    laboratory studies on control of semivolatile radionu-
    clides in gaseous effluents and of operating experience
    in that area. Includes references and a list of parti-
    cipants.

Co-Ordinated Research and Environmental Surveillance Pro-
    gramme Related to Sea Disposal of Radioactive Waste.
    Washington, D.C.: Nuclear Energy Agency, Organisation
    for Economic Co-Operation and Development, 1984.

    This report details sea disposal operations of packaged
    low-level radioactive wastes which are carried out
    under the provisions of the Convention on the
    Prevention of Marine Pollution by Dumping of Wastes and
    Other Matter, also referred to as the London Dumping
    Convention. Periodically under review, this is a 1983
    progress report.

Criteria for Underground Disposal of Solid Radioactive
    Wastes. Vienna: International Atomic Energy Agency,
    1983.

This report describes the basic requirements and criteria for underground disposal of solid radioactive
wastes. It is intended to serve both implementing and
regulatory bodies when an underground disposal system
is being developed. Includes references and a bibliography.

Current Practices and Options for Confinement of Uranium
    Mill Tailings. Vienna: International Atomic Energy
    Agency, 1981.

Uranium mill tailings are the solid residues and the
associated liquids remaining after uranium has been
extracted from an ore. This report presents an overview of the current practices for disposal of mill
tailings and appropriate site selection which can be
considered with the objective of minimizing the potential radiological risks of such an operation now and in
the future. Includes a list of references and a glossary of terms used.

Cutting Techniques as Related to Decommissioning of Nuclear
    Facilities. Washington, D.C.: Nuclear Energy Agency,
    Organisation for Economic Co-Operation and Development,
    1981.

This report deals with the segmenting of large reactor
vessels and their components as well as questions of
practicality and feasibility, and the demolition of
activated or contaminated concrete. Illustrated with a
list of references.

Decommissioning of Nuclear Facilities: Decontamination,
    Disassembly and Waste Management. Vienna: Inter-
    national Atomic Energy Agency, 1983.

The purpose of this report is to provide an information
base on the considerations important to decommissioning, the methods available for decontamination and
dissasembly of a nuclear facility, the management of
the resulting radioactive wastes, and the areas of
decommissioning methodology where improvements might be
made. Includes a list of references, a bibliography
and a glossary of terms used.

Decontamination Methods as Related to Decommissioning of
    Nuclear Facilities. Washington, D.C.: Nuclear Energy
    Agency, Organisation for Economic Co-Operation and
    Development, 1981.

This report discusses objectives of a decontamination
program relative to decommissioning, factors and logic
influencing program selection, and current chemical,

mechanical and electromechanical decontamination
processes. Illustrated with a list of references.

Design, Construction, Operation, Shutdown and Surveillance
of Repositories for Solid Radioactive Wastes in Shallow
Ground. Vienna: International Atomic Energy Agency,
1984.

Shallow-ground disposal of radioactive wastes has been
practiced for many years in several countries using a
variety of procedures to dispose of different types and
quantities of wastes. This report is intended for
administrative and technical authorities and specialists who are considering the use of shallow-ground
disposal for selected radioactive wastes. Illustrated,
with a list of references.

Disposal of Low- and Intermediate-Level Solid Radioactive
Wastes in Rock Cavities: A Guidebook. Vienna: International Atomic Energy Agency, 1983.

Disposal of radioactive wastes underground after appropriate conditioning is generally considered to be a
feasible method of providing the necessary protection
for man and the environment. This report is intended
to serve as a source of general information for those
considering disposal of low- and intermediate-level
radioactive wastes in rock cavities. It provides guidelines to be used by regulatory bodies and implementing
organizations. Includes a list of references.

Disposal of Radioactive Grouts into Hydraulically Fractured
Shale. Vienna: International Atomic Energy Agency,
1983.

This report describes a method for liquid waste
disposal in which radioactive waste effluents in the
form of a slurry containing hydraulic binders (grouts)
are injected by means of fracturing into a deep
underground formation considered to be isolated from
the surface. The composition of the grout is chosen so
that the slurry solidifies in situ. Illustrated with
references and a glossary of terms used.

Disposal of Radioactive Waste: An Overview of the Principles
Involved. Washington, D.C.: Nuclear Energy Agency,
Organisation for Economic Co-Operation and Development,
1982.

Disposal is the final step in waste management and may
be simply defined as a method of dealing with wastes
for which there is no intention of retrieval. This
report is intended to provide a review of the philoso-

phies underlying the current technical approach to the disposal of radioactive waste.

The Environmental and Biological Behaviour of Plutonium and Some Other Transuranium Elements. Washington, D.C.: Nuclear Energy Agency, Organisation for Economic Co-Operation and Development, 1981.

The production and the need to dispose of transuranium elements, particularly plutonium, have been considered in the the last decades as serious drawbacks to the development of nuclear energy. This report is intended to provide in a concise and simple form the basic information about plutonium and other transuranics in order to assess the importance of health problems posed to man by this production as a result of the developments of nuclear energy. Includes a glossary of terms and list of references.

Environmental Assessment Methodologies for Sea Dumping of Radioactive Wastes. Vienna: International Atomic Energy Agency, 1984.

Dumping of wastes in the ocean is currently carried out under the provisions of, and constraints imposed by, the Convention on the Prevention of Marine Pollution by Dumping of Wastes and Other Matter, London, 1972, commonly known as the London Dumping Convention (LDC). This report presents guidelines for the preparation and evaluation of environmental assessments relevant to the dumping of radioactive wastes in the ocean. Includes a list of references.

Epidemiological Studies of Groups With Occupational Exposure to Radiation. Washington, D.C.: Nuclear Energy Agency, Organisation for Economic Co-Operation and Development, 1985.

The possible carcinogenic effects of exposure to ionizing radiation, past, present or future, are currently of great public concern. This report is concerned with studies of persons exposed to ionizing radiation at work. The purpose of this report is to provide a simple description of the different types of epidemiological studies, of how they are carried out and of their potential, limitations and problems. Includes a list of references.

Generic Models and Parameters for Assessing the Environmental Transfer of Radionuclides from Routine Releases: Exposures of Critical Groups. Vienna: International Atomic Energy Agency, 1982.

This report is intended for national regulatory bodies and technical and administrative personnel responsible for performing environmental impact analyses. In particular, for generic assessments of doses to most exposed individuals from routine releases of radioactive effluents to atmospheric and aquatic environments. Includes a list of references.

Geological Disposal of Radioactive Waste: A Review of the Current Technical Understanding of Geochemical Processes Related to Geological Disposal of Radioactive Waste. Washington, D.C.: Nuclear Energy Agency, Organisation for Economic Co-Operation and Development, 1982.

The safe disposal of radioactive waste into deep geological formations will depend on a technical understanding of the chemical and physical phenomena that influence the behavior of radioactive elements in the geosphere. This report reviews current information about the geochemical processes of importance in safety assessments.

Geological Disposal of Radioactive Waste: Research in the OECD Area. Washington, D.C.: Nuclear Energy Agency, Organisation for Economic Co-Operation and Development, 1982.

This report presents brief descriptions of research activities in the OECD area, including those supported by the Commission of European Communities and the NEA, which contribute to the development of safe options for the disposal of radioactive wastes deep underground. It covers principally geological disposal options for high-level wastes from nuclear fuel reprocessing and spent nuclear fuel. Illustrated with bibliography.

Guide to the Safe Handling of Radioactive Wastes at Nuclear Power Plants. Vienna: International Atomic Energy Agency, 1980.

This report defines the aims and objectives of waste management at nuclear power plants. It is intended to provide guidance to regulatory authorities, designers and plant operators. Topics include: design considerations of waste management systems, sources and characteristics of wastes, transport of wastes, monitoring systems, safety analysis and a review of future trends in waste management. Illustrated with a list of references and a bibliography.

Guidebook on Spent Fuel Storage. Vienna: International Atomic Energy Agency, 1984.

Spent fuel storage has been identified as an important
independent step within the nuclear fuel cycle. This
report is an attempt to provide a summary of the
experience and information available in many areas
related to spent fuel storage. Illustrated with a list
of references and a glossary of terms used.

Handling and Storage of Conditioned High-Level Wastes.
    Vienna: International Atomic Energy Agency, 1983.

This report deals with aspects of the management
(handling and storage) of conditioned (immobilized and
packaged) high-level wastes from the reprocessing of
spent nuclear fuel. This report is intended for
authorities and specialists responsible for or involved
in the planning of facilities for the handling, transport and storage of conditioned high-level waste.

Handling of Tritium-Bearing Wastes. Vienna: International
    Atomic Energy Agency, 1981.

The generation of nuclear power and reprocessing of
nuclear fuel results in the production of tritium and
the possible need to control the release of tritium-
contaminated effluents. This report contains basic
information on tritium and tritium disposal as well as
seven papers presented at an IAEA Technical Committee
Meeting held in December 1978. Illustrated, with a
list of references.

In Situ Experiments in Granite Associated With the Disposal
    of Radioactive Waste. Washington, D.C.: Nuclear Energy
    Agency, Organisation for Economic Co-Operation and
    Development, 1985.

This document reports on the results and conclusions of
Phase I of the International Stripa Project for the
safe disposal of radioactive wastes in suitable
geologic formations, together with the preliminary
findings of Phase II. Illustrated with references.

Interface Questions in Nuclear Health and Safety.
    Washington, D.C.: Nuclear Energy Agency, Organisation
    for Economic Co-Operation and Development, 1985.

This report addresses the problems that arise in
interfacing the different disciplines associated with
nuclear power. Effective management of the potential
hazards connected with nuclear power programs requires
that the fields of radiation protection, nuclear safety
and health, and radioactive waste management all gain a
better understanding of one another.

International Co-Operation for Safe Radioactive Waste Management. Washington, D.C.: Nuclear Energy Agency, Organisation for Economic Co-Operation and Development, 1983.

The Nuclear Energy Agency supports national waste management efforts through a program of international co-operation. The NEA promotes the transfer of experience between Member countries, assists in research and development, and contributes to the development of an international understanding and guidance on current issues of waste management. This is a report outlining these activities. Illustrated.

The International Stripa Project: Background and Research Results. Washington, D.C.: Nuclear Energy Agency, Organisation for Economic Co-Operation and Development, 1983.

The International Stripa Project concerns research into the feasibility and safety of disposal of highly radioactive wastes from nuclear power generation, deep underground in crystalline rock. This report summarizes the objectives and preliminary results of experimental work performed within the framework of the Stripa Project. Illustrated.

Ionizing Radiation: Sources and Biological Effects. New York: United Nations Scientific Committee on the Effects of Atomic Radiation (UNSCEAR), 1982.

Contains the text of the 1982 UNSCEAR report to the General Assembly and annexes that support the conclusions of their study. Contains numerous charts, tables, graphs and illustrations. Each annex is individually and extensively referenced.

Isotope Techniques in the Hydrogeological Assessment of Potential Sites for the Disposal of High-Level Radioactive Wastes. Vienna: International Atomic Energy Agency, 1983.

Of particular interest for the disposal of high-level radioactive wastes is the behavior or performance of hydrological systems over periods of up to one hundred thousand years. This report describes the present state of development and the applications of various isotope techniques used in studying groundwater origin and flow, with emphasis on low-permeability rock systems. Illustrated with a list of references and a bibliography.

Long-Term Management of High-Level Radioactive Waste: The
    Meaning of a Demonstration. Washington, D.C.: Nuclear
    Energy Agency, Organisation for Economic Co-Operation
    and Development, 1983.

    The "demonstration" of the safe management of high-
    level radioactive waste is often regarded as a pre-
    requisite for the further development of nuclear
    energy. It is therefore essential to be clear about
    both the meaning of the term "demonstration" and the
    practical means to satisfy this request, particularly
    with respect to long-term aspects of waste disposal.
    This document attempts to respond to these points and
    to define the background against which the issues
    involved should be considered. Illustrated.

Long-Term Management of Radioactive Waste: Legal, Adminis-
    trative, and Financial Aspects. Washington, D.C.:
    Nuclear Energy Agency, Organisation for Economic Co-
    Operation and Development, 1984.

    This is an initial analysis, at an international level,
    of the institutional aspects of the long-term manage-
    ment of radioactive waste. It describes how legisla-
    tion and regulatory controls, financial methods, and
    the nuclear third party liability regime may be adopted
    so as to help ensure the long-term safety of the tech-
    nical containment systems for radioactive waste.
    Illustrated with a bibliography.

Long-Term Radiological Aspects of Management of Wastes from
    Uranium Mining and Milling. Washington, D.C.: Nuclear
    Energy Agency, Organisation for Economic Co-Operation
    and Development, 1985.

    This report presents several examples and their advan-
    tages, disadvantages and limitations of the application
    of the International Commission of Radiological Protec-
    tion methodology for the optimization of radiation
    protection to wastes from uranium mining and milling.
    Illustrated.

Management of Radioactive Waste at the Oak Ridge National
    Laboratory: A Technical Review. Washington, D.C.:
    National Research Council, 1985.

    This is a report of the findings and conclusions by the
    Board on Radioactive Waste Management after a two-year
    study of the management of radioactive wastes at the
    Oak Ridge National Laboratory (ORNL). It includes
    sections on waste management plans, practices, and
    programs at ORNL, environmental safety, facility
    decontamination and decommissioning and alternative

methods for long-term management of radioactive waste.
Illustrated with references and a glossary of terms
used.

Management of Radioactive Wastes Produced by Users of Radioactive Materials. Vienna: International Atomic Energy
   Agency, 1985.

   This report is intended to provide guidance for regulatory, administrative and technical authorities who are
   responsible for or are involved in planning, approving,
   executing and reviewing national waste management programs related to the safe use of radioactive materials
   and the subsequent disposal of radioactive wastes produced. Includes a list of references.

Mannano, F., ed. Management and Disposal of Alpha-Contaminated Wastes: A Survey of Current Practices, Strategies,
   and R and D Activities in Some EC Countries and the
   USA. Washington, D.C.: National Technical Information
   Service, 1983.

   This report is intended to give the necessary background for the critical review of waste management
   practices so far applied at the Ispra Establishment in
   Italy, as well as for their possible modifications
   according to more up-to-date management schemes.

Metrology and Monitoring of Radon, Thoron, and Their
   Daughter Products. Washington, D.C.: Nuclear Energy
   Agency, Organisation for Economic Co-Operation and
   Development, 1985.

   An important aspect in the control of radiation hazards
   from exposure to radon, thoron and their daughter
   products in mines and dwellings depends on accurate and
   reliable measurement of these radionuclides. This
   report reviews the principles and techniques of
   metrology and monitoring, and gives guidance on the
   selection and use of measurement methods and equipment
   for the various conditions of exposure of workers and
   members of the public. Illustrated with references.

Nuclear Aerosols in Reactor Safety: Supplementary Report.
   Washington, D.C.: Nuclear Energy Agency Group of
   Experts on Nuclear Aerosols in Reactor Safety, 1985.

   This book is a self contained summary of the original
   1979 report. The main emphasis is placed on accidents
   involving Light Water Reactors and extensive reactor
   core damage (i.e. Three Mile Island). Includes
   references.

Nuclear Power and Public Opinion. Washington, D.C.: Nuclear
    Energy Agency, Organisation for Economic Co-Operation
    and Development, 1984.

    This report examines the different experiences of the
    Nuclear Energy Agency Member countries and emphasizes
    basic approaches and practices aimed at winning greater
    public acceptances of nuclear power. Illustrated, with
    a list of references.

The Oceanographic and Radiological Basis for the Definition
    of High-Level Wastes Unsuitable for Dumping at Sea.
    Vienna: International Atomic Energy Agency, 1984.

    This report presents two potential models providing an
    approach to the definition of wastes unsuitable for
    dumping at sea. The models are based on ensuring that
    the radiation dose equivelants received by individuals
    as a consequence of dumping do not exceed recommended
    limits. Includes a list of references and a glossary
    of terms used.

Performance Assessment for Underground Radioactive Waste
    Disposal Systems. Vienna: International Atomic Energy
    Agency, 1985.

    In the development and utilization of nuclear energy,
    work is proceeding on concepts and strategies for
    disposal of radioactive wastes. This report focuses on
    performance assessment of subsystems within the total
    waste disposal system. It is intended for use by
    authorities and specialists responsible for or involved
    in planning, making and reviewing performance assess-
    ments of underground disposal systems for radioactive
    wastes. Includes a list of references.

Planning for Off-Site Response to Radiation Accidents in
    Nuclear Facilities. Vienna: International Atomic
    Energy Agency, 1981.

    Nuclear facilities are designed and constructed with
    consideration for safety. Their sophisticated safety
    systems are designed to protect not only the nuclear
    facility and its personnel, but also the public and the
    environment. The purpose of this report is to give
    guidance to those who are responsible for the
    protection of the public in the event of an accident
    occuring at a land-based nuclear facility. Includes a
    bibliography.

Radiological Significance and Management of Tritium,
    Carbon-14, Krypton-85 and Iodine-129 Arising From the
    Nuclear Fuel Cycle. Washington, D.C.: Nuclear Energy
    Agency, Organisation for Economic Co-Operation and
    Development, 1980.

    This report has been prepared by an NEA Group of
    Experts and is an analysis of the potential public
    health significance as well as the management of four
    long-lived radionuclides arising from operations in the
    nuclear fuel cycle: tritium, carbon-14, krypton-85 and
    iodine-129. Includes a list of references and a
    glossary of terms.

Regulations for the Safe Transport of Radioactive Material-
    1985 Edition. Vienna: International Atomic Energy
    Agency, 1985.

    This is the revised version of the Agency's "Regula-
    tions for the Safe Transport of Radioactive Material"
    as approved by the Board of Governors in September
    1984. This edition supercedes all previous editions of
    this report. Includes sections on test procedures,
    requirements for packages, and general principles and
    provisions.

Review of the Continued Suitability of the Dumping Site for
    Radioactive Waste in the North East Atlantic. Washing-
    ton, D.C.: Nuclear Energy Agency, Organisation for
    Economic Co-Operation and Development, 1980.

    This report contains the results of a 1979 review of
    the North-East Atlantic region disposal area conducted
    by an international group of oceanographic and radia-
    tion protection experts convened for this purpose by
    the NEA. Illustrated, with a list of references.

Safety Analysis Methodologies for Radioactive Waste
    Repositories in Shallow Ground. Vienna: International
    Atomic Energy Agency, 1984.

    Safe management and disposal of radioactive wastes from
    the various parts of the nuclear fuel cycle and from
    the use of radioisotopes for various purposes are
    important aspects of nuclear technology development.
    This report is intended for authorities and specialists
    responsible for or involved in planning, performing,
    and/or reviewing safety assessments of underground
    radioactive waste repositories. Includes a list of
    references.

Safety Assessment for the Underground Disposal of Radio-
    active Wastes. Vienna: International Atomic Energy
    Agency, 1981.

Countries in which electricity is produced by nuclear
energy are faced with adopting appropriate systems for
the management and disposal of radioactive wastes.
This report is intended for authorities and specialists
responsible for or involved in planning, performing
and reviewing safety assessments of underground
radioactive waste repositories.   Includes a glossary of
terms used and a bibliography.

Seabed Disposal of High-Level Radioactive Waste.   Washington, D.C.: Nuclear Energy Agency, Organisation for
Economic Co-Operation and Development, 1984.

This report presents the status of the work by the
Seabed Working Group of the OECD Nuclear Energy Agency
to examine the concept of disposing of high-level
radioactive wastes by burial in the continuously depositional sedimentary geologic formations of the deep
ocean flat.

Shallow Ground Disposal of Radioactive Wastes: A Guidebook.
Vienna: International Atomic Energy Agency, 1981.

This report is intended for those considering the use
of shallow ground disposal for selected radioactive
wastes and provides guidelines to cover the needs and
interests of both developed and developing countries
for the use of both regulating and implementing organizations.   Subjects include: generic and regulatory
activities and safety assessments, investigation and
selection of repository sites, waste acceptance criteria, design and construction of repositories, and operation, shut-down and surveillance of repositories.
Includes references and a glossary of terms used.

Site Investigations, Design, Construction, Operation, Shut-
down and Surveillance of Repositories for Low- and
Intermediate-Level Radioactive Wastes in Rock Cavities.
Vienna: International Atomic Energy Agency, 1984.

The safe disposal of radioactive waste, which is the
final step in the waste management operation chain, is
of vital importance for ensuring the long-term
protection of man and his environment.   This report is
meant for administrative and technical authorities as
well as specialists planning the disposal of low- and
intermediate-level radioactive wastes in rock cavities.
Illustrated, with a list of references.

Site Investigations for Repositories for Solid Radioactive
Wastes in Deep Continental Geologic Formations.
Vienna: International Atomic Energy Agency, 1982.

The potential impacts of radioactive wastes on man and
the environment should be kept acceptably low. This
report is intended for administrative and technical
authorities responsible for or involved in planning,
approving, executing and reviewing national waste
management programs. Includes a list of references, a
bibliography and a glossary of terms used.

Site Investigations for Repositories for Solid Radioactive
    Wastes in Shallow Ground. Vienna: International Atomic
    Energy Agency, 1982.

This report is a companion document to "Shallow Ground
Disposal of Radioactive Wastes: A Guidebook." It is
intended for administrative and technical authorities
responsible for or involved in planning, approving,
executing and reviewing national waste management programs. Includes a list of references, a bibliography
and a glossary of terms used.

Storage With Surveillance Versus Immediate Decommissioning
    for Nuclear Reactors. Washington, D.C.: Nuclear Energy
    Agency, Organisation for Economic Co-Operation and
    Development, 1985.

This report reviews the current debate over whether the
decommissioning of nuclear reactors should be immediate
or delayed for decades or centuries. It also reports
the consensus of a group of experts on the principle
criteria to be considered in the selection of decommissioning options. Illustrated with references.

Summary of Nuclear Power and Nuclear Cycle Data in OECD
    Member Countries. Washington, D.C.: Nuclear Energy
    Agency, Organisation for Economic Co-Operation and
    Development, 1984.

A questionnaire on electricity generation, nuclear
power, and fuel cycle data is distributed annually to
OECD Member Countries. In the most recent questionnaire, countries were asked to provide statistics for
1983 and the most likely projections up to the year
2000. The replies to this questionnaire are presented
in this summary.

Technical Appraisal of the Current Situation in the Field of
    Radioactive Waste Management. Washington, D.C.: Nuclear
    Energy Agency, Organisation for Economic Co-Operation
    and Development, 1985.

This report presents the collective view of the Radioactive Waste Management Committee on the main scientific and technical issues in the field of radioactive
waste management, particularly of disposal and the

associated long term safety aspects. It is addressed
to all persons interested in an up-to-date and objective assessment of the scientific and technical
achievements in this field.

Testing and Monitoring of Off-Gas Cleanup Systems at Nuclear
Facilities. Vienna: International Atomic Energy
Agency, 1984.

Releases from nuclear facilities to the atmosphere
during normal operations usually account for a
relatively small part of the total radiation exposure
of the public. This report describes the methods
currently employed, especially in nuclear power plants,
for testing and monitoring the effectiveness of the
cleanup systems installed to limit the emission of
radioactive particulate aerosols, gases and vapors to
the environment. Includes a list of references.

Treatment of Low- and Intermediate-Level Liquid Radioactive
Wastes. Vienna: International Atomic Energy Agency,
1984.

Liquid wastes are generated from the various processes
used in the nuclear fuel cycle and in the applications
of radionuclides in research institutions, hospitals
and industry. This report is addressed to regulatory
authorities and specialists who are responsible for or
involved in the design and operation of low- and
intermediate-level waste management facilities.
Illustrated, with a list of references.

Treatment of Low- and Intermediate-Level Solid Radioactive
Waste. Vienna: International Atomic Energy Agency,
1983.

Like almost any other industrial activity, the generation of electricity by nuclear fission, the operation
of different facilities in the nuclear fuel cycle and
the application of radionuclides in medicine, research,
industry and other fields, are attended by the production of waste materials. This report is aimed at
compiling the experience gained in treating low- and
intermediate-level solid wastes, one of the major
waste sources in nuclear technology. Illustrated with
a bibliography.

Underground Disposal of Radioactive Wastes: Basic Guidance.
Vienna: International Atomic Energy Agency, 1981.

This report gives background guidance on the wide variety of potential waste types and forms, and on the
differing technical and environmental aspects of the
disposal options. Includes a glossary of terms used.

Uranium Mill Tailings Management. Washington, D.C.: Nuclear
    Energy Agency, Organisation for Economic Co-Operation
    and Development, 1982.

The ultimate objective in the management of wastes from
uranium mining and milling is to dispose of them in
such a manner as to provide for the protection of man
and the environment from all potential effects of both
radioactive and non-radioactive contaminants in the
tailings. This report is the proceedings of two workshops which focused on the long term aspects of mill
tailings.

Proceedings

Advances in Ceramics. Vol. 8. Nuclear Waste Management.
    Chicago, Illinois, April 24-27, 1983. Columbus, OH: The
    American Ceramics Society, 1984.

Contains 81 papers presented at the Second International Symposium on Ceramics in Nuclear Waste Management
dealing with all aspects of nuclear waste management in
which ceramics, particularly glass and Synoc, play a
leading role. Included are papers on leachability,
radiation effects on glass, waste forms, and waste form
processing. Includes an author and a subject index.

Barney, G. Scott, et al., eds. Geochemical Behavior of
    Disposed Radioactive Waste. ACS Symposium Series 246.
    Washington, D.C.: American Chemical Society, 1984.

Examines the issues involved in radioactive waste
disposal and the health hazards at underground waste
sites. Assesses the chemical and physical behavior of
wastes from the nuclear fuel cycle, nuclear weapons
testing and from medical and research activities.
Includes an author and a subject index.

Blasewitz, J. M., et al., eds. The Treatment and Handling
    of Radioactive Wastes. Richlad, Washington, April 19,
    1982. New York: Springer-Verlag, 1983.

Contains over 100 papers presented at the American
Nuclear Society Topical Meeting on "The Treatment and
Handling of Radioactive Wastes", covering the treatment
and associated handling of high-level, transuranic and
low-level liquid and solid wastes on an international
scale.

Cecille, L. and R. Simon, eds. The Acid Digestion Process
   for Radioactive Waste. Radioactive Waste Management,
   Volume 11. New York: Harwood, 1983.

This book is the proceedings of a seminar focused on
the acid digestion process for the treatment of alpha
combustible solid waste. Detailed performance figures
(i.e. waste sorting and shredding, wet combustion and
plutonium recovery) for the principal sub-assemblies of
the Alona Demonstration Facility at Eurochemic, Mol,
Belgium are presented as determined on the basis of
inactive test runs and lab-scale action experiments.
Illustrated.

Conditioning of Radioactive Wastes for Storage and Disposal.
   Utrecht, The Netherlands, June 21-25, 1982. Vienna:
   International Atomic Energy Agency, et al., 1983.

Contains the proceedings of an international symposium
on the conditioning of radioactive wastes for storage
and disposal. Major topics included: the technical
bases for conditioning, the immobilization processes,
packaging methods and materials, evaluation of conditioned wastes, quality assurance and cost/benefit
considerations. Includes references and an author
index.

Decommissioning Requirements in the Design of Nuclear
   Facilities. Washington, D.C.: Nuclear Energy Agency,
   Organisation for Economic Co-Operation and Development,
   1980.

This report contains the proceedings of a meeting held
to discuss the views of experts from various disciplines (civil and mechanical design, decontamination,
radiological protection, waste management, and operations) toward those considerations that can be taken at
the design phase of a nuclear facility and which could
have an important influence during decommissioning.
Illustrated with a list of references.

The Dose Limitation System in the Nuclear Fuel Cycle and in
   Radiation Protection. Madrid, Spain, October 19-23,
   1981. Vienna: International Atomic Energy Agency,
   1982.

Contains the proceedings of the International Symposium
on the Application of the Dose Limitation System in
Nuclear Fuel Cycle Facilities and other Radiation Practices held in 1981. The Symposium dealt with: the
application of the dose limitation system to the design
of nuclear facilites, to radioactive waste management
practices, to aspects of monitoring the exposure of

individuals and populations, and to emergency planning
recovery operations. Includes references and an author
index.

Executive Conference on Emergency Preparedness. San
Antonio, Texas, February 10-13, 1980. La Grange Park,
IL: American Nuclear Society, 1980.

Contains 25 papers presented at the ANS Executive
Conference on Emergency Preparedness which focued on
first-hand experience with recent incidents of national
importance and progress in preparedness planning in
local, state and federal government and the nuclear
industry. Includes an author index and a list of
attendees.

Executive Conference on State, Federal, Nuclear Interface.
Monterey, California, February 1-4, 1981. La Grange
Park, IL: American Nuclear Society, 1981.

Contains over 20 papers presented at the ANS Executive
Conference on State, Federal, Nuclear Interface covering cooperation and controversy between state and
federal government and the nuclear industry. Emphasis
is placed upon spent fuel storage, high-level waste
management, emergency preparedness and low-level waste
management. Includes an author index and a list of
attendees.

Fuel Reprocessing and Waste Management. Jackson, Wyoming,
August 26-29, 1984. La Grange Park, IL: American
Nuclear Society, 1984.

Includes 90 papers, two-thirds by authors from outside
the United States, on all aspects of reprocessing,
including: fuel storage and reprocessing and waste
processing. Although spent commercial reactor fuel is
not currently being reprocessed in the U.S., research
and development activities are continuing with vitrifi-
cation of high-level waste contributing the bulk of the
U.S. papers presented.

Hebel, W. and G. Cottone, eds. Management and Modes for
Iodine-129. Radioactive Waste Management, Volume 7.
New York: Harwood, 1982.

This book contains the proceedings of a meeting on
management modes for Iodine-129 held in 1981. Topics
included: where iodine-129 comes from, analysis and
monitoring, removal from liquid and gaseous effluents,
methods of immobilization and the radiological impact
of iodine-129 in case of possible releases and
different disposal options. Includes references.

Helsel, W. and G. Cottone, eds. Conditioning and Storage of
    Spent Fuel Element Hulls. Brussels, Belgium, January
    19, 1982. Radioactive Waste Management, Volume 9. New
    York: Harwood, 1982.

    Contains the proceedings of a meeting held in the scope
    of the R&D Programme of the European Communities on
    Radioactive Waste Management and Storage. The meeting
    was convened to collect the available knowledge and to
    assess the state of research and development towards
    technical solutions of managing spent fuel element
    hulls. Includes references.

Lyon, W. S., ed. Analytical Chemistry in Nuclear Technology.
    Lancaster, PA: Technomic, 1982.

    This is the proceedings of the 25th Conference on
    Analytical Chemistry in Energy Technology. Includes 45
    reports prepared by specialists in nuclear materials
    analysis. Analytical methods, equipment and results as
    they relate to heavy metals, nuclear waste, etc. are
    discussed. Illustrated with references and a subject
    index.

Management of Alpha-Contaminated Wastes. Vienna: Inter-
    national Atomic Energy Agency, 1981.

    This report contains the proceedings of a meeting held
    in 1980 discussing the management of wastes generated
    by spent fuel reprocessing and mixed oxide fuel
    fabrication plants as well as the potential for wastes
    generated by fast breeder reactors. Illustrated, with
    references and author index.

Management of Gaseous Wastes from Nuclear Facilities.
    Vienna: International Atomic Energy Agency, 1980.

    During the course of the nuclear fuel cycle, different
    radioactive wastes and effluents are generated. Mini-
    mizing the release of these airborne radionuclides into
    the environment is important to assure the protection
    of man and his environment. This report contains the
    proceedings of a meeting dealing with this problem.
    Illustrated, with references and an author index.

Navratil, James D. and Wallace W. Schulz, eds. Actinide
    Recovery from Waste and Low-Grade Sources. Radioactive
    Waste Management, Volume 6. New York: Harwood, 1982.

    This book contains a series of papers presented at the
    International Symposium on Actinide Recovery from Waste
    and Low-Grade Sources held in 1981. They provide an
    in-depth coverage on the following topics: uranium
    recovery from oceans and phosphoric acid; recovery of

actinides from solid and liquid wastes; plutonium scrap recovery technology; and other new developments in actinide recovery processes. Includes references and a subject index.

Near-Field Phenomena in Geologic Repositories for Radioactive Waste. Seattle, Washington, August 31-September 3, 1981. Washington, D.C.: Nuclear Energy Agency, Organisation for Economic Co-Operation and Development, 1981.

The disposal of radioactive waste in underground repositories may itself create various phenomena in the local host rock. It is essential to check that such phenomena do not affect the ability of the geologic formation to isolate the radioactive material for very long periods of time. These proceedings contain 28 papers contributed to an NEA workshop on this topic. Illustrated.

Post, Roy G. and M. E. Wacks, eds. Waste Management 75, 78, 79, 80, 81, 82, 83, 84, 85... Tucson, AZ: The Arizona Board of Regents, 1975, et al. Distributed by the American Nuclear Society.

A long running, annual series of symposia dedicated to presenting papers dealing with all aspects of radioactive waste management and the nuclear fuel cycle from an international viewpoint. Each year, the symposium has grown larger with upwards of 200 papers being presented in 1985. Illustrated with an author index.

Proceedings of an International Conference on Radioactive Waste Management. Seattle, Washington, May 16-20, 1983. Vienna: International Atomic Energy Agency, 1984.

Contains in five volumes 146 papers presented at the Symposium. Major topics include: waste management policy and its implementation, waste handling, treatment and conditioning at nuclear facilities, storage and disposal of radioactive waste, environmental and safety assessment of waste management systems, and radioactive releases into the environment from nuclear operations. Not all papers are in English. Contains references and an author index.

Proceedings of the International Conference on Robotics and Remote Handling in the Nuclear Industry. Toronto, Canada, September 23-27, 1984. Toronto: Canadian Nuclear Society, 1984.

Contains 39 papers presented on the methods and reasons why robots and remote handling techniques are used in

all aspects of the nuclear industry: mining, reactor
inspections, power generation and radioactive waste
management. Illustrated.

Proceedings of the Workshop on Geological Disposal of Radio-
    active Waste. Insitu Experiments in Granite. Stockholm,
    Sweden, October 25, 1982. Washington, D.C.: Organisa-
    tion for Economic Co-Operation and Development, 1983.

    In the development of safe methods for disposal of
    highly radioactive wastes from nuclear power genera-
    tion, the International Stripa Project is a key element
    in the in-site experimental programs in granite. These
    proceedings provide an overview of conclusions that can
    be drawn from on-going research at the Stripa mine in
    Sweden and at similar facilities in the OECD area.

Radionuclide Release Scenarios for Geologic Repositories.
    Paris, France, September 8-12, 1980. Washington, D.C.:
    Nuclear Energy Agency, Organisation for Economic Co-
    Operation and Development, 1981.

    The safety of radioactive waste disposal in geologic
    formations cannot be verified experimentally. Safety
    analysis provides the only means to ensure that all
    risks associated with the waste repositories are accep-
    tably low. This is a collection of 15 papers that
    attempt to define radionuclide release scenarios on an
    international scale. Illustrated, with references.

Remote Handling in Nuclear Facilities. Washington, D.C.:
    Nuclear Energy Agency, Organisation for Economic Co-
    Operation and Development, 1985.

    Remote handling equipment, from a simple crane to
    modern robot, plays an important role in all nuclear
    fuel cycle processes. These proceedings provide a
    review of the present status of research and devlopment
    and applications of remote handling in the nuclear fuel
    cycle. Includes references.

Scientific Basis for Nuclear Waste Management IX...
    Pittsburgh: Materials Research Society, 1979, et al.

    Held annually as an international symposium since 1978,
    these annual meetings take place in the United States
    as well as Europe. The reports presented in the series
    give the results of research and development activities
    at universities, government laboratories and private
    industry. Topics range from repository characteriza-
    tion and waste form production to product and perfor-
    mance assessment. Illustrated with an author and a
    subject index.

Siting of Radioactive Waste Repositories in Geological
    Formations. Paris, France, May 19-22, 1981. Washing-
    ton, D.C.: Nuclear Energy Agency, Organisation for
    Economic Co-Operation and Development, 1981.

    The process of siting radioactive waste repositories in
    deep geological formations is a complex one that
    includes a number of phases. The objectives of this
    workshop were: to review the approaches used in NEA
    Member countries for the siting of repositories, to
    review the technical means that can be used to obtain
    information that is required to determine whether a
    particular location is suitable as a disposal site, and
    to discuss possible approaches for effectively
    communicating to the public the results of the site
    characterization process. Illustrated, with
    references.

Organizations

American Nuclear Society
555 N. Kensington Avenue
La Grange Park, IL 60525

    Founded in 1954, ANS includes physicists, chemists,
    engineers, administrators, etc. with professional exper-
    ience in nuclear science and engineering. Its goal is
    to advance science and engineering in the nuclear
    industry by sponsoring research and technical meetings
    as well as by working with governments and any other
    agency dealing with nuclear issues.

Atomic Industrial Forum
4101 Wisconsin Avenue
Bethesda, MD 20814

    Founded in 1953, AIF is a collection of industrial,
    R&D, educational and governmental groups dedicated to
    the use of nuclear power for peaceful uses, and to
    increasing public understanding of nuclear energy.
    Sponsors topical workshops and conferences.

Board on Radioactive Waste Management
National Academy of Sciences
2101 Constitution Avenue
Washington, D.C. 20418

    Founded in 1955, the Board's purpose is to advise the
    federal government on long-range radioactive waste
    management plans and programs. Assesses the adequacy
    of present and projected technology and standards in

meeting long-range health, safety and environmental
aspects of radioactive waste management.

International Atomic Energy Agency
Vienna International Center
P.O. Box 100
Wagamerstrasse 5, Vienna
Austria

Founded in 1957, IAEA is composed of over 100 nation
states dedicated to the research and development of
practical applications of nuclear energy for peaceful
uses. Establishes health and safety standards for each
state to follow. Sponsors technical reports and meetings throughout the world.

National Campaign for Radioactive Waste Safety
P.O. Box 4524
105 Stanford, SE
Albuquerque, NM 87106

A part of the Southwest Research and Information Center,
this group opposes the federal government's current
efforts at radioactive waste disposal. Provides information to citizens groups around the country and maintains an active speakers bureau. They also act as a
clearinghouse for information on radioactive waste
management.

Nuclear Information and Resources Service
1346 Connecticut Avenue, NW, 4th Floor
Washington, D.C. 20036

Purpose is to assist individuals and organizations
interested in and concerned about nuclear issues. Provides information, advice, materials, and speakers to
people trying to stop nuclear activities. Promotes all
alternatives to nuclear power.

Abstracting and Indexing and Online Services

DOE ENERGY data bases. Produced by the Technical Information
    Center, Department of Energy. Available from Dialog
    Information Services.

DOE Energy is a multidisciplinary file containing
references to the world's scientific and technical
literature on energy from 1974 to the present. All
unclassified information processed at the Technical
Information Center, Department of Energy is included in
the data base. DOE ENERGY corresponds in part to

several abstracting journals, including <u>INIS Atomindex</u>
and <u>Energy Research Abstracts</u>. Approximately 50% of
the records in the data base are not announced in the
print formats. Abstracts are included for records
added from 1976 to the present.

<u>Energy Research Abstracts</u>. Washington, D.C.: Superintendent
of Documents, Government Printing Office. Bimonthly,
published since 1976.

ERA provides abstracting and indexing coverage of all
scientific and technical reports, journal articles,
conference papers and proceedings, books, patents,
theses, and monographs originated by the Department of
Energy, its laboratories, energy centers and contractors. ERA also covers other energy information prepared in report form by federal and state government
organizations, foreign governments, and domestic and
foreign universities and research organizations. Each
issue includes a file number correlation index, a
report number index, a contract number index, a subject
index, a personal author index and a corporate author
index.

<u>INIS Atomindex</u>. Vienna: International Atomic Energy Agency.
Bimonthly, published since 1970.

Atomindex is a computer produced bibliography issued by
the International Nuclear Information System (INIS).
It contains bibliographic descriptions and abstracts or
descriptions for books, research reports, patents,
journal articles, conference papers, etc. Each issue
includes a personal author index, a corporate entry
index, a report, standard and patent number index, a
subject index, and a conference index. The references
are in English with the titles also given in the
original language.

<u>Waste Management Research Abstracts (New Series)</u>. Vienna:
International Atomic Energy Agency. Irregular,
published since 1982.

WMRA collects and disseminates information on research
in progress in the field of radioactive waste management. Abstracts intended for inclusion in this publication were sent by IAEA member states, the Commission of
the European Communities and by the Nuclear Energy
Agency, Organisation for Economic Co-Operation and
Development and have been printed in the language
(English, French, Russian or Spanish) and in the form
of submittal without any changes.

10.
Laws and Regulations

Mary Frances Miller

Introduction

Over the past 20 years, there has been an explosion of
legislation and regulation addressing the problems of toxic
and hazardous materials. This proliferation of information
can be traced to the impetus provided by a revived social
consciousness based on the realization that man's impact on
our environment requires a farsighted managerial approach if
our legacy to future generations will be one of enlightened
response.

The purpose of this chapter is to provide a listing of
information sources which orbit the central issue of toxic
and hazardous materials. With this analogy, each individual
listing should be viewed as the nucleus of its own self
directed information storehouse.

Conceptual awareness of the necessity for legislative action
to control all phases of the manufacture, distribution,
disposal and ultimate disposition of these materials culmin-
ated in the environmental social movements of the 1960's.
By the 1970's actual effectual legislation and regulation
had reached deluge proportions. During this period a multitude of new environmental statutes had been passed and
existing legislation amended to oversee the manufacture,
storage, testing, use, transportation and disposal of toxic
and hazardous materials.

Some of the major legislation relative to toxic and hazardous materials heretofore enacted include:

1. Clean Air Act 1963
2. Clean Water Act 1977 (formerly Federal Water
     Pollution Control Act)
3. Comprehensive Environmental Response,
     Compensation, and Liability Act 1980
4. Federal Insecticide, Fungicide, and Rodenticide Act 1975
5. Hazardous Materials Transportation Act 1975

6. Marine Protection, Research and Sanctuaries
   Act 1972
7. National Environmental Policy Act (NEPA) 1969
8. Occupational Safety and Health Act 1970
9. Resource Conservation and Recovery Act 1976
10. Safe Drinking Water Act 1974
11. Toxic Substances Control Act 1976

1. <u>Clean Air Act</u> of 1963 provided for increased federal
   regulatory implementation. States were required to
   attain stringent standards of air quality within a specified time frame. Under the direction of the Environmental Protection Agency national ambient air quality
   standards were established for air pollutants causing
   danger to the public health and welfare.

2. <u>Clean Water Act</u> of 1977 enacted a regulatory program
   for dealing with water pollution. Substantial federal
   funding was provided for construction and operation of
   public water treatment facilities. The act encouraged
   state control over water pollution problems. The year
   1985 was set to mark the elimination of pollutants
   discharged into the nation's waters and subsequently
   restoring and maintaining the integrity of the nation's
   waters.

3. <u>Comprehensive Environmental Response, Compensation, and
   Liability Act</u> (SUPERFUND) 1980 provided federal funding
   to deal with hazardous substances, pollutants or contaminants released into the air, water or land. The EPA
   has the authority to order parties responsible for
   noxious releases to do clean up work if they don't
   voluntarily take such action. If the responsible
   parties fail to comply with the order, the EPA can use
   federal funds to clean up the site and recover expenses
   from responsible parties later.

4. <u>Federal Insecticide, Fungicide, and Rodenticide Act</u> of
   1975 regulates the marketing, registration and labelling
   of all pesticides in the U.S.

5. <u>Hazardous Materials Transportation Act</u> 1975 established
   regulation oversight and review policies improving the
   public's protection regarding the dangers inherent in
   the transportation of hazardous materials.

6. <u>Marine Protection, Research and Sanctuaries Act</u> 1972
   (Ocean Dumping Act) was enacted to regulate ocean
   dumping of materials which would have an adverse effect
   on the marine ecosystem. This act also gave authority
   to the Secretary of Commerce to designate marine
   sanctuaries and to regulate their activities.

7. <u>The National Environmental Policy Act</u> (NEPA), passed in
   1969, marked a beginning of a new consciousness of the
   quality of life. With the implementation of NEPA came
   the responsibility for all agencies of the federal

government to recognize and address the short-term and
long-range effects of environmental decisions on the
quality of the human environment. These considerations
imposed on federal agencies take the form of environmental impact statements and must be prepared for all
actions federally recommended or proposals for legislation that significantly affect the quality of the human
environment.

The concerns to be addressed in each environmental
impact statement are:

a) the environmental impact of the proposed action
b) any adverse environmental effects which cannot be
   avoided should the proposal be implemented
c) alternatives to the proposed action
d) the relationship between local short-term uses of
   man's environment and the maintenance and
   enhancement of long-term productivity
e) any irreversible and irretrievable commitments of
   resources which would be involved in the proposed
   action should it be implemented

NEPA was also responsible for the establishment of the
Council on Environmental Quality which serves as an
advisory group to the President on environmental issues.

8. Occupational Safety and Health Act of 1970 officially
   recognized the federal government's role in assuring
   safe and healthy working conditions for the entire labor
   force. This act allows for the continuous monitoring
   and enforcement of the safety standards developed under
   its authority. In addition, private industry along with
   state, and local government agencies are encouraged to
   participate in the programs conducted through the aus-
   pices of OSHA, promoting information, research, education and training in the field of occupational safety
   and health.

9. Resource Conservation and Recovery Act of 1976 provides
   technical and financial assistance to state and local
   governments and interstate agencies for the development
   of solid waste management. The Act prohibited future
   open dumping on land and provided for the promulgation
   of guidelines for solid waste collection, transport,
   separation, recovery and disposal practices.

10. Safe Drinking Water Act of 1974 specifies standards for
    public drinking water supplies. The states are allowed
    to enforce the standards of acceptable levels of contam-
    inants in drinking water.

11. Toxic Substances Control Act 1976 is aimed at the manu-
    facturing and processing of toxic substances. It strives
    to keep substances that are unreasonable risks to the
    public health and environment off the market. It is
    concerned with the control of new chemical substances
    and new uses of existing chemical substances.

The above represent a selection of major pertinent legislation and a synopsis of each.   Full text of all federal
legislation is contained in the United States Code.   The
U.S. Code is a multi-volume set, published every six years,
containing full text of all laws enforced.   It is codified
into 50 titles or subjects.   Some of the most appropriate
titles for researching toxic and hazardous material legislation are:

Title 7 - Agriculture
Title 15 - Commerce and Trade
Title 16 - Conservation
Title 21 - Food and Drugs
Title 27 - Intoxicating Liquors
Title 29 - Labor (Chapter 15) Occupational Safety & Health
Title 30 - Mineral Lands and Mining
Title 33 - Navigation and Navigable Waters
Title 42 - The Public Health and Welfare
Title 43 - Public Lands
Title 45 - Railroads
Title 46 - Shipping
Title 49 - Transportation

The U.S. Code also has a General Index which contains an
Index to Acts by Popular Name. Once a year the U.S. Statutes
at Large, a chronological compilation of laws passed by
Congress, are published.

Once Congress has passed a law it is up to the agency
charged with the implementation to develop the regulations.
There are two major sources of federal administrative
regulations: 1) Federal Register and 2) Code of Federal Regulations.   The Federal Register is published daily, except
weekends and official holidays. It contains all rules, regulations and amendments promulgated by the various government
agencies which have the force of law.   The Code of Federal
Regulations is analogous to the U.S. Code in that it is a
subject arrangement of the Federal Register. It is divided
into 50 titles similar to the U.S. Code.

Loose Leaf Services

Due to the nature of the law and the constant changes and
revisions that take place, the loose leaf is an efficient
format to keep material up to date.   Included in this section are both traditional loose leaf services and treatises
published in loose leaf form.

Air Pollution Control.   Bureau of National Affairs. Periodic
    supplements, published since 1980.

A reference manual consisting of one binder and biweekly bulletin. Provides guidance on complying with
state and federal air pollution standards. Coverage is
in nontechnical terminology and includes practical features such as glossaries, conversion equations and sample forms. Also contains information on trends in air
pollution control developments, policies, strategies
and techniques.

Air Pollution: Federal Law and Analysis. Callaghan & Co.
Periodic supplements, published since 1983.

Asbestos - Federal & State Regulations. Asbestos Information. Periodic supplements, published since 1978.

Cal-OSHA Reporter. Sten-O-Press. Weekly, published since
1973.

This looseleaf contains short articles on OSHA regulations and other environmental laws. Decisions of the
Cal-OSH Appeals Board are indexed by code section
number, name, and subject.

Chemical Regulation Reporter.   Bureau of National Affairs.
Weekly supplements, and reports, published since 1977.

A weekly service providing text of regulations and
policies, RPAR notices and rules and guidelines on
chemical testing, manufacture, transporting hazardous
materials, administration and use of pesticides and new
and existing chemicals. Issued in five binders including separate binders for Index to Government Regulations and Hazardous Materials Transportation.

Chemical Substances Control. Bureau of National Affairs.
Periodic supplements, published since 1980.

A one-binder quick-reference manual. Provides practical guidance for complying with law and rules governing
pesticides, toxic, hazardous and other chemical substances. Coverage spans from pre-manufacture and testing through disposal. Includes international requirements and handling emergency situations.

Coal Law and Regulation. Matthew Bender & Co., Inc.   Periodic supplements, published since 1983.

A five-volume treatise covering issues such as mine
health and safety, environmental impacts and the Clean
Air Act. Includes step-by-step analysis of procedures
required by state and federal regulations.

Employment Safety And Health Guide. Commerce Clearing House.
   Weekly supplements and newsletter, published since
   1971.

   Explains the practical applications of the 1977 Mine
   Safety and Health Act and the Occupational Safety and
   Health Act in nontechnical language. Weekly reports
   provide coverage of OSHA proposals, laws, regulations,
   decisions, rulings, standards and research studies.

Environment Reporter. Bureau of National Affairs. Weekly
   supplements and reports, published since 1970.

   A 17-volume weekly loose-leaf service covering the
   major environmental issues. Contains full text of
   federal and state laws and regulations. A comprehensive reference source with a cumulative digest index of
   cases covered since 1970.

Environmental Law Reporter. Environmental Law Institute.
   Monthly supplements, published since 1970.

   A comprehensive reporter service consisting of five
   loose leaf binders: 1) ELR Indexes; 2) Volume Year;
   3) Pending Litigation; 4) Statutes; 5) Regulations.

   The Pending Litigation volume summarizes developments
   and ongoing court actions providing a timely and unique
   service. Both the Statutes and Regulations binders
   contain full text of selected federal statutes and
   regulations as well as international agreements and
   Executive Orders. Another special feature is the
   Annual Bibliography, a comprehensive listing of law
   school reviews and legal journal articles dealing with
   environmental law and policy.

Federal Regulation of the Chemical Industry. Shepard's/
   McGraw-Hill. Annual supplements, published since 1980.

   A practical reference source explaining the law and
   federal regulations governing the chemical industry.
   Examines each area affected by federal regulations and
   advises how to comply with the requirements.

Food Drug Cosmetic Law Reporter. Commerce Clearing House,
   Inc. Weekly supplements, published since 1938.

   A six-volume set offering full text of state and federal statutes, regulations and court decisions relating
   to food, drug and cosmetic law problems. Included in
   the subject content are additives and adulteration of
   food, pesticides in foodstuffs, packaging and labeling
   of food, drugs and cosmetics.

Hazardous Materials Guide. J. J. Keller Associates, Inc.
    Semiannual supplements, published since 1976.

    A one-volume reference and operational tool designed to
    answer questions about shipping, handling and transpor-
    tation of hazardous materials. Provides complete Hazar-
    dous Materials Regulations as issued by the MTB and ex-
    plains how to comply with each portion of the rules and
    how to use the commodity list of materials.

Hazardous Materials Transportation. Bureau of National
    Affairs. Monthly supplements, published since 1977.

    See the chapter on Transportation of Hazardous
    Materials for complete description.

Hazardous Waste Regulatory Guide. J. J. Keller Associates,
    Inc. Periodic supplements, published since 1982.

    A one-volume, easy to use guide detailing information
    on identifying, monitoring, treating and disposing of
    hazardous wastes as defined by the state agency respon-
    sible for hazardous waste administration.

Hazardous Wastes Handbook. 5th ed. Government Institute.
    Supplements planned, published since 1982.

    A one-volume reference source in non-legal terminology.
    Provides a step-by-step approach to the EPA hazardous
    waste regulations. Analyzes the impact of RCRA and
    offers suggestions on efficient and cost-effective
    compliance with the regulations.

Hazardous Wastes Management Guide. J. J. Keller Associates,
    Inc. Semiannual supplements, published since 1980.

    A comprehensive guide outlining EPA's regulations con-
    cerning the identification, monitoring, treatment and
    disposal of hazardous wastes. Includes EPA regula-
    tions, a state regulations synopsis and reference
    sources.

Index to Government Regulation. Bureau of National Affairs.
    Weekly supplementation, published since 1979.

    A one-volume reference source covering chemicals feder-
    ally regulated. Contains citations to the Code of
    Federal Regulations and Federal Register announcements.
    Pertinent regulations to chemicals are accessible by
    generic name, industrial use name or Chemical Abstract
    Service Registry Number.

International Environment Reporter. Bureau of National
    Affairs. Monthly supplements, published since 1978.

    Reports pollution control and environmental protection
    policies of 13 countries including Japan, Canada and
    members of the European Economic Community. Features
    programs, policies and directories of the United
    Nations and other international organizations.

International Hazardous Materials Transport Manual. Bureau
    of National Affairs. Monthly supplements, published
    since 1983.

    See the chapter on Transportation of Hazardous
    Materials for complete description.

Nuclear Regulation Reports.   Commerce Clearing House, Inc.
    Weekly supplements, published since 1975.

    A two-volume loose-leaf service publishing rules for
    development, production, licensing, processing and use
    of nuclear energy and its byproducts and waste disposal.

OSHA Compliance Guide. Commerce Clearing House, Inc. Monthly
    supplements, published since 1977.

    A one-volume source which thoroughly explains federal
    and state statutory provisions, standards, regulations
    and interpretations. A separate monthly newsletter
    highlights current changes and views notes about pend-
    ing and upcoming developments.

Occupational Exposure Guide. J. J. Keller Associates, Inc.
    Semiannual supplements, published since 1979.

    A one-volume reference source thoroughly covering the
    identification, classification and regulation of work-
    place hazards caused by exposure to hazardous sub-
    stances. Contains explanations of OSHA rules for
    record-keeping, access to medical records, exposure
    limits, labeling and signs and employee hazard notifi-
    cation. Also includes reference data on suspected
    carcinogens, enforcement officials, and compliance
    products and services.

Occupational Safety and Health Act.   Matthew Bender & Co.,
    Inc. Annual supplements, published since 1977.

    Two volumes containing texts of OSHA, Senatorial Commit-
    tee Reports, OSHRC Rules of Procedure, citations to the
    statute, regulations, cases and literature of OSHA.

Occupational Safety and Health Reporter. Bureau of National
    Affairs. Weekly supplements, published since 1971.

    A comprehensive reference source offering practical
    guidance for compliance with changing requirements in
    the workplace. Includes coverage of the private sector
    and state and federal governments.

Pesticides Guides. J. J. Keller Associates, Inc. Semi-
    annual supplements, published since 1979.

    A one-volume reference tool providing comprehensive
    coverage of the EPA regulations governing pesticides.
    Included in the guide are the rules for registering
    pesticides with the EPA, label requirements, permits
    for experimental use and recordkeeping mandates, com-
    plete regulations as issued by the EPA, required forms,
    pesticide listings and program administration proce-
    dures.

Pollution Control Guide. Commerce Clearing House, Inc.
    Weekly supplements and newsletter, published since
    1973.

    A seven-volume set which provides coverage on statutes,
    regulations and court decisions on toxic substances,
    air and water pollution, radiation, pesticides, solid
    waste disposal and noise. A weekly newsletter high-
    lights current and new developments.

Sewage Treatment Construction Grants Manual. Bureau of
    National Affairs. Monthly supplements, published
    since 1976.

    A guide to acquiring and administering federal con-
    struction grants. Provides full text of official EPA
    handbooks, federal statutes, pertinent regulations, EPA
    policy and directives and guidelines. Kept current
    with monthly reports.

Toxic Substances Control Guide. J. J. Keller Associates,
    Inc. Semiannual supplements, published since 1979.

    A reference tool dealing with the safe and legal hand-
    ling of toxic substances. Provides guidelines for
    compliance with the Toxic Substances Control Act regu-
    lations. EPA-issued regulations and practical reference
    information on government interagency controls on toxic
    substances are also included.

Treatise On Environmental Law. Matthew Bender & Co., Inc.
    Annual supplements, published since 1973.

A five-volume treatise covering such environmental
issues as air and water pollution, radiation, pesticide
pollution, toxic substances and toxic waste, fertilizer
and feed lot pollution and the National Environemntal
Policy Act. Analyzes federal and state statutes, explains how to comply with regulations and cites cases.

Water Pollution Control. Bureau of National Affairs. Bi-
    weekly supplements and bulletin, published since
    1979.

    A one-volume quick reference source. Explains clearly
    how to comply cost-effectively and efficiently with
    federal, state and local water pollution control re-
    quirements.

Journals and Law Reviews

Asbestos Litigation Reporter. Andrews Publications, Inc.
    Bimonthly, published since 1979.

Atomic Energy Law Journal. Invictus Publications Co.
    Quarterly, published since 1959.

Boston College Environmental Affairs Law Review. Boston
    College Law School. Quarterly, published since 1971.
    (Title varies; v. 1-6, 1971-77, Environmental Affairs)

CELA Newsletter. Canadian Environmental Law Association.
    Bimonthly, published since 1976.

Canadian Environmental Law Reports. Canadian Environmental
    Law Research Foundation. Bimonthly, published since
    1972.

Chemical and Radiation Waste Litigation Reporter. Law
    Reporters. Monthly, published since 1981.

Columbia Journal of Environmental Law. Columbia University,
    School of Law. Semiannually, published since 1974.

EDF Letter. Environmental Defense Fund. Bimonthly, pub-
    lished since 1969.

EIS Annual Review. Information Resources Press. Annually, published since 1978.

EIS: Digests of Environmental Impact Statements. Information Resources Press. Monthly, published since 1977.

ELI Associates Newsletter. Environmental Law Institute. Bimonthly, published since 1981.

Ecology Law Quarterly. University of California, Berkeley School of Law. Quarterly, published since 1971.

Environment Regulation Handbook. EIC Intelligence. Monthly, published since 1977.

Environmental Forum. Environmental Law Institute. Monthly, published since 1982.

Environmental Law. Lewis & Clark College, Northwestern School of Law. Quarterly, published since 1970.

Environmental Law. American Bar Association, Standing Committee on Environmental Law. Quarterly, published since 1972.

Environmental Law Newsletter. State Bar of Texas. Quarterly, published since 1970.

Environmental Protection Report. National Association of Attorneys General. Monthly, published since 1980.

Food, Drug, Cosmetic Law Journal. Academic Press, Inc. Quarterly, published since 1946.

From the State Capitals - Waste Disposal and Pollution Control. Wakeman Walworth, Inc. Monthly, published since 1945.

Harvard Environmental Law Review. Harvard University Law School. Semiannually, published since 1976.

Hazardous Materials/Hazardous Waste Legal Reporter. Business and Legal Reports. Weekly, published since 1983.

Hazardous Waste Litigation Reporter. Andrews Publications, Inc. Bimonthly, published since 1980.

Inside E.P.A. Weekly Report. Inside Washington Publishers. Weekly, published since 1980.

Journal of Planning and Environmental Law. Sweet and Maxwell. Monthly, published since 1948.

Land Use and Environment Law Review. Clark Boardman Co. Annually, published since 1970. (Title varies; v. 1-8, 1970-77, _Environment Law Review_.)

New York Sea Grant Law and Policy Journal. New York Sea Grant Institute. Annually, published since 1976.

Occupational Health and Safety Law. Business Law Reporting Services. Monthly, published since 1977.

Outlook Environmental Law and Technology Journal. Temple University, School of Law. Annually, published since 1982.

Stanford Environmental Law Annual. Stanford Environmental Law Society. Annually, published since 1978.

State Regulation Report; Toxic Substances & Hazardous Waste. Business Publishers, Inc. Bimonthly, published since 1981.

Toxic Chemicals Litigation Reporter. Andrews Publications, Inc. Biweekly, published since 1983.

UCLA Journal of Environmental Law and Policy. UCLA School of Law, Environmental Law Society. Semiannually, published since 1980.

Virginia Journal of Natural Resources Law. University of Virginia School of Law. Semiannually, published since 1980.

Data Bases

There are two data bases which provide full text legal
research of federal and state statutes, cases and some
regulations. The high cost of these specialized services
frequently restricts user accessibility to law school
libraries, law firms and county libraries.

LEXIS
Mead Data Central
200 Park Avenue
New York, NY 10017   (212) 883-8560

WESTLAW
West Publishing Company
50 W. Kellogg Boulevard
P.O. Box 3526
St. Paul, MN 55165   (618) 228-2500

National and International Organizations

Center for Study of Responsive Law
P.O. Box 19367
Washington, D.C. 20036   (202) 387-8030

> An independent nonprofit organization concerned with,
> among other issues, toxic and hazardous waste and the
> response of government and business institutions to the
> needs of citizens and consumers. Conferences held
> intermittently. Publishes research results.

Citizens for a Better Environment
59 E. Van Buren
Suite 2610
Chicago, IL 60605   (312) 939-1984
                                  Research   (312) 939-1530
Other offices:
111 King Street, Madison, WI 53703
(608) 251-2804

536 W. Wisconsin Ave. #502, Milwaukee, WI 53203
(414) 271-7475

381 Park Ave. S, New York, NY 10004 c/o INFORM
(212) 689-4040

88 First Street, San Francisco, CA 94105
(415) 777-1984

A nonprofit environmentalists group active in research
and litigation on air and water pollution problems,
energy issues, and toxic substances control. Publishes
CBE Environmental Review.

Common Cause
Box 220
Washington, D.C. 20044   (202) 833-1200

Strives to advance state and federal government operations. Its current emphasis is on the Clean Air Act.

Energy Law Institute
4995 Glenway Avenue
Cincinnati, OH 45238   (513) 921-1853

Attorneys and interested individuals who act as a
clearinghouse and consulting agency regarding energy
law issues. Will provide assistance to state and
local government agencies.

Environmental Action
1346 Connecticut Ave., NW
Suite 731
Washington, D.C. 20036   (202) 833-1845

Citizens' lobbying organization working to promote
environmental legislation.

Environmental Action Foundation
724 Dupont Circle Building
Washington, D.C. 20036   (202) 296-7570

This national nonprofit public interest organization
specializes in providing technical and organizational
know-how to local activist groups. Its activities
include waste and toxic materials projects, utility
projects, participation in Congressional hearings and
educational activities.

Environmental Defense Fund, Inc.
444 Park Avenue S.
New York, NY 10016   (212) 686-4191

A national public interest group seeking to protect the
well-being of both human life and wildlife, through
educational programs and selective litigation in opposition to policies adversly affecting the quality of
air and water.

Environmental Law Institute
1346 Connecticut Ave., NW
Suite 600
Washington, D.C. 20036   (202) 452-9600

Founded in 1970, this non-profit institute is a national center for policy research in the fields of environmental protection, natural resources and pollution
control. The Institute has been especially active in
publishing, research and education in areas of air and
water pollution control, toxic substances and hazardous
wastes and environmental economics.

The ELI Library is a specialized collection of thousands of books, periodicals, reports and data bases,
professionally staffed and available by appointment.

The educational branch of the Institute conducts conferences, seminars and training courses. Each year the
Institute holds programs on environmental law, air and
water pollution law, and toxic substances and hazardous
wastes with the co-sponsorship of the American Law
Institute and American Bar Association. The ELI
"Roundtable" conducts programs on toxic torts, regulatory reform, and the Toxic Substances Control Act.

The Institute created the ELI Associate Program to
provide a national network of environmental professionals a forum for sharing a dedication to the improvement
of research data, policy analysis, and public discussion of environmental issues.

Food and Drug Law Institute
1200 New Hampshire Avenue, NW
Suite 380
Washington, D.C. 20036   (202) 833-1601

An organization comprised of representatives from the
food and drug manufacturers sector. Its goal is to
disseminate information about new laws and regulations
throughout the food and drug manufacturing industry.

Government Institutes, Inc.
966 Hungerford Drive, #24
Rockville, MD 20850   (301) 251-9250

Founded in 1973, provides continuing education in the
fields of energy and environment. Offers books,
courses, and in-house seminars geared to the professional. The Institute aims to provide the professional
with information needed to understand and comply with
laws and regulations in a practical and cost effective
manner.

Harvard Environmental Law Society
Harvard Law School
Cambridge, MA 02138   (617) 495-3125

Concerned law students representing the vanguard of
active research for many issues concerning the environment. Sample projects include toxic wastes, land-use
planning, nuclear power and wilderness preservation.
Provides advice, research and assistance services in
the formulation of proposed legislation and grass roots
lobbying.

Institute of International Law
22 Avenue William-Favre
CH-1207 Geneva
Switzerland

An appointed membership body of professionals in the
field of international law. A major issue within this
group is the promotion of international pollution control laws.

International Council of Environmental Law
Adenauerallee 214
D-5300 Bonn
Federal Republic of Germany

An international organization of professionals interested in exchanging a wide variety of information in
the field of environmental law, policy and administration. Publications include: References, 20/yr.;
Directory, semiannual; and a quarterly periodical
Environmental Policy and Law, English and French
editions.

International Nuclear Law Association
Square de Meeus 29
B-1000 Brussels
Belgium

Specifically interested in addressing the legal ramifications associated with the peaceful development of the
nuclear power industry. Ultimately concerned with the
protection of life and the environment as related to
the use of nuclear energy.

Methyl Chloride Industry Association
1075 Central Park Avenue
Scarsdale, NY 10583   (914) 725-1492

Founded in 1981, this association has primary interest
in industry needs regarding government regulation of
methyl chloride (chloromethane). Prepares regulations
and comments and represents the industry at hearings.

Natural Resources Defense Council, Inc.
122 East 42nd Street
New York, NY 10168   (212) 949-0049

The membership of this nonprofit organization includes
a complement of lawyers, scientists and various professionals dedicated to protecting our natural resources
and insuring the quality of life in the United States.
Specifically interested in the subjects of air and
water pollution, nuclear safety, public lands management, transportation regulations, noise pollution, and
environmental carcinogens. This group encourages a
close watch on governmental actions and court decisions
while monitoring scientific research projects and maintaining public awareness of these issues.

Natural Rights Center
156 Drakes Lane
Summertown, TN 38483   (615) 964-3992

A volunteer membership organization specifically concerned with the protection of wilderness areas and
human environment. Has resolved to address the issue
of industrial "cost benefit analysis" vs. a consideration of the effects these processes will have on the
well being of future generations. Has participated in
litigation aimed at insuring constitutional protection
regarding the enforcement of environmental laws.

Federal Regulatory Agencies

Agency for Toxic Substances and Disease Registry
1600 Clifton Road, NE
Atlanta, GA 30333   (404) 452-4111

In cooperation with state, federal and local officials,
the mission of the agency is to provide leadership and
direction to programs relating to the protection of the
public and workers from exposure to hazardous substances in storage sites, released in fires or explosions, and transportation accidents. ATSDR maintains a
listing of restricted areas contaminated by toxic
substances and develops procedures for determining the
public health risk with respect to the proximity of
toxic or hazardous materials.

Consumer Product Safety Commission
1111 Eighteenth Street, NW
Washington, D.C. 20207   (202) 634-7740

The Consumer Product Safety Commission has among its
responsibilities the challenge of safeguarding the
public from the risks of injury associated with
potentially hazardous consumer products.

Department of Justice
Land and Natural Resources Division
Tenth Street & Pennsylvania Avenue, NW
Washington, D.C. 20530   (202) 633-2701

Included among the various duties of the Land and
Natural Resources Division is the responsibility for
the conduct of law suits, both federal and state,
relating to the protection of the American environment.
Civil and criminal actions encompassing air, water and
noise pollution, hazardous waste, and wildlife laws may
be represented by this agency.

Environmental Protection Agency
401 M Street, SW
Washington, D.C. 20460   (202) 382-2090

The purpose of the Environmental Protection Agency is
to control and find solutions for ending pollution in
the areas of air, water, solid waste, pesticides,
radiation and toxic substances.

Federal Maritime Commission
1100 L Street, NW
Washington, D.C. 20574   (202) 523-5707

Food and Drug Administration
5600 Fishers Lane
Rockville, MD 20857   (301) 443-3380
                                      or (301) 443-4177

The Food and Drug Administration is assigned the task
of protecting the national health against impure and
unsafe foods, drugs and cosmetics.

Interstate Commerce Commission
12th Street & Constitution Avenue, NW
Washington, D.C. 20423   (202) 275-7252
    or
Office of Compliance and Consumer Assistance
                                          (202) 275-0860

Mine Safety & Health Administration
Department of Labor
4015 Wilson Boulevard, Room 601
Arlington, VA 22203   (703) 235-1452

It is the reponsibility of the Mine Safety and Health
Administration to develop and regulate mine safety and
health standards aimed at preventing accidents and
occupational diseases in the mining industry.

National Oceanic & Atmospheric Administration
14th Street & Constitution Avenue, NW
Washington, D.C. 20230   (202) 377-4190

Nuclear Regulatory Commission
1717 H Street, NW
Washington, D.C. 20555   (301) 492-7715

Occupational Safety & Health Administration
Department of Labor
200 Constitution Avenue, NW
Washington, D.C. 20210   (202) 523-8151

OSHA develops and promulgates occupational safety and
health standards. Its policy includes the issuance of
minimum standard and regulations and it has the
authority to conduct investigations and inspections to
determine compliance with its mandates. Its enforcement procedures include citations and penalties for
noncompliance offenders.

Special Libraries & Information Centers

College of William and Mary
Marshall-Wythe Law Library
Williamsburg, VA 23185   (804) 253-4680

Special Collection: Environmental Law

Holdings: 120,075 volumes
          46,919 volumes in microform
          167 audiovisual items

Subscriptions: 3792 journals and serials
               10 newspapers

Services: Interlibrary loans; copying; LEXIS

Library open to the public

Department of Justice
Land & Natural Resources Division Library
10th & Pennsylvania Ave., NW
Room 2333
Washington, D.C. 20530   (202) 633-2768

Subjects: Civil cases regarding lands, titles, water
rights, Indian claims, hazardous waste, public works,
pollution control, marine resources, fish and wildlife,
environment.

Special collections: Legislative histories (55)

Holdings: 13,300 volumes
            900 records and briefs

Subscriptions: 160 journals and serials

Services: Interlibrary loans

Computerized Information Services: JURIS, DIALOG, SDC,
LEXIS, New York Times Information service, Legi-Slate,
Dow Jones News Retrieval

Environmental Law Institute - Library
1346 Connecticut Avenue, NW
Suite 620
Washington, D.C. 20036   (202) 452-9600

Special Collection: Federal Environmental Impact
                    Statement (9,000)

Holdings: 5,000 volumes

Subscriptions: 255 journals and serials
               5 newspapers
               5 vertical file drawers of clippings,
               pamphlets and unpublished papers
               150 computer data bases

Library open to the public by appointment

Environmental Protection Agency
National Enforcement Investigations Library
Denver Federal Center
Building 53, Box 25227
Denver, CO 80225   (303) 234-5765

Subjects: Air pollution, enforcement of EPA guidelines
            on water pollution, pesticides, toxic sub-
            stances and effluent.

Holdings: 1,500 books
            150 periodical titles
            8,000 reports

Services: Reference, reproduction services, inter-
            library loan.

Computerized Information Services: Access to more than
150 computerized data bases through Lockheed DIALOG,
SDC, ORBIT, NLM, RECON, New York Times Information

Bank, Dow Jones News Retrieval System, WESTLAW, and
NIH-EPA Chemical Information System.

Library Services are primarily for enforcement
personnel.

Nuclear Regulatory Commission
Law Library
STOP 555
Washington, D.C. 20555   (301) 492-7584

Library Location:
7735 Old Georgetown Road
Bethesda, MD 20014

> Subjects: Law - nuclear energy, environmental,
>   administrative.

> Special Collections: Publications of the Joint
>   Committee on Atomic Energy
>   1945-1975 (complete set);
>   AEC and NRC Reports 1956 to
>   present

> Holdings: 10,000 books
>   1,500 bound volumes
>   500 technical reports
>   10 drawers of Federal Register,
>   Congressional Record and law journals in
>   microform

> Subscriptions: 113 journals and serials

> Services: Interlibrary loan

> Computerized Information Services: LEXIS, TERA
>   (internal data base)

> Special Catalogs: Administrative Law Article File
>   (card and loose-leaf)

> Special Indexes: Significant Federal Court Case File -
>   AEC and NRC issues (loose leaf)

Press Relations Staff
Associate Commissioner for Legislation and Information
Food and Drug Administration
5600 Fishers Lane, Room 15B-42
Rockville, MD 20857   (301) 443-3285

> Special Collection: Federal regulations concerning
>   efficacy, labeling, manufacture,
>   packaging, safety, toxicity,
>   adverse or side effects, and con-
>   taminants of food, drugs, cosme-
>   tics, biological products, diagnos-

tic products, medical devices, and
ionizing and nonionizing radiation-
emitting products and substances.

Holdings: Access to current information on press
releases, recalls, FDA Consumer indexes, drug
approval lists, Federal Register summaries,
and import detention summaries. Publications
include the FDA Consumer (monthly), enforcement reports, press releases, fact sheets,
reprints, informational and educational
materials. A list of publications is available. Services provided for the media only.

Stanford Environmental Law Society
2575 Sand Hill Road
Menlo Park, CA 94305   (415) 854-3300

Subjects: Water law, resource management, toxic waste
disposal, land use, beverage container laws, smoking
regulations

Holdings: Figures not available

Library open to the public

Research Centers

Congressional Information Service, Inc.
4520 East-West Highway, Suite 800
Bethesda, MD 20814   (301) 654-1550

CIS attempts to provide access to all the publications
of Congress, with the exception of the Congressional
Record. Among the types of Congressional publications
that CIS makes available are hearings, committee
prints, House and Senate Reports, House and Senate
Documents, Special Publications, and Public Laws.
Access to documents is gained through the CIS/ Index, a
monthly service, which catalogs, abstracts, and indexes
Congressional publications issued during the previous
month.

University of Colorado
Rocky Mountain Mineral Law Foundation
Fleming Law Building, B405
Boulder, CO 80309   (303) 449-0943

Research interests include: environmental law, mineral
and natural resource law.

11.
Transportation of Hazardous Materials

James K. Webster

Introduction

The transporting of hazardous, toxic, and dangerous
materials is a subject that impinges on several other
chapters in this book. Most wastes, whether hazardous or
not, have to be transported to a disposal site. There are
more and more local, state, federal and international
regulations that govern the vehicles and containers that
these materials are shipped in. And inevitably, there will
be accidents and spills that will require emergency measures
and cleanup.

In this chapter, I have attempted to bring together all of
those publications and other resources that deal predominately with transportation, but the reader should be aware
that Chapter 10, Laws and Regulations, and Chapter 12,
Accidents, Spills and Cleanup will include information about
transportation as it relates to that chapter topic.

There are a number of items mentioned in this chapter that
deserve special comment. First of all, the reader should
note the three-volume set of the Code of Federal Regulations
that deals with hazardous materials transportation. Many
other publications cited here are derived from and based on
these regulations. Perhaps the most current information on
this subject may be found in Conference on Recent Advances
in Hazardous Materials Transportation Research: An International Exchange. This TRB special report is scheduled for
publication in April 1986. Also, the entry in Books,
Reports & Monographs that describes the Proceedings of the
7th International Symposium on the Transportation of
Dangerous Goods by Sea and Inland Waterways includes some
valuable information about the 9th Symposium in this series,
to be held in the Netherlands in April 1987.

The loose-leaf services produced by J. J. Keller and the
Bureau of National Affairs should be particularly useful to
those people who have a continuing concern with this

subject. The subscription costs are high, but the continuous updating features make them especially attractive in some situations.

For those who have computer search services available, the TRIS data base is the best source for bibliographic literature searches.

Most of the items mentioned in the sections on Audiovisual Materials and Seminars & Workshops originated as or from training programs, but they could be very useful as sources of information under some circumstances.

A number of government organizations, associations, and research centers are listed. The four components of Department of Transportation are all important sources; many of their recent publications are described elsewhere in the chapter. Among the associations mentioned, the **Association of American Railraods, Bureau of Explosives** and the **Transportation Research Board** are deeply involved in hazardous materials transportation, and have produced a number of publications and other materials.

Finally, the reader should note the somewhat unusual sources listed in the Companies & Industries section. J. J. Keller & Associates and the Union Pacific System both have much to offer in the way of publications and information.

Books, Reports, & Documents

This section lists printed monographic material--books, documents, conference proceedings, etc. Annotations and supplementary information are provided whenever possible.

Bierlein, Lawrence. **Hazardous Materials: A Guide for State and Local Officials**. Washington, D.C.: Department of Transportation Report No. DOT-I-82-2 (PB82-201203), Feb. 1982.

> Intended as a guide to the federal regulatory program on the transportation of hazardous materials for State and local officials. This guide begins by defining the differences between hazardous materials, substances and wastes. The roles of the various DOT agencies involved with regulatory and enforcement activities are outlined, as well as their relationship with other government entities. A large section is devoted to the regulatory process, current regulations, the regulated parties, and their basic obligations. Also included are sections on classification, packaging, package markings, labeling, documentation, and placarding.

Bureau of Explosives Tariff. BOE-6000-D. Washington, D.C.:
     Association of American Railroads, Bureau of Explo-
     sives, May 1984.

     This publication not only fulfills an ICC requirement
     that carrier rules be published in tariff format, but
     permits the Bureau to maintain the DOT hazardous
     materials regulations on a more up-to-date basis than
     is possible with the Code of Federal Regulations.
     Available bound or as a looseleaf service.

Code of Federal Regulations, Title 49: Transportation.
     Washington, D.C.: Government Printing Office, 1983.

     The DOT hazardous materials regulations are published
     in 3 volumes described below.

     Parts 100-177 (S/N 022-003-95266-7)
          100 Office of Transportation Security-
              Cargo security advisory standards
          106 Rulemaking procedures
          107 Hazardous materials program procedures
          171 General information, regulations, and
              definitions
          172 Hazardous materials table and hazardous
              materials communications regulations
          173 Shippers-General requirements for shipments and
              packagings
          174 Carriage by rail
          175 Carriage by aircraft
          176 Carriage by vessel
          177 Carriage by public highway

     Parts 178-199 (S/N 022-033-95267-5)
          178 Shipping container specifications
          179 Specifications for tank cars
          190 Pipeline Safety Program Procedures
          191 Transportation of natural and other gas by
              pipeline; reports of leaks
          192 Transportation of natural and other gas by
              pipeline; Minimum Federal Safety Standards
          193 Liquefied natural gas facilities; Federal Safety
              Standards
          195 Transportation of liquids by pipeline

     Parts 300-399 (S/N 050-001-00279-5)
          Including: Federal Motor Carrier Safety Regulations.

Community Teamwork: Working Together to Promote Hazardous
     Materials Transportation Safety; A Guide for Local
     Officials. Washington, D.C.: Department of Transporta-
     tion, Research & Special Programs Administration,
     Materials Transportation Bureau, May 1983.

     This Guide is designed to provide ideas on how to
     develop a hazardous materials transportation safety

program at the most economical cost. Examples are
presented of different state and local agencies sharing
the cost of providing labor, equipment and materials.
Ways in which private industry have supported state/
local safety programs are illustrated. Publications
and other sources of information are also identified.

Conference on Recent Advances in Hazardous Materials Transportation Research: An International Exchange. Lake
Buena Vista, Florida, November 10-13, 1985. State of
the Art Report 3. Washington, D.C.: National Research
Council, Transportation Research Board, 1986.

Cosponsored by the Federal Highway Administration and
the Federal Emergency Management Agency. Topics covered
include: vehicle design, packages and containers,
vehicle operation, information technology, management
systems, routing, risk assessment, hazardous waste,
release effects, emergency response, emergency preparedness, statistical analysis, planning, techniques,
research methods, materials technology, accident
investigation, and intermodal aspects.

Emergency Action Guides. Washington, D.C.: Association of
American Railroads, Bureau of Explosives, Oct. 1984.

The railroad industry's newest emergency response
publication. Covers 134 chemicals that comprise 98% of
the volume of hazardous materials carried by U.S. railroads. Data reported for each chemical include:
General Hazards--threshold odor concentration, STEL,
TLV, conditions to avoid; Health Hazards--public
health, skin, eye contact, inhalation, ingestion; Fire
Hazards--lower and upper flammable limits, behavior in
fire, hazardous combustion products; Explosive Hazards;
Protective Clothing and Equipment; First Aid; Fire
Response; Spill Response; Air Spills; Land Spills;
Water Spills; NFPA Hazard Diamond. Other sections
include: (1) a brief overview of the chemical emergency
preplanning process, (2) sources of additional aid, (3)
chemical regulation, (4) identification and classification, including cross references for U.N. identification numbers and the AAR's 49 STCC codes numbering
series, and (5) a glossary of technical terms used in
the guide.

Emergency Handling of Hazardous Materials in Surface
Transportation. Washington, D.C.: Association of
American Railroads, Bureau of Explosives, 1981.

Gives commodity specific information on over 2400
hazardous materials, including the basic properties of
the chemical and the recommended way to deal with it in
the early stages of an emergency. Also lists emergency

environmental mitigation procedures for each EPA-named
hazardous substance.

This publication is also available as a microcomputer
software program, using either CP/M or MS-DOS operating
system, minimum 64K RAM and hard disk storage. For
information contact:

> Customized Business Control, Inc.
> 2827 Truman Drive
> Hatfield, PA 19440
> (215) 362-2123

Emergency Procedures for Ships Carrying Dangerous Goods.
London, England: International Maritime Organization,
1981. (Available from UNIPUB).

Emergency Response to Hazardous Materials in Transportation,
Self-Study Guide. Publication SPP-70. Quincy, MA:
National Fire Protection Association, 1982.

Explosives Motor Vehicles Terminals. Standard NFPA 498.
Quincy, MA: National Fire Protection Association, 1982.

Guidance Manual for Operators of Small Gas Systems.
Washington, D.C.: Department of Transportation,
Research & Special Programs Administration, Materials
Transportation Bureau, 1982.

Provides a broad and general overview of compliance
responsibilities under federal pipeline safety
regulations. Designed for the nontechnically trained
person who operates a master meter system, a liquified
petroleum gas system, a small municipal system or a
small independent system.

A Guide to the Federal Hazardous Materials Transportation
Regulatory Program. Washington, D.C.: Department of
Transportation Report No. DOT-I-83-12, Jan. 1983.

Prepared by the DOT's Materials Transportation Bureau
to promote awareness of the Federal effort to improve
the safety of hazardous materials transportation.
Includes definitions of regulated materials, an
historical overview, and the relationship of the
federal program to state and local efforts. Provides
an explanation of the requirements imposed by the
regulations on shippers and carriers of hazardous
materials.

Handling Hazardous Materials. Alexandria, VA: American
Trucking Associations, Jan. 1985.

Summarizes DOT requirements and gives standard operating procedures for the systematic handling of hazardous freight, bills and equipment. Lists duties of all personnel in handling hazardous material freight. Illustrates placards and labels with instructions for use and application.

Hazardous Materials Transportation - A Legislator's Guide. Publication EN-84-1. Denver, CO: National Conference of State Legislatures, 1984.

Provides an overview of the hazardous materials transportation system including the federal and state regulatory framework and emergency response mechanisms. Highlights existing state programs designed to improve the system's safety and efficiency. Presents private industry and federal sources of assistance to states and localities.

IATA Dangerous Goods Regulations. 24th ed., effective Dec. 31, 1982. Montreal, Canada: International Air Transport Association, 1983.

Improving Transportation of Hazardous Materials Through Risk Assessment and Routing. Transportation Research Record 1020. Washington, D.C.: National Research Council, Transportation Research Board, 1985.

Contains five papers:

"Risk Assessment of Transporting Hazardous Material: Route Analysis and Hazard Management"

"Assessing the Risk and Safety in the Transportation of Hazardous Materials"

"Economic Evaluation of Routing Strategies for Hazardous Road Shipments"

"Routing Models for the Transportation of Hazardous Materials - State Level Enhancements and Modifications"

"Cost-effectiveness Analysis of Transportation Strategies for Nuclear Waste Repository Sites"

International Civil Aviation Organization, Air Navigation Commission, Dangerous Goods Panel. Technical Instructions for the Safe Transport of Dangerous Goods by Air. International Regulations Publishing & Distributing Organization, 1983.

Also available in French, Spanish, and Russian. Distributed in the U.S. by Labelmaster, 5724 Pulaski Road, Chicago, Illinois 60646.

International Maritime Dangerous Goods Code. Looseleaf for
     updating. London, England: International Maritime
     Organization, 1981. (Available from UNIPUB).

A Layman's Guide to Radioactive Materials Transportation.
     Publication EN-83-4. Denver, CO: National Conference
     of State Legislatures, 1983.

     Presents an overview of the sources of the radioactive
     materials that are transported in interstate commerce.
     The safety record of the shipments is reviewed along
     with the federal and state regulations. Appendices
     include maps with the locations of nuclear power
     plants, approved fuel routes, a review of states' Good
     Samaritan laws, as well as other pertinent information.

Manufacture, Transportation, Storage and Use of Explosive
     Materials. Standard NFPA 495. Quincy, MA: National
     Fire Protection Association, 1982.

Mostyn, Harri P. Packaging Dangerous Goods for Transport.
     4 volumes. Surry, England: Aurigny, Ltd., 1985.

     Contents: Guide to Class 3 inflammable liquids; Guide
     to Class 6.1 poisons; Guide to Class 8 corrosives;
     General guide.

1984 Emergency Response Guidebook: Guidebook for Hazardous
     Materials Incidents. DOT P 5800.3. Washington, D.C.:
     Department of Transportation, Research & Special
     Programs Administration, Materials Transportation
     Bureau, 1984.

     This guidebook was developed for use by firefighters,
     police and other emergency services personnel as a
     guide for initial actions to be taken to protect
     themselves during an incident involving hazardous
     materials. Primarily designed for use at an incident
     occurring on a highway or railroad.

Proceedings of the 7th International Symposium on the Trans-
     portation of Dangerous Goods by Sea and Inland Water-
     ways. Vancouver, British Columbia, Canada, September
     27-30, 1982, ICHCA, P.O. Box 2366, Station "D", Ottawa,
     Ontario, Canada K1P 5WP, 1983.

     This is a continuation of a series of such meetings
     which began with the first hosted by the Dutch and the
     American National Academy of Sciences Committee on
     Hazardous Materials at the Nieuwe Doelen, Rotterdam,
     The Netherlands, May 1968.

The two volumes include papers on the following topics:

Carriage of Dangerous Goods in Bulk (9 papers)
Total Inter-Modality -- A Viable Concept? (6 papers)
Carriage of Dangerous Goods in Specialized Marine
    Systems and Remote Areas (6 papers)
Emergency Response and Associated Training (7 papers)
Pollution Control (6 papers)
Dangerous Goods in Ports (7 papers)

Of special interest is the attention to training of
personnel, a subject which has received relatively
little attention in the past. For example, the
Maritime Training Section, Dangerous Goods, of IMO in
London now has a syllabus for the various courses
recommended for the seagoing and on-shore personnel
handling hazardous cargoes, including emergency
procedures in case of spill or other incident.

The 8th Symposium was held in Havana, Cuba, September
24-27, 1984. The proceedings were not published. The
9th symposium is scheduled to be held April 13-17, 1987
in Rotterdam, Netherlands. For details, contact International Maritime Organization, 4 Albert Embankment,
London SE1 7SR, U.K.

Public Technology, Inc. <u>Transportation of Hazardous</u>
    <u>Materials</u>. Washington, D.C.: Department of Transporta-
    tion Report No. DOT-I-80-46 (PB82-154071), Sept. 1980.

Covers approaches to dealing with incidents involving
hazardous materials in transport with the legal and
safety issues as a key concern. Planning and training
in response to vehicles transporting hazardous materials is addressed. The report also includes summaries
of exemplary state local programs, and describes
federal activity on the topic.

<u>Recommendations on the Safe Transport, Handling and Storage</u>
    <u>of Dangerous Substances in Port Areas</u>. London,
    England: International Maritime Organization, 1981.
    (Available from UNIPUB).

<u>Risk Assessment Processes for Hazardous Materials Transpor-</u>
    <u>tation</u>. National Cooperative Highway Research Program,
    Synthesis of Highway Practice 103. Washington, D.C.:
    National Research Council, Transportation Research
    Board, 1983.

Provides an overview of the use of risk assessment with
regard to transporting hazardous materials (including
hazardous wastes). Although the synthesis addresses
risk assessment for use in rule making, it focuses on
the needs, and means, to prevent, reduce, mitigate, and

respond to hazardous material transport risks from the perspective of local jurisdictions.

State Statutes and Regulations on Radioactive Materials Transportation. Publication EN-83-3. Denver, CO: National Conference of State Legislatures, 1983.

Outlines the state statutes and regulations that apply to radioactive materials transportation. Local regulations are also included.

Technical Instructions for the Safe Transport of Dangerous Goods by Air. Montreal, Canada: International Civil Aviation Organization, 1983.

Toward a Federal/State/Local Partnership in Hazardous Materials Transportation Safety. Washington, D.C.: Department of Transportation, Research & Special Programs Administration, Materials Transportation Bureau, Sept. 1982.

This document is an update and refinement of an MTB internal strategy paper originally prepared in August 1981. It establishes MTB's approach for creating an intergovernmental partnership for the resolution of the problem of hazardous materials transportation.

Transfrontier Movements of Hazardous Wastes: Legal and Institutional Aspects. Washington, D.C.: Organization for Economic Cooperation and Development, 1985.

This collection of papers, presented at an OECD Seminar, addresses a number of legal and institutional issues which arise in protecting the environment against potential risks created by hazardous waste. Particular attention is paid to the responsibilities of the parties concerned, their liability and insurance matters.

Transport of Dangerous Goods; Recommendations of the Committee of Experts on the Transport of Dangerous Goods. 3rd rev. ed. ST/SG/AC.10/1/Rev.3. New York: United Nations, 1984.

The recommendations cover principles of classification and definition of classes, listing of the principal dangerous goods, general packing requirements, testing procedures, marking, labelling or placarding, and shipping documents.

Transportation of Hazardous Materials: Planning and Accident Analysis. Transportation Research Record 977. Washington, D.C.: National Research Council, Transportation Research Board, 1984.

Contains six papers:

"Planning for a Transportation-related Hazardous Material Spill in a Municipal Watershed"

"Hazardous Materials: Developing Transportation Safety Programs on a Limited Budget"

"Risk of Multiple Small-package Spills of Hazardous Substances"

"Estimating the Release Rates and Costs of Transporting Hazardous Waste"

"Chemical Spill Response Information System of the Association of American Railroads"

"A Survey of Foreign Hazardous Materials Transportation Safety Research Since 1978"

Transportation of Hazardous Materials: Toward a National Strategy. 2 volumes. Special Report 197. Washington, D.C.: National Research Council, Transportation Research Board, 1983.

Contains the proceedings of a conference of the Steering Committee to Develop a National Strategy for the Transportation of Hazardous Materials and Hazardous Wastes in the 1980's, held in Williamsburg, Virginia, February 17-20, 1981. Volume 1 includes an overview of the problem and a discussion of Committee findings and recommendations. Volume 2 contains 12 resource papers and the reports of the conference workshop rapporteurs.

Transporting Hazardous Waste. Alexandria, VA: American Trucking Associations, April 1984.

A complete guide to the federal and state requirements for transporting hazardous waste. Identifies each state agency responsible for regulating the transportation of hazardous waste, as well as providing a source of contact. All state manifest, registration and permit requirements are provided to assist with compliance efforts.

Work Zone Safety, Maintenance Management and Equipment, and Transportation of Hazardous Materials. Transportation Research Record 833. Washington, D.C.: National Research Council, Transportation Research Board, 1981.

Contains ten papers on three different topics presented at the 1981 TRB annual meeting. The one paper of interest is entitled: "Regulation of the Movement of Hazardous Cargoes on Highways (Abridgment)".

Periodicals

Hazardous Cargo Bulletin. Intapress Publishing, Ltd.
    Monthly, published since 1980.

Published in London. Principal subscribers are shippers, carriers, and other professionals involved in the transport of chemicals, oils and gases. Covers regulations, safety, training courses, publications, upcoming conferences, tankers, terminals, port news, and packaging. Subscription includes a separately published quarterly booklist.

Hazardous Materials Transportation. Cahners Publishing Co.
    Bimonthly, published since 1978.

Aimed at shippers, carriers, regulators, counsel and packaging manufacturers. Emphasis is on state, local, and national rules and regulations. Also covers trends in transportation policy, packaging and handling developments, and information on changes in classifications, markings, permits, shipping documents, etc.

Not to be confused with the BNA loose-leaf service with the same title, described in the next section.

Toxic Materials Transport. Business Publishers, Inc.
    Biweekly, published since 1980.

Focuses on legislative and judicial issues affecting the hazardous materials transportation field, especially recent and continuing federal preemption and state routing restriction controversies. Reports on how products must be packaged and labeled, and how they may be transported. Also covers accident response, routing requirements, and technological developments.

Loose-Leaf Services

Hazardous Materials Guide: Shipping, Materials Handling and
    Transportation. Neenah, WI: J. J. Keller & Associates.

A quick reference source providing complete Hazardous
Materials Regulations as issued by the Materials
Transportation Bureau. A useful reference and
operational tool for shippers and transporters of
hazardous materials. The 900 page guide explains how
to comply with each portion of the rules, including
documentation, storage, labeling and packaging,
placarding, and how to use the commodity list of materials. Also included are DOT rules issued under Title
49, CFR, Parts 106, 107, 171 through 179, and Part 397.
Also includes reference data on enforcement agencies,
containers, data sources, exemptions, products and
services. Looseleaf, updated periodically.

Hazardous Materials Transportation. Washington, D.C.:
    Bureau of National Affairs.

Consists of two reference binders which contain the
full text of the rules and regulations that govern the
shipment of hazardous materials within the U.S. by
rail, air, ship, highway and pipeline. Also provides a
comprehensive table of rules for almost 10,000 materials, cross-referenced by United Nations identification
number and generic name. Kept up-to-date by a monthly
bulletin.

Should not be confused with a periodical with the same
name, described in the previous section.

International Hazardous Materials Transportation Manual.
    Washington, D.C.: Bureau of National Affairs.

The reference manual includes information on air and
sea transport rules for more than 3000 commonly carried
hazardous materials. In addition, the manual's compliance guides provide explanations of requirements at
each stage of shipping for every type of hazardous substance, and help to locate rules by mode of transport
and country. The monthly updating bulletin reports on
proposed changes in international standards, enforcement activity, and compliance efforts.

Special Articles

"DOT's Alan Roberts Assesses Current and Future Trends for Transportation of Hazardous Materials". In Hazardous Materials & Waste Management 1, No. 5 (Sept.-Oct. 1983): 18-21.

A candid interview with the Associate Director for Hazardous Materials Regulation of the DOT's Materials Transportation Bureau. Roberts is a leading expert on DOT regulations, having spent 16 years with the Department working on them.

Phillips, Gregory T. "Hazardous Materials Transportation: A Local Approach". In In The Public Interest: A Review of Law and Society IV No.'s 1 & 2 (Fall 1983/Spring 1984): 26-34.

This article focuses on the role of localities - cities, towns, and counties - in regulating the transportation of hazardous materials within their boundaries. It discusses the historic basis of local safety regulation - the police power - and an evaluation of specific local safety measures that have been enacted. Limited to highway transportation by truck. Rail, air, and sea modes are not covered.

In The Public Interest is published annually by The Center for Public Interest Law, Faculty of Law and Jurisprudence, O'Brian Hall, State University of New York at Buffalo, Buffalo, NY 14260. Copies of this issue may be available from the Center; copies of the article are available from the author of this chapter.

Wood, William S. "Transportation of Hazardous Materials". In Safety and Accident Prevention in Chemical Operations. 2nd ed., edited by H. H. Fawcett and W. S. Wood. New York: John Wiley & Sons, 1982: 681-710.

A good summary article. Covers emergency services, railroad transportation, label and placard requirements, tank-truck operations, water and pipeline transport, loading and unloading operations, and regulating organizations. Includes an extensive bibliography.

Data Bases & Information Systems

Expert Information Systems Ltd. (EXIS)
EXIS 1
38 Tavistock Street
London WC2E 7PB
England   01-240 0837

EXIS 1 is a computerized information retrieval system
which covers the transportation, storage, and handling
of hazardous cargoes and chemicals. The system
provides online access to data bases containing information on hazardous materials regulations, properties,
and commercial aspects. The system is designed as a
series of separately accessible modules dealing with
particular areas of tranport, emergency response, or
technical applications. EXIS modules are being
released as they are completed; initially available are
the IMO Module, the ICAO Regulations Module, and the
Materials Information Module. In the future the
following modules are expected to be added: Chemdata
(U.K. Chemical Emergency Response); Hazardous Cargo
Contacts; U.K. Blue Book National Maritime Regulations;
ADR, on European road transport; RID, on European rail
transport; ADN, on European inland waterways transport;
and other modules covering national and port regulations and bulk chemicals information.

EXIS 1 is a menu-driven system that can be accessed
online by computer terminals, microcomputers, or word
processors over public telephone lines, the national
PSS packet-switching system, or the International
Packet Switching Service (IPSS) data network. The
system can also be accessed using a conventional telex
machine and can be loaded onto many types of client
computers.

EXIS 1 is intended for use by chemical manufacturers
and shippers, shipowners and operators, road haulers,
rail operators, tank container operators, cargo
brokers, stevedores, port authorities, terminal operators, emergency services, packaging manufacturers,
handling equipment manufacturers, government departments, trade associations, transport consultants, and
forwarders.

TRIS (Transportation Research Information Services)
Producer: Transportation Research Board
Time Span: 1968 to present
Coverage: International
Updated: 750 abstracts monthly

Contains citations, with abstracts, to literature on
transportation, including air, highway, rail, maritime,
mass transit, and other transportation modes. Also

includes resumes of data from recent of ongoing
research projects on transportation-related subjects.
Coverage includes about 93,000 records and 3900 resumes
of recent or ongoing highway research projects from the
Highway Research Information Service (HRIS); about
25,000 abstracts from the Maritime Research Information
Service (MRIS); about 21,000 abstracts from the
Railroad Research Information Service (RRIS); about
4000 abstracts from the Air Tranportation Research
Information Service (ATRIS); about 33,000 abstracts
from the National Highway Traffic Safety Administration; over 31,000 government report abstracts from the
National Technical Information Service (NTIS) that are
not included in other files, over 1400 abstracts of
statistical reports and time series in the
Transportation Statistical Reference File; about 1000
abstracts of non-U.S. documents from the Department of
Trasportation, Transportation Systems Center; and more
than 13,400 records from the Urban Mass Transportation
Research Information Service (UMTRIS).

Audiovisual Materials

Handling Hazardous Materials Slide Program.   Alexandria, VA:
American Trucking Associations, 1982.   Slide/tape
programs.

A comprehensive slide presentation designed to
complement a company's hazardous materials training
program.   This five-part series provides an in-depth
review of the DOT hazardous materials regulations for
shipping papers, labeling and package mailing, loading
and storage, and placarding of vehicles.

Hazardous Materials Transportation Emergencies.   Quincy, MA:
National Fire Protection Association, 1982-83.   Slide/
tape programs.

Consists of four units, available together or
separately:

Unit 1: Assessing the Problem.
Unit 2: Identifying Hazardous Products.
Unit 3: Responding to the Incident.
Unit 4: Resolving the Incident.

Safe Transport.   Washington, D.C.: Materials Transportation
Bureau, 198_.   Motion Picture, 16mm, sound, color, 21
min.

An overview of the Dept. of Transportation and hazardous material transportation requirements.   Topics

include: definitions and hazard classes; packaging; marking, including identification numbers, labeling, placarding, shipping papers, emergency response, CHEMTREC and National Response Center, interface DOT/ EPA hazardous waste/hazardous substances; regulation; and an overview of pipeline safety. Also available on 3/4 inch videotape.

Transportation of Hazardous Materials - Awareness and Recognition. Washington, D.C.: Forest Service, 1983. 112 color slides, audiocassette, 102-page script, 26- page student workbook, 45 min.

Contains a non-technical introduction to the complete course (described below) as well as a broad overview of the CFR Title 49 Hazardous Materials Regulations. Designed primarily for upper management who oversee the handling or transporting of hazardous materials by truck, rail, ship or plane. Distributed by the National AudioVisual Center.

Transportation of Hazardous Materials - Awareness, Recognition and Working Knowledge. Washington, D.C.: Forest Service, 1983. 435 color slides, 4 audiocassettes, 102-page script, 298-page student workbook, 6-8 hours.

The complete course includes the program described above, plus additional technical and detailed information. Topics include definition, identification, classification and separation of hazardous materials; labeling, stowing and transporting of these materials. Distributed by the National AudioVisual Center.

Seminars & Workshops

Several organizations offer periodic courses on the subject of transportation of hazardous materials. The information given below is based on recent literature from the company or agency. For upcoming schedules and registration information, contact them at the address or phone number indicated.

Hazardous Materials Transportation Regulations Seminar.

Unz & Co. offers specialized seminars designed to help attendees comply with current and new regulations on domestic and international transportation. The two-day programs cover domestic transport during Day 1 and international transport on Day 2. In 1984, six seminars were scheduled in cities such as Boston, Houston, Newark, and Baltimore. Registration Fees in 1984 were $499 for two days and $265 for one day.

Unz & Co.
190 Baldwin Avenue
P.O. Box 308
Jersey City, NJ 07303
(800) 631-3098

Hazardous Shipment Workshop: Applying U.S./International Regulations.

Described as "A program for operating and managerial
personnel responsible for domestic or international
movement of hazardous materials, substances and
wastes." This course is similar in format to the pre-
ceding one: two days, with the first day for domestic
transportation, the second day for international. In
1984, this workshop was offered at the World Trade
Center in New York City. Fees were $570 for two days,
$360 for one day.

A special feature of this course is that it is available for in-house presentation or at a nearby location
arranged by the attendees.

World Trade Institute
One World Trade Center, 55W
New York, NY 10048
(212) 466-4044

Intermodal Transportation of Hazardous Materials, Course for Industry.

This one-week course is designed to provide a basic
working knowledge of the Hazardous Materials
Transportation Regulations contained in Title 49, Code
of Federal Regulation, Parts 100-178. Instructors are
from the Transportation Safety Insitute and the modal
administrations of DOT. In 1985, this course was held
in Oklahoma City in April and June. The registration
fee was $250 for the 5-day program.

Hazardous Materials Safety Division, DMA-62
Transportation Safety Institute
6500 South MacArthur Boulevard
Oklahoma City, OK 73125
(405) 686-4824

Government Organizations

Four components of the Department of Transportation and an independent federal board are the primary governmental sources of information on tranportation of hazardous materials. Each of these is described below.

Coast Guard
Marine Technical and Hazardous Materials Division
2100 Second Street, SW
Washington, D.C. 20593   (202) 426-1577

> Relevant Areas of Interest: Ship safety standards; safe water transport of chemicals and cargoes; foreign chemical tankship certification; incinerator vessels; liquified gas ship standards; occupational safety and health for merchant mariners; control of vapor hazards; cargo containment and handling; environmental dispersion modeling of chemical releases; water pollution from oil and chemicals.
>
> Publications: Chemical Data Guide for Bulk Shipment by Water (CIM 16616.6); Regulations in Title 46 Code of Federal Regulations; Navigation and Vessel Inspection Circulars; safety pamphlets; technical notes; Environmental Response Newsletter (annual).
>
> Information Services: Answers inquiries; makes referrals to other sources of information; issues guidance.
>
> See also the description of the Coast Guard's Chemical Hazards Response Information System (CHRIS) in Chapter 12, Accidents, Spills and Cleanup.

Federal Highway Administration
Bureau of Motor Carrier Safety
400 Seventh Street, SW
Washington, D.C. 20590   (202) 426-1790

> Relevant Areas of Interest: Movement on the nation's highways of dangerous cargos such as hazardous wastes, explosives, flammables and other volatile materials.
>
> Publications: Motor Carrier Accident Investigations (irregular); Accidents of Motor Carriers of Property (annual); On Guard (periodic safety bulletin). The Bureau maintains a mailing list for those interested in its publications.
>
> Information Services: Answers inquiries; makes referrals; provides consulting and training services to state organizations.

Federal Railroad Administration
Office of Safety
400 Seventh Street, SW
Washington, D.C. 20590   (202) 755-9260

Relevant Areas of Interest: Railroad safety
regulations, transportation of explosives and other
hazardous materials, reporting and investigation of
rail accidents.

Publications: Rail-Highway Grade-Crossing Accidents and
Accident Bulletin (both annual); Railroad Accident
Investigation Reports (irregular); employee fatality
accident reports; annual report.

Information Services: Answers inquiries; distributes
publications; makes referrals to other sources of
information.

National Transportation Safety Board
800 Independence Avenue, SW
Washington, D.C. 20594   (202) 382-6735

Relevant Areas of Interest: Accidents involving all
forms of transportation--civil aviation, railroad,
pipeline, highway, marine--especially those involving a
fatality or substantial property damage; studies and
investigations that result in recommendations for
transportation safety and accident prevention; evaluation of the adequacy of safeguards and procedures
concerning the transportation of hazardous materials
and the performance of other government agencies
charged with ensuring the safe transportation of such
materials.

Publications: Special reports; accident investigation
reports; annual reports.

Information Services: Publications are provided free of
charge to the following categories of subscribers -
federal, state, or local transportation agencies;
international transportation organizations or foreign
governments; educational institutions or public
libraries; nonprofit public safety organizations; and
the news media.

A Public Reference Room of the Safety Board is available for record inspection or copying.

Research and Special Programs Administration
Materials Transportation Bureau
400 Seventh Street, SW
Washington, D.C. 20590   (202) 755-9260

Relevant Areas of Interest: Transportation of hazardous
materials (explosives and other dangerous articles),

including transportation by carrier by rail, highway,
air, and water; preparation of shipments, including
classification, packaging, and labeling requirements;
specifications for packaging and packages; liquids and
gases pipeline transportation safety and regulation;
pipeline components and facilities.

Publications: Code of Federal Regulations, Title 49-
Transportation, Parts 100-199; Special reports;
Hazardous Materials Newsletter; Guides for the inspection of shipments by motor vehicle or at freight facilities.

Information Services: Consults with and advises
shippers, carriers, governmental agencies, and the
general public on all phases of shipping and transportation of hazardous materials and informs participants
on the formulation, promulgation, and administration of
the hazardous materials and pipeline safety regulations.

Associations, Organizations & Research Centers

American Petroleum Institute
1220 L Street, NW
Washington, D.C. 20005   (202) 682-8000

Membership includes producers, refiners, marketers, and
transporters of petroleum and allied products. API is
very interested in transportation of hazardous materials since petroleum products constitute the greatest
tonnage of regulated materials transported. Provides
extensive information services and maintains a library
of several thousand volumes.

American Trucking Associations
2200 Mill Road
Alexandria, VA 22314   (703) 838-1849

A federation of 51 state trucking associations and 13
national conferences of trucking companies. Maintains
a 20,000-volume library. Relevant publications and
slide programs are described in previous sections of
this chapter.

Association of American Railroads
Bureau of Explosives
1920 L Street, NW
Washington, D.C. 20036   (202) 835-9500

Membership consists of railroads, steamship companies,
motor carriers, manufacturers, and shippers of hazardous materials and container manufacturers. Conducts
educational programs and research in safe transportation and storage of explosives and other hazardous
materials. Provides emergency procedures in the
handling of hazardous articles. A number of their
publications are listed elsewhere in this chapter.

Chemical Manufacturers Associations
2501 M Street, NW
Washington, D.C. 20037   (202) 887-1100

Founded in 1872 as the Manufacturing Chemists
Association. Its membership consists of 200 manufacturers of basic chemicals who sell a substantial
portion of their production to others. Administers
research in areas of broad import to chemical manufacturing, such as air and water pollution control, and
other special research programs. Conducts committee
studies, workshops, and technical symposia.

CMA's major contribution to transportation of hazardous
materials is CHEMTREC--the Chemical Transportation
Emergency Center. CHEMTREC provides 24-hour guidance
to emergency services on handling transportation
accidents involving chemicals. In ten years, CHEMTREC
received over 155,000 calls and gave assistance in
21,700 transportation emergencies. The toll-free
number for CHEMTREC is (800) 424-9300.

Council for Safe Transportation of Hazardous Articles
685 Third Avenue
New York, NY 10017   (212) 878-5204

Inter-industry federation of shipper trade associations, traffic conferences, and sustaining companies.
Supports reasonable and necessary government safety
regulations to achieve safe and efficient distribution
of small-package articles regulated under the
Department of Transportation and other agencies for
hazardous material transportation, specializing in unit
of one gallon or less domestically and 55 gallons or
less internationally. Attempts to shape regulations
and keep participants advised of new rule-making
proposals and amendments relevant to regulated package
freight distribution and for hazardous substances and
wastes. Sponsors public meetings on topical issues,
and an annual seminar. Maintains library of regulations and documents.

Hazardous Materials Advisory Council
1012 14th Street, NW, Suite 907
Washington, D.C. 20005   (202) 783-7460

Membership includes 300 shippers, carriers, and container manufacturers of hazardous materials, substances and wastes; shipper and carrier associations. Works to promote safe transportation of these materials; provides assistance in answering regulatory questions, guidance to appropriate governmental resources, and advice in establishing corporate compliance and safety programs. Conducts seminars on hazardous domestic and international material packaging and transporting. Presents the George L. Wilson Memorial Award for outstanding achievements in hazardous materials tranportation safety by a person or company. HMAC's publications include a monthly _Courier_ and an annual directory.

National Fire Protection Association
Batterymarch Park
Quincy, MA 02269   (617) 770-3000

Membership drawn from the fire service, business and industry, health care, educational and other institutions, and individuals in the fields of insurance, government, architecture, and engineering. Develops, publishes, and disseminates standards, prepared by approximately 175 technical committees, intended to minimize the possibility and effects of fire and explosion; conducts fire safety education programs for the general public. Provides information on fire protection, prevention, and suppression; compiles annual statistics on causes and occupancies of fires, large-loss fires, fire deaths, and fire fighter casualties. Provides field service by specialists on electricity, flammable liquids and gases, and marine fire problems. Maintains a library of more than 50,000 books, reports, periodicals, audiovisual materials and microforms. Several publications and audiovisual items are cited elsewhere in this chapter.

National Safety Council
444 N. Michigan Avenue
Chicago, IL 60611   (312) 527-4800

Three of the 29 industrial sections of the National Safety Council are devoted to safety in cargo transportation. These are the marine, railroad, and commercial vehicle sections. Each of these prepare draft safety data sheets and other training materials aimed at improving employee and public safety. Movies and slide presentations on transportation safety that have been produced by government, industry, and carriers are distributed or listed through the Council. See Chapter 1 for more information on NSC.

Transportation Research Board
National Research Council
2101 Constitution Avenue, NW
Washington, D.C. 20418   (202) 334-3220

TRB is a major source of information on all aspects of
transportation. Some of their recent publications have
been listed in a previous section of this chapter.

In particular, Committee A3C10 deals with transportation of hazardous materials. This committee is concerned with the movement of hazardous materials
including conditions and forces encountered during
transportation; type and extent of hazards associated
with material type of class; preservation and packaging
to prevent cargos' damage or decomposition during
transportation and handling; laws controlling and legal
liabilities pertaining to hazardous materials movements; signs and labels to indicate the hazardous
nature of the cargo; and the knowledge of incidents
occurring during transportation and handling movement
risks.

Transportation Safety Institute
6500 South MacArthur Boulevard
Oklahoma City, OK 73125   (405) 686-2153

Established in 1971 by DOT. Develops and conducts
training programs for federal, state, and local government, and secondarily for industry representatives and
international students. Publications include a course
catalog and handout material.

Virginia Polytechnic Institute and State
  University, Safety Projects Office
167 Whittemore
Blacksburg, VA 24061   (703) 961-5635

Performs research for U.S. Government and industry in a
number of safety-related fields, including hazardous
materials transportation. Research results are published in technical reports and journals.

Companies & Industries

J. J. Keller & Associates
145 W. Wisconsin Avenue
Neenah, WI 54956   (800) 558-5011

A commercial publishing, servicing, and consulting
corporation specializing in federal and state regulatory requirements governing the motor carrier industry,

distribution, toxic substances control, hazardous
wastes, etc. Publishes 20 guides and manuals, over 50
handbooks and booklets, 3 periodicals, hundreds of
forms, labels, placards and other transportation-
related materials.

The Hazardous Materials Guide is described elsewhere in
this chapter, and a number of Keller publications are
covered in other parts of this book.

Labelmaster
5724 N. Pulaski Road
Chicago, IL 60646   (800) 621-5808

Provides forms, labels, placards, publications, etc.
Distributing agent for INTEREG - the International
Regulations & Distributing Organization. Several of
the publications are described in earlier parts of this
chapter.

Union Pacific System
1416 Dodge Street
Omaha, NE 68179   (402) 271-3313

Union Pacific has developed a comprehensive program for
handling hazardous materials and dealing with problems
and incidents. The Hazardous Material Training
Department has developed a number of publications:

Defining Your Hazardous Material Problem (Dec. 1983)
Hazardous Material First Responder Course (No date)
Hazardous Materials Incident Management (Nov. 1982)
Instructions for Handling Hazardous Materials (Oct.
  1983)
Recognizing and Identifying Hazardous Materials (Oct.
  1983).

A Hazardous Material Training master plan is expected
to be available soon.

12.
Accidents, Spills and Cleanup

Margaret R. Wells

Introduction

Spills, leaks, explosions, fires, and numerous other incidents can occur at any time between the initial manufacture
and final disposal of hazardous substances. The December
1984 poisonous gas leak in Bhopal, India, which killed 2500
people, is a tragic example of the devastating effects of
such occurrences. Reports of chemical spills, leaks, and
other accidents appear regularly in the media and often
result in human tragedy, environmental damage, and corporate
liability.

Heightened public and industry awareness of the widespread
dangers posed by hazardous materials emergencies is reflec-
ted in the proliferation of popular and technical literature
on the subject. Risk assessment, hazard management, and
remedial and emergency response are regular topics of books,
periodical articles, conferences, and related information
sources.

The interdisciplinary nature of hazardous substance emergencies literature illustrates the breadth of concern. Chemistry, technology, fire science, safety, and environmental
science are representative subject areas covered in the literature of hazardous materials research. Industry and emergency response personnel must maintain an ongoing awareness
of new developments in these fields in an effort to respond
promptly and effectively to all types of hazardous materials
emergencies.

The sources described in this chapter discuss many aspects
of hazardous materials emergencies planning, prevention, and
response. Additional sources cited throughout this book,
such as materials on transportation, disposal, and environmental effects, are also relevant and should be consulted
when appropriate.

Books

The Library of Congress classification system supplies headings for many subjects on accidents and emergencies involving hazardous materials. "Chemicals -- Safety Measures," "Hazardous Substances -- Accidents," and "Hazardous Substances -- Transportation -- Accidents" are particularly useful subject headings. The following books provide valuable information on hazardous substance accidents.

Bennett, Gary F., et al. Hazardous Materials Spills Handbook. New York: McGraw-Hill, 1982.

Regulations, risk assessment, prevention, response plans, and other information on hazardous materials spills are covered. Suggestions for further reading and case histories are also included.

Cashman, John R. Hazardous Materials Emergencies: Response and Control. Lancaster, PA: Technomic, 1983.

This lucid, comprehensive reference source covers all aspects of emergency preparedness and response for spills, explosions, fires, fumes, contamination, and other incidents. Information on planning, training, and equipping response teams is covered. Also included are case histories and extensive lists of research sources.

Cormack, D. Response to Oil and Chemical Marine Pollution. New York: Elsevier, 1983.

A discussion of new techniques and equipment for response to oil and chemical spills in water. Various cleanup methods are evaluated.

Ecology and Environment, Inc. and Whitman, Requardt, and Associates. Toxic Substance Storage Tank Containment. Pollution Technology Review, No. 116. Park Ridge, NJ: Noyes, 1985.

Toxic substance storage tanks must be rigorously designed and maintained to avoid dangerous spills and leaks. This book treats design, maintenance, inspection, and emergency procedures for efficient storage tank use. Standards and codes, the necessity of regular facility evaluation, and corrective measures are covered. Spill control information is also included.

Emergency Response Guide. Alexandria, VA: American Trucking Associations, 1984.

Developed by DOT, this guide helps the user make critical decisions in the event of an accident, fire, or
spill involving hazardous materials. All DOT regulated
materials are listed, with reference to a guide that
specifies emergency action to be taken in the event of
a hazardous materials incident.

Fawcett, Howard H. and William S. Wood. Safety and Accident
    Prevention in Chemical Operations. 2nd ed. New York:
    Wiley, 1982.

    Covers a broad spectrum of information on chemical
    accidents, including laws, personnel safety, and fire
    response. Loss prevention checklist and information on
    formulating a community response plan are included.

Fire and Emergency Response Resource Directory. Quincy, MA:
    National Fire Protection Association, 1984.

    Lists individuals, organizations, and other resources
    for emergency response personnel. Radiological and
    hazardous substance emergencies are included.

Green, A. E., ed. High Risk Safety Technology. New York:
    Wiley, 1982.

    A discussion of risk assessment and emergency prepared-
    ness in selected industries. Chemical and radiation
    accidents, fire, explosions, and other incidents are
    covered.

Guswa, J. H., et al. Groundwater Contamination and Emergency
    Response Guide. Pollution Technology Review, No. 111.
    Park Ridge, NJ: Noyes, 1984.

    Part III of this book discusses emergency response
    methodologies for groundwater contamination incidents.
    Part I details initial response in cases of suspected
    contamination. Part II serves as a reference manual
    for monitoring groundwater contamination.

Handbook of Hazardous Waste Regulation, Volume II: How to
    Protect Employees During Environmental Incident
    Response. Madison, CT: Bureau of Law and Business,
    Inc., 1985- .

    A collection of EPA safety guides intended for any
    individual or organization involved in hazardous
    materials incident response. Detailed treatments of
    employee training, safety, contingency planning, and
    emergency response dominate this loose-leaf manual.
    Charts and check-off sheets are also included.

Hazardous Wastes and Environmental Emergencies: Management, Prevention, Cleanup, Control. Silver Spring, MD: Hazardous Materials Control Research Institute, 1984.

Practical information on the management of hazardous materials emergencies, such as spill control and clean- up, site remediation, and waterway decontamination.

Industrial and Hazardous Waste Management Firms, 1984. Minneapolis: Environmental Information, Ltd., 1984.

Includes a list of firms offering remedial response services for spills and cleanup.

Lagadec, Patrick. Major Technological Risk. Elmsford, NY: Pergamon, 1982.

Describes and analyzes five major disasters occurring between 1974 and 1979, including the history and after- math of each incident. A list of the locations, dates, products involved, and damages for similar accidents from 1921 to 1979 is an informative feature.

Manual for Spills of Hazardous Materials. Prepared by the Technical Services Branch, Environment Canada. Ottawa, Canada: Canadian Government Publishing Center, 1984.

Designed for emergency response personnel, this manual covers control, containment, and other emergency mea- sures for responding to chemical spills. Quantitative data on hazardous chemicals is also included.

Saxena, Jitendra, ed. Hazard Assessment of Chemicals: Current Developments. Vol. 2. New York: Academic, 1983.

Contains five subject reviews and two case studies of chemical spills. Environmental effects and cleanup processes are described.

Shaver, Deborah K. and Robert L. Berkovitz. Post-accident Procedures for Chemicals and Propellants. Pollution Technology Review, No. 109. Park Ridge, NJ: Noyes, 1984.

All aspects of emergency response to hazardous mater- ials transportation accidents involving 28 widely used chemicals and propellants are discussed. Initial trans- portation precautions, accident reporting, damage assessment, and cleanup measures are some examples of the topics treated. Appendices list chemical proper- ties, additional information sources, and other relevant materials.

Smith, Al J., Jr. Managing Hazardous Substances Accidents.
    New York: McGraw-Hill, 1981.

    A concise discussion of developing and implementing
    relief systems for hazardous materials emergencies.
    The causes and types of accidents, assistance agencies
    and groups, cleanup technology, and other information
    for response personnel are covered.

Terrien, Ernest J. Hazardous Materials and Natural Disaster
    Emergencies: Incident Action Guidebook. Lancaster,
    PA: Technomic, 1984.

    A concise guidebook providing detailed instructions for
    emergency response personnel. Checklists, procedures,
    diagrams, forms, and other materials are appropriate
    for training and actual response.

Weaver, L. Albert. Techniques for Hazardous Chemical and
    Waste Spill Control. 2nd ed. Raleigh, NC: L. A.
    Weaver, 1983. Available from Lab Safety Supply Co.

    Emphasizing safety in handling hazardous substances,
    this manual reviews regulatory guidelines, chemical
    reactions, and suggested methods for reducing the
    probability of spills. Emergency response measures and
    contingency planning are also discussed.

Additional handbooks, manuals, encyclopedias, and other
reference materials describing chemical processes and safety
measures also provide valuable information for emergency
response. Selected examples of these materials are listed
below.

Alliance of American Insurers. Handbook of Hazardous
    Materials: Fire, Safety, Health. 2nd ed. Schaumburg,
    IL: The Alliance, 1983.

Goldfarb, Alan S., et al. Organic Chemicals Manufacturing
    Hazards. Ann Arbor, MI: Ann Arbor Science, 1981.

International Technical Information Institute. Toxic and
    Hazardous Industrial Chemicals Safety Manual for
    Handling and Disposal with Toxicity and Hazard Data.
    Tokyo: The International Technical Information
    Institute, 1982.

National Fire Protection Association. Fire Protection Guide
    on Hazardous Materials. 8th ed. Boston: The Associa-
    tion, 1984.

Periodicals

Dangerous Properties of Industrial Materials Report. Van
Nostrand Reinhold. Bimonthly, published since 1980.

Reports on the physical properties of widely used
industrial materials, including data on toxicity and
flammability.

Emergency Preparedness News. Business Publishers, Inc.
Biweekly, published since 1977.

Contingency planning, crisis management, and relief
from man-made disasters, including chemical spills, are
covered.

Fire Command. (Formerly Fire Service Today.) National Fire
Protection Association. Monthly, published since 1933.

A magazine designed for professional firefighters which
reports on various types of fires and containment and
management methods. Information on explosions and
leaks involving hazardous materials, case histories,
new products, and a calendar of upcoming events for
firefighters are included.

Fire Engineering. Technical Publishing Co. Monthly,
published since 1877.

Covers information on firefighting tactics for all
types of fire emergencies, including those involving
hazardous materials. "Fire Report" analyzes selected
case studies.

The Safety Practitioner. Paramount. Monthly, published
since 1983.

The "official journal of the British Institution of
Occupational Safety and Health" (title page). The
"Hazard Data Bank" section highlights a single
substance each month, discussing properties, uses, fire
hazards, leakage procedures, and other information for
emergency response and safety. The majority of the
publication focuses on occupational safety with
occasional articles on emergency equipment and response
measures.

Abstracts and Indexes

Fire Technology Abstracts. Federal Emergency Management
    Agency. (Available from the Superintendent of
    Documents.) Bimonthly, published since 1976.

Covers international fire literature, including
materials on fire prevention and hazard identification.
Books, conference proceedings, and articles from 600
periodicals are indexed by author and subject. Indexes
are cumulated annually.

Proceedings

Control of Hazardous Materials Spills - 1980. Silver
    Spring, MD: Hazardous Materials Control Research Insti-
    tute, 1980.

Features over 90 papers on methods of spill cleanup
management. Case histories, personnel safety, training, site investigation, remedial action, and other
topics are treated.

Dabberdt, Walter F., ed. Atmospheric Dispersion of
    Hazardous/Toxic Materials from Transport Accidents.
    Proceedings related to the course given at the Inter-
    national Center for Transportation Studies (ICTS),
    Amalfi, Italy, September 20-24, 1982. With the
    cooperation of SRI International and the Bureau of
    National Affairs, Inc. New York: Elsevier, 1984.

A collection of nine conference papers discussing atmospheric dispersion principles for emergency response and
materials transport planners. Processes, measurement,
and modeling techniques for assessing the severity of
hazardous materials accidents are covered. Response
planning and understanding the atmospheric dispersion
of substances are emphasized.

Hazardous Materials Spills Conference Proceedings. Sponsored
    by the Bureau of Explosives, et al. Rockville, MD:
    Government Institutes, Inc., 1982- .

Heavy Gas Dispersion Trials at Thorney Island. Proceedings
    of a Symposium held at the University of Sheffield,
    Great Britain, April 3-5, 1984. J. McQuaid, guest
    editor. New York: Elsevier, 1985.

Reports on a major research project resulting from the
June 1, 1984 explosion at a chemical plant in Flixbor-
ough. The flammable gas cloud dispersing from the
explosion site prompted research on the atmospheric
dispersion of hazardous substances.

New England Legal Foundation, ed. Managing Hazardous Wastes:
Proceedings on the Practical Aspects of Hazardous Waste
Regulation. Hartford, Connecticut, November 20, 1981.
Boston: New England Legal Foundation, 1982.

Focuses on regulations and industry compliance in
hazardous waste management. "Spill Emergencies," by
John Cuthbertson, "Inspections," by Timothy G. Rogers,
and "Abandoned Sites," by William Blatchley are particularly relevant.

Oil and Hazardous Materials Spills and Dump Conference -
1980. Silver Spring, MD: Hazardous Materials Control
Research Institute, 1980.

Information on identifying, controlling, and implementing cleanup procedures for spills and uncontrolled
hazardous waste sites is featured.

Rodricks, Joseph V. and Robert G. Tardiff, eds. Assessment
and Management of Chemical Risks. Based on a Symposium
Sponsored by the American Chemical Society, Kansas
City, Missouri, September 12-17, 1982. Washington,
D.C.: American Chemical Society, 1984.

Papers by M. Granger Morgan, "Use of Risk Assessment
and Safety Evaluation," and Virgil O. Wodicka on "The
Need for Risk Assessment of Chemicals in Corporate
Decision Making" are relevant to hazardous materials
emergencies.

Reports and Documents

The proliferation of documents and reports generated by U.S.
government agencies and independent laboratories involved in
government-sponsored research includes analyses of transportation accidents, new response methods for all types of
hazardous materials emergencies, and other research on emergency response. Regular monitoring of government indexes
and computerized data bases covering reports and documents
is essential for maintaining a continuing awareness of these
materials. The Monthly Catalog of United States Government
Publications and its online equivalent, the GPO Monthly
Catalog Data Base, cover publications available from the
Government Printing Office and other U.S. government

agencies. Government Reports Announcement and Index and the
NTIS Bibliographic Database include reports available from
the National Technical Information Service (NTIS). Examples
of selected reports relevant to hazardous materials emergencies follows. Numerous reports are also available for
incidents involving specific substances, transportation
methods, and emergency situations.

Esposito, M. P., et al. Guide for Decontaminating Buildings,
    Structures, and Equipment at Superfund Sites (Final
    Report Sep 83-Nov 84). Report #EPA/600/2-85/028.
    Sponsor: Battelle Columbus Labs., Ohio; Environmental
    Protection Agency, Cincinnati, OH. Hazardous Waste
    Engineering Lab. 1985.

Green, A. J. Annotated Bibliography for Cleanup of Hazar-
    dous Waste Disposal Sites. Report #WES/MP/EL-82-7.
    Army Engineer Waterways Experiment Station, Vicksburg,
    MS. Environmental Lab. 1982.

Guide to Developing Contingency Plans for Hazardous Chemical
    Emergencies. SuDocs No. HE 20.7008:C42. Prepared for
    Centers for Disease Control, Center for Environmental
    Health; prepared by James E. Powers, et al. Palo Alto,
    CA: Center for Planning and Research, Inc., 1981.

Hansen, D. J. Potential Effects of Oil Spills and Other
    Chemical Pollutants on Marine Mammals Occurring in
    Alaskan Waters. Anchorage, AK: Minerals Management
    Service, Alaska Outer Continental Shelf Office, 1985.

    Describes and assesses the potential effects of oil
    spills and other contaminants on marine mammals that
    occur in Alaskan waters. The report focuses primarily
    on the potential direct and indirect effects of oil
    spills on marine mammals and addresses both short-term
    effects that may occur at the time of contact with oil,
    and long-term effects that may occur long after contact
    with oil. The report also briefly reviews the litera-
    ture on the potential effects of other contaminants
    such as heavy metals, DDT and PCB's on marine mammals.

Kot, C. A., et al. Hazardous Materials Accidents Near
    Nuclear Power Plants: An Evaluation of Analyses and
    Approaches. Report #ANL-83-53. Argonne National Lab.,
    Illinois. Sponsor: Nuclear Regulatory Research Commis-
    sion, Washington, D.C., Office of Nuclear Regulatory
    Research. 1983.

Marshall, M. D. _Capture-and-Containment Systems for Hazardous Materials Spills on Land_. Report #EPA-600/2-84-084. MSA Research Corp., Evans City, Pennsylvania. Sponsor: Municipal Environmental Research Lab., Cincinnati, OH. 1984.

Melvold, R. W. and L. T. McCarthy, Jr. _Emergency Response Procedures for Control of Hazardous Substance Releases_. Report #EPA-600/D-84-023. Rockwell International, Newbury Park, CA. Sponsor: Municipal Environmental Research Lab., Edison, NJ. 1983.

National Transportation Safety Board, Washington, D.C. Bureau of Accident Investigation. _Hazardous Materials Investigation Report - Release of Hazardous Waste Acid from Cargo Tank Truck, Orange County, Florida, March 6, 1984_. Report #NTSB/HZM-85/01. 1985.

Oil Spills: Environmental Effects. VA: National Technical Information Service, 1983.

Considers the environmental effects of oil spills on marine animals and fish.

Schneider, George R. _Removal of Water Soluble Hazardous Materials Spills from Waterways by Activated Carbons_. Report #EPA-600/S2-81-195. Cincinnati, OH: Environmental Protection Agency, Municipal Environmental Research Laboratory. Center for Environmental Research Information (distributor), 1981. SuDocs EP1.89/2.W29/8.

Special Investigation Report - Railroad Yard Safety: Hazardous Materials and Emergency Preparedness. National Transportation Safety Board. Report #NTSB/SIR-85/02. 1985.

Unterberg, W., et al. _Manual and Training Course for Prevention of Spills of Hazardous Substances at Fixed Facilities_. Report #EPA-600/D-84-073. Rockwell International, Newbury Park, California. Sponsor: Municipal Environmental Research Lab., Cincinnati, OH. 1984.

Data Bases

Chemical Hazards in Industry
Producer: Royal Society of Chemistry (Nottingham, England)
Time Span: January 1984 to date
Coverage: 200 primary journals
Updated: Monthly

Scans 200 major journals for any information on chemical hazards, including explosions, safety in handling and storage, and waste management.

Hazard Assessment Computer System (HACS)
Producer: Coast Guard, Office of Marine and Environment
           Systems, Environmental Response Division
Time Span: Current information
Coverage: Industry surveys, experiments, selected
           literature
Updated: Information not available

Online counterpart to Chemical Hazards Response Information System (CHRIS). (See Coast Guard listing under "Government Organizations" for detailed information.) Includes planning, assessment and response measures for chemical spills in water.

Hazardline
Producer: Occupational Health Services, Inc.
Time Span: Current information
Coverage: Books, journals, court decisions, and U.S.
           government agency information on over 3,000
           chemical substances
Updated: Monthly

Records on individual chemicals cover extensive safety information, including response procedures for leaks, spills, fire, and evacuation. Reactions, exposure levels, and other health materials are also covered.

Oil and Hazardous Materials Technical Assistance Data System (OHMS-TADS)
Producer: Environmental Protection Agency, Oil and
           Special Materials Control Division, Office of
           Water Program Operations
Time Span: 1950 to present, with some material published in
           the early 1900's
Coverage: Unknown
Updated: Monthly, 6000 records per year

Contains data gathered from published literature on approximately 1200 materials that have been designated oil or hazardous materials. Provides technical support

for dealing with potential or actual dangers resulting
from the discharge of oil or hazardous substances.
Emphasis is placed on the effects of these substances
on water quality.

Oil Spill Data Base
Producer: Center for Short-Lived Phenomena
Time Span: 1967 to present
Coverage: Reports of oil spills in water
Updated: Annually

  Information on ocean oil spills, including descriptions
  of the substances, amounts, causes, and cleanup procedures.

Pollution Incident Reporting System (PIRS)
Producer: Coast Guard, Office of Marine and
           Environment Systems, Environmental Response
           Division
Time Span: Not available
Coverage: Pollution incident reports
Updated: As incident reports are filed

  Information on marine pollution occurrences, including
  cleanup information.

Spill Prevention Control and Countermeasure System (SPCCS)
Producer: Environmental Protection Agency, Office of
           Solid Waste and Emergency Response
Time Span: Current information
Coverage: EPA compliance reports; spill reports
Updated: Weekly

  Includes reports of oil and hazardous materials spills.
  Used by the EPA to monitor compliance.

Audiovisual Materials

Chemical Hazard Control II. Berkeley, CA: Biology Media,
    Inc., 1981. Videocassette, 25 min.

  Discusses safety measures for personnel involved in the
  handling, disposal, and storage of dangerous laboratory
  chemicals. Procedures for responding to fires and
  spills are included.

First on the Scene. Washington, D.C.: Chemical Manufac-
    turers Association, 1985. (Available for purchase or
    loan from the Association.)

Designed for all initial response personnel, this
training program teaches response and injury avoidance
methods. The videotape stresses methods of approaching
the scene, identifying the material, stabilizing the
scene, obtaining assistance, and entering the site.
Accompanied by a discussion guide and brochure.

Handling Hazardous Chemicals Safely. Morristown, NJ: Allied
    Corp., 1982. (Distributed by BNA Communications, Inc.)
    16mm film, sound, color, 16 min.

    Safety in chemical handling and procedures for emer-
    gency situations are covered in this training film.
    Accompanied by a workbook and a discussion leader's
    guide.

If You Only Knew. London: Millbank Films, 1982. 16mm
    film, sound, color, 24 min.

    Emphasis on safety information for employees handling
    dangerous chemicals. A leader's guide for discussion
    is included.

Associations and Other Groups

Academy of Hazard Control Management
5010A Nicholson Lane
Rockville, MD 20852   (301) 469-9448

    A coalition of certified hazard control managers in all
    areas of safety and health, including environmental
    concerns. Publishes a newsletter.

American Society for Testing and Materials
1916 Race Street
Philadelphia, PA 19103   (215) 299-5400

    Committee F-20 on Spill Control Systems includes Sec-
    tion F-20.20, the Division on Hazardous Substances
    Spills, and Section F-20.21, the Division on Initial
    Response Actions.

American Society of Safety Engineers
850 Gusse Highway
Park Ridge, IL 60068   (312) 692-4121

    Individuals concerned with industrial safety and acci-
    dent prevention. Annual conference. Publishes Profes-
    sional Safety.

Association of American Railroads
Bureau of Explosives
1920 L Street, NW
Washington, D.C. 20036   (202) 835-9500

Provides information on emergency response measures for
the handling of hazardous materials. Main concern is
the safe transportation and handling of explosives and
other dangerous substances. Also described in Chapter
11, "Transportation of Hazardous Materials."

Chemical Manufacturers Association
2501 M Street, NW
Washington, D.C. 20037   (202) 887-1100

Primary service is CHEMICAL TRANSPORTATION EMERGENCY
CENTER (CHEMTREC), which provides information on emer-
gency services for chemical transportation accidents.
CHEMTREC is described in more detail in Chapter 11,
"Transportation of Hazardous Materials."

National Fire Protection Association
Batterymarch Park
Quincy, MA 02269   (617) 770-3000

Provides field service by specialists in all types of
fire emergencies. Publishes Fire Command; produces
audiovisual programs on fire safety and prevention.
See Chapter 11, "Transportation of Hazardous Materials"
for more information on NFPA.

Spill Control Association of America
17117 W. Nine Mile Road, Suite 1040
Southfield, MI 48075   (313) 552-0500

An organization of companies and individuals involved
in the manufacture and supply of spill control
equipment. Focus is on new developments in the spill
control industry. Publishes a newsletter; sponsors
quarterly conventions.

Government Organizations

Coast Guard
Office of Marine Environment and Systems
Environmental Response Division
Chemical Hazards Response Information System (CHRIS)
G-WER-2
2100 2nd Street, SW
Washington, D.C. 20593   (202) 426-9568

Collects data on chemical hazards, published in a series of CHRIS Manuals -- <u>A Condensed Guide to Chemical
Hazards</u>, <u>Hazardous Chemical Data</u>, <u>Hazard Assessment
Handbook</u>, and <u>Response Methods Handbook</u>. Provides
information on chemical emergencies in water, response
measures, and equipment and resources available to
response teams. (See also National Response Center
entry under "Government Organizations" and Hazard
Assessment Computer System in "Data Bases.")

<u>Emergency Response Division</u>
Office of Emergency and Remedial Response
Environmental Protection Agency
401 M Street, SW
Washington, D.C. 20460   (202) 245-3045

Responsibilities include impact assessment of spills,
emergency notification, and establishing reportable
quantity standards for over 1,000 hazardous chemicals.

<u>Federal Emergency Management Agency</u>
1725 Eye Street, NW
Washington, D.C. 20472   (202) 287-0390

Conducts disaster planning for hazardous materials
emergencies, including nuclear accidents.

<u>National Response Center</u>
Environmental Response Division
Office of Marine Environment and Systems
Coast Guard
400 7th Street, SW, Room 7402
Washington, D.C. 20590   (202) 426-1192
            Toll-free for reporting spills (800) 424-8802

Provides emergency advice and information for oil and
hazardous materials spills emergency response.

<u>National Response Team for Oil and Hazardous Materials
     Spills</u>
Office of Emergency and Remedial Response
Environmental Protection Agency
401 M Street, SW
Washington, D.C. 20460   (202) 382-2180

Establishes national policy on spill response and
preparedness. Also assists regional forces at spill
sites.

<u>National Transportation Safety Board</u>
800 Independence Avenue, SW
Washington, D.C. 20594   (202) 382-6600

Bureau of Accident Investigation and Hazardous Materials Division conduct research on hazardous materials accidents. See the chapter on Transportation of Hazardous Materials for additional information.

Office of Solid Waste and Emergency Response
Environmental Protection Agency
401 M Street, SW
Washington, D.C. 20460   (202) 382-4610

Regulates hazardous wastes produced or deposited in non-controlled sites.

Oil and Special Materials Control Division
Office of Water Program Operations
Environmental Protection Agency
401 M Street, SW
Washington, D.C. 20460   (202) 245-3048

Develops programs and standards for the control of hazardous materials and oil spills.

Libraries and Information Centers

Major science and engineering libraries offer many relevant holdings on hazardous substances. In addition to these collections, the following libraries and information centers collect materials on various aspects of hazardous substance accidents, industrial and fire safety, and other pertinent topics.

Environmental Research & Technology Inc.
Life Sciences Information Center
Box 2105
Fort Collins, CO 80522   (303) 493-8878

Special collection on oil spills.

National Fire Protection Association
Charles S. Morgan Library
Batterymarch Park
Quincy, MA 92269   (617) 328-9290

All aspects of fire prevention and protection, fire services management, and flammability of materials.

New York State Department of State - Fire Academy Library
600 College Avenue
Box K
Montour Falls, NY 14865   (607) 535-7136

    Fire protection and prevention and other aspects of
    fire response and management. Special collection on
    hazardous materials.

Spill Control Association of America - Library
17117 W. Nine Mile Road, Suite 1040
Southfield, MI 48075   (315) 552-0500

    Oil and hazardous substance spills statistics. Special
    collection on history, containment techniques, and
    research on oil and hazardous materials spills.

University of California, Santa Barbara
Sciences/Engineering Library
Santa Barbara, CA 93106   (805) 961-2765

    Special Collection -- the Oil Spill Information Center
    Archives -- contains information on the Santa Barbara
    oil spill of 1969.

Woodward-Clyde Consultants
Environmental Systems Division, Library
3 Embarcadero Center, Suite 700
San Francisco, CA 94111   (415) 956-7070

    Information on the environment, including pollution and
    environmental impact.

Research Centers and Industrial Laboratories

Numerous organizations affiliated with the government, universities, and industry conduct investigations involving
hazardous substance emergencies. The number of these organizations far exceeds the limits of this chapter and a search
of the standard sources listing these centers is advisable.
The following research centers and industrial laboratories
are representative of the multitude of relevant organizations.

Bowdoin College
Marine Station
Brunswick, ME 04011

Center for Radiological Protection
Interdisciplinary Programs Office
Georgia Institute of Technology
Atlanta, GA 30332

Disaster Research Center
Ohio State University
128 Derby Hall
154 N. Oval Mall
Columbus, OH 43210

Environmental Hazards Management Institute
45 Pleasant Street
Box 283
Portsmouth, NH 03801

Oil and Hazardous Materials Simulated Environmental Test
    Tank (OHMSETT) (EPA affiliate)
Raritan Center GS Depot, Building 109
Woodbrige Avenue
Edison, NJ 80017

Oil and Hazardous Materials Spills Branch (OHMSB) (EPA
    affiliate)
GSA Depot, Building 209
Woodbridge Avenue
Edison, NJ 08817

Pennsylvania State University
Office of Hazardous and Toxic Waste Management
Land & Water Resources Research Building
University Park, PA 16802

Syracuse Research Corp.
Merrill Lane
Syracuse, NY 13210

Texas A&M University
Environmental Engineering Division
College Station, TX 77843

University of Toronto
Institute for Environmental Studies
Haultain Building
170 College Street
Toronto, Ontario M5S 1A4
Canada

Publishers and Vendors

(Telephone numbers are included whenever possible)

Academic Press, Inc.
111 Fifth Avenue
New York, NY 10003
(800) 321-5068

Air Force Aerospace Medical
   Research Laboratory
Aerospace Medical Division
Air Force Systems Command
Wright-Patterson Air Force
   Base, OH 45433

Air Pollution Control
   Association
P.O. Box 2861
Pittsburgh, PA 15230
(412) 621-1090

Air Pollution Control
   Association
Florida Section
Gainesville, FL
(Complete address unavailable)

Alliance of American Insurers
20 N. Wacker Drive
Chicago, IL 60606
(312) 490-8500

American Chemical Society
1155 16th Street, NW
Washington, D.C. 20036
(202) 872-4600

American Conference of
   Governmental Industrial
   Hygienists
Building D-5
6500 Glenway Avenue
Cincinnati, OH 45211
(513) 661-7881

American Geophysical Union
2000 Florida Avenue
Washington, D.C. 20009
(202) 462-6903

American Industrial Hygiene
   Association
475 Wolf Ledges Parkway
Akron, OH 44311-1087
(216) 762-7294

American Institute of Chemical
   Engineers
345 East 47th Street
New York, NY 10017
(212) 705-7338

American Institute of
   Professional Geologists
7828 Vance Drive, Suite 103
Arvada, CO 80003
(303) 431-0831

American Medical Association
Order Dept. OP-128
P.O. Box 821
Monroe, WI 53566
(312) 645-5000

American Nuclear Society
555 N. Kensington Avenue
LaGrange Park, IL 60525
(312) 352-6611

American Public Health
  Association
1015 15th Street, NW
Washington, D.C. 20005
(202) 789-5660

American Society of Agronomy
677 South Segoe Road
Madison, WI 53711
(608) 274-1212

American Society of Civil
  Engineers
345 East 47th Street
New York, NY 10017
(212) 705-7496

American Society for Metals
Metals Abstracts Trust
9639 Kinsman Road
Metals Park, OH 44073
(216) 338-5151

American Society of Safety
  Engineers
850 Busse Highway
Park Ridge, IL 60068
(312) 692-4121

American Society for Testing
  and Materials
1916 Race Street
Philadelphia, PA 19103
(215) 299-5400

American Water Resources
  Association
5410 Grosvenor Lane, Suite 220
Bethesda, MD 20814
(301) 493-8600

American Water Works
  Association
Washington Office
704 National Press Building
Washington, D.C. 20045
National Office (303) 794-7711

Ann Arbor Science Publishers
c/o Butterworth
  Publishers, Inc.
10 Tower Office Park
Woburn, MA 01801
(617) 935-9361

Annual Reviews, Inc.
4139 El Camino Way
Palo Alto, CA 94306
(415) 493-4400

J. W. Arrowsmith Ltd.
Wetherby, West Yorkshire
England
(Complete address unavailable)

Appleton-Century-Crofts
25 Van Zant Street
East Norwalk, CT 06855
(203) 838-4400

Arizona Center for Occupa-
  tional Safety and Health
University of Arizona
Health Sciences Center
Division of Biomedical
  Communications
Tucson, AZ 85721
(602) 626-4824

Army Engineer Waterways
  Experiment Station
P.O. Box 631
Vicksburg, MS 39180
(601) 636-3111

Aspen Systems Corporation
1600 Research Boulevard
Rockville, MD 20850
(301) 251-5000

Association of American
  Railroads
Bureau of Explosives
1920 L Street, NW
Washington, D.C. 20036
(202) 835-9500

Atomic Industrial Forum, Inc.
7101 Wisconsin Avenue
Bethesda, MD 20814
(301) 654-9260

Aurigny, Ltd.
P.O. Box 12
Guildford, Surrey GU4 7PL
England

Bailliere Tindall
1 Vincent Square
London SW1P 2PN
England

Ballinger Publishing Co.
54 Church Street
Cambridge, MA 02138
(617) 492-0670

Basic Books, Inc.
10 East 53rd Street
New York, NY 10022
(212) 207-7292

Battelle Press
505 King Avenue
Columbus, OH 43201
(614) 424-6393

Baywood Publishing Co., Inc.
120 Marine Street
P.O. Box D
Farmingdale, NY 11735
(516) 293-7130

Biology Media
918 Parker Street
P.O. Box 10205
Berkeley, CA 94710
(415) 524-5929

Biomedical Publications
P.O. Box 495
Davis, CA 95617
(916) 756-8453

BioSciences Information
  Service (BIOSIS)
2100 Arch Street
Philadelphia, PA 19103-1399
(800) 523-4806

BNA Communications, Inc.
9439 Key West Avenue
Rockville, MD 20850
(301) 948-0540

R.R. Bowker Company
205 East 42nd Street
New York, NY 10017
(212) 916-1600

BRS (Bibliographic Retrieval
  Services, Inc.)
1200 Route 7
Latham, NY 12110
(800) 833-4707

Bureau of Law and
  Business, Inc.
64 Wall Street
Madison, CT 06443-9988
(203) 245-7448

Bureau of National
  Affairs, Inc.
1231 25th Street, NW
Washington, D.C. 20037
(202) 452-4379

Business Publishers, Inc.
951 Pershing Drive
Silver Spring, MD 20910
(301) 587-6300

Business Research Publications
817 Broadway
New York, NY 10003
(718) 673-4700

Cahners Publishing Co.
P.O. Box 716
Back Bay Annex
Boston, MA 02117
(617) 536-7780

California, Toxic Waste
    Assessment Group
Governor's Office of
    Appropriate Technology
Sacramento, CA
(Complete address unavailable)

Cambridge Scientific Abstracts
5161 River Road
Bethesda, MD 20816
(301) 951-1403

Cambridge University Press
32 E. 57th Street
New York, NY 10022
(212) 688-8888

Canadian Advisory Council on
    the Status of Women
Box 1541 Station B
Ottawa, Ontario K1P 5R5
Canada

Canadian Government Publishing
    Centre
Ottawa, Ontario K1A 0S9
Canada
(819) 997-2560

Canadian Institute of Public
    Health Inspection
P.O. Box 130
Etobicoke, Ontario M9C 4V2
Canada

Canadian Nuclear Society
111 Elizabeth St., 11th Floor
Toronto, Ontario M5G 1P7
Canada
(416) 977-6152

Capital Planning Information
6 Castle Street
Edinburgh EH2 3AT
Scotland

Cash Crop Farming
    Publications, Ltd.
Industrial Magazine Division
222 Argyle Avenue
Delhi, Ontario N4B 2Y2
Canada
(519) 582-2510

John R. Cashman
P.O. Box 204
Barre, VT 05641
(802) 479-2307

Center for Environmental
    Research Information
c/o Environmental Protection
    Agency
401 M Street, SW
Washington, D.C. 20460

Center for Occupational
    Hazards
5 Beekman Street
New York, NY 10038
(212) 227-6220

Center for Short-Lived
    Phenomena
138 Mt. Auburn Street
Cambridge, MA 02138
(617) 492-3310

Chapman & Hall
11 New Fetter Lane
London EC4P 4EE
England
NY City (212) 244-3336

Chemcontrol A/S
Dagmarhus
1553 Copenhagen V
Denmark

Chemical Abstracts Service
P.O. Box 3012
Columbus, OH 43210
(800) 848-6533

Chemical Manufacturers
   Association
2501 M Street, NW
Washington, D.C. 20037
(202) 887-1100

Coast Guard
Office of Marine Environment
   and Systems
Environmental Response
   Division
2100 Second Street, SW
Washington, D.C. 20593
(202) 426-2010

Commerce Clearing House
4025 W. Peterson Avenue
Chicago, IL 60646
(312) 583-8500

Communication Channels, Inc.
6255 Barfield Road
Atlanta, GA 30328
(404) 256-9800

CompuServe, Inc.
5000 Arlington Centre Blvd.
Columbus, OH 43220
(800) 848-8990

Computer Sciences Corp.
CIS User Support Group
      See
NIH/EPA Chemical Information
   System (CIS) User Support
   Group

Congressional Information
   Service, Inc.
4520 East-West Highway
Bethesda, MD 20814-3389
(301) 654-1550

Conservation Foundation
1717 Massachusetts Avenue, NW
Washington, D.C. 20036
(202) 797-4300

Corpus Information
   Services, Ltd.
1450 Don Mills Road
Don Mills, Ontario M3B 2X7
Canada
(416) 445-7101

Council of Europe, U.S.
   Publications Sales Agent
Manhattan Publishing Co.
80 Brook Street
P.O. Box 650
Croton, NY 10520
(914) 271-5194

Council on Economic
   Priorities
84 Fifth Avenue
New York, NY 10011
(212) 420-1133

CRC Press, Inc.
2000 Corporate Blvd., NW
Boca Raton, FL 33431
(305) 994-0555

Cuadra/Elsevier
P.O. Box 1672, Grand Central
   Station
New York, NY 10163
(212) 916-1010

Walter de Gruyter, Inc.
200 Saw Mill River Road
Hawthorne, NY 10532
(914) 747-0110

Marcel Dekker, Inc.
270 Madison Avenue
New York, NY 10016
(212) 696-9000

Department of Energy
Technical Information Center
Box 62
Oak Ridge, TN 37830
(615) 576-6299

Department of the Interior,
Geological Survey
Reston, VA 22092
(703) 860-7444

Department of Transportation
Research & Special Programs
  Administration
Materials Transportation
  Bureau
400 Seventh Avenue, SW
Washington, D.C. 20590
(202) 755-9260

DIALOG Information
  Services, Inc.
3460 Hillview Avenue
Palo Alto, CA 94304
(800) 227-1927

DIMDI
Weisshausstrasse 27
5000 Koln 41 (Sulz)
West Germany

Direct Cinema, Ltd.
P.O. Box 69589
Los Angeles, CA 90069
(213) 656-4700

Dokukagaku-kai
1st Floor, Gakkai Center Bldg.
4-16, Yayoi 2-Chome, Bunkyo-ku
Tokyo 113
Japan

Duke University Press
6697 College Station
Durham, NC 27708
(919) 684-2173

Dustri Verlag
Dr. Karl Feistle
Bahnkofstrasse 9, Postfach 49
D-8024 Munchen-Deisenhofen
Germany

EQES, Inc.
799 Roosevelt Road
Building 6, Suite 104
Glen Ellyn, IL 60137
(312) 858-6161

EIC/Intelligence, Inc.
48 W. 38th Street
New York, NY 10018
(212) 944-8500

Electric Power Research
  Institute
P.O. Box 10412
Palo Alto, CA 94303
(415) 855-2000

Elsevier Science Publishing
  Co., Inc.
52 Vanderbilt Avenue
New York, NY 10017
(212) 867-9040

Engineering Index, Inc.
345 E. 47th Street
New York, NY 10017
(212) 705-7615

Environews, Inc.
1097 National Press Building
Washington, D.C. 20045
(202) 347-3868

Environment Information
  Center
          See
EIC/Intelligence

Environmental Information Ltd.
7400 Metro Blvd., Suite 400
Minneapolis, MN 55435

Environmental Studies
  Institute
International Academy at Santa
  Barbara
Riviera Campus, Alameda
  Padre Serra
Santa Barbara, CA 93103
(805) 965-5010

ESA/Quest
European Space Agency
Information Retrieval Service
ESRIN, Via Galileo Galilei,
  C.P. 64
I-00044 Frascati, Rome
Italy

European Weed Research
  Society (EWRS)
c/o Biologische,
Bundesanstalt für Land-und
  Forstwirtschaft,
Messeweg 11/12, D-3300
Braunschweig
Federal Republic of Germany

Excerpta Medica
Box 211
1000 AE
Amsterdam, Netherlands

Exchange Publications
2550 M Street, NW, Suite 620
Washington, D.C. 20037

Executive Enterprises, Inc.
33 W. 60th Street
New York, NY 10023
(212) 489-2671

F & S Press
106 Fulton Street
New York, NY 10038
(212) 233-1080

Stuart Finley, Inc.
3428 Mansfield Road
Falls Church, VA 22041

Finnish Institute of
  Occupational Health
Haartmanninkatu 1
SF-00290
Helsinki 29 Finland

Flournoy Publishers, Inc.
1845 West Morse
Chicago, IL 60626
(312) 761-3955

Freed Publishing Co.
Box 1144, FDR Station
New York, NY 10022

Gale Research Co.
Book Tower
Detroit, MI 48226
(313) 961-2242

Garland STPM Press
136 Madison Avenue
New York, NY 10016
(212) 686-7492

Geological Survey
Water Resources Division
420 National Center
12201 Sunrise Valley Drive
Reston, VA 22092
(703) 860-6031

Gordon & Breach Science
  Publishers, Inc.
One Park Avenue
New York, NY 10016
(212) 689-0360

Government Institutes, Inc.
966 Hungerford Drive, #24
Rockville, MD 20850
(301) 251-9250

Government Printing Office
  (G.P.O.)
North Capitol and H
  Streets, NW
Washington, D.C. 20401
(202) 275-2051

Willard Grant Press
Statler Office Building
20 Providence Street
Boston, MA 02116
(617) 482-9399

Grune & Stratton
c/o Academic Press
111 Fifth Avenue
12th Floor
New York, NY 10003
(212) 741-4888

Harwood Academic Publishers
P.O. Box 786 Cooper Station
New York, NY 10276
(212) 206-8900

Hazardous Materials Control
  Research Institute
9300 Columbia Boulevard
Silver Spring, MD 20910
(301) 587-9390

Hazardous Materials
  Publishing Co.
243 West Main Street
Kutztown, PA 19530

Haztrain
P.O. Box 2206
La Plata, MD 20646

Heldref Publications
4000 Albemarle Street, NW
Washington, D.C. 20016
(202) 362-6445

Hemisphere Publishing Corp.
79 Madison Avenue
New York, NY 10016
(212) 725-1999

Hutchinson Ross Publishing
  Co., distributed by Van
  Nostrand Reinhold Co., Inc.
135 W. 50th Street
New York, NY 10020
(212) 265-8700

Idaho Department of Health
  and Welfare, Division of
  Environment
450 West State Street
Boise, ID 83720
(208) 334-4061

Indiana University
Audio-Visual Center
Bloomington, IN 47405
(812) 335-8087

Information Handling Services
15 Inverness Way E.
Englewood, CO 80150
(303) 790-0600

Institute of Environmental
  Sciences
940 E. Northwest Highway
Mt. Prospect, IL 60056
(312) 255-1561

Institute for Scientific
  Information
3501 Market Street
University City Science
  Center
Philadelphia, PA 19104
(800) 523-1850

Institute for Toxic Waste
  Management
New Jersey Institute of
  Technology
323 High Street
Newark, NJ 07102

Institute of Water Pollution
  Control
Ledson House
53 London Road
Maidstone, Kent ME16 8JH
England

Institute of Water Research
Michigan State University
334 Natural Resources Bldg.
East Lansing, MI 48824
(517) 353-3742

Institution of Chemical
   Engineers
165-171 Railway Terrace
Rugby CV21 3HQ
England

Intapress Publishing, Ltd.
38 Tavistock Street
London WC2E 7PB
England

Interdok
P.O. Box 326
Harrison, NY 10528
(914) 835-3506

International Air Transport
   Association
2000 Peel Street, W.
Montreal, Quebec H3A 2R4
Canada
(514) 844-6311

International Association of
   Hydrological Sciences
Institute of Hydrology
Maclean Building
Crowmarsh Gifford
Wallingford, Berkshire
   OX10 8BB
England

International Atomic Energy
   Agency Publications
Available from UNIPUB
205 W. 42nd Street
New York, NY 10017
(212) 916-1659

International City Management
   Association
1120 G Street, NW
Washington, D.C. 20005
(202) 626-4600

International Civil Aviation
   Organization
1000 Sherbrooke Street, W
P.O. Box 400
Montreal, Quebec H3A 2R2
Canada
(514) 285-8219

International Film Bureau
332 S. Michigan Avenue
Chicago, IL 60604
(312) 427-4545

International Labor Office
Washington Branch
1750 New York Avenue, NW
Washington, D.C. 20006
(202) 376-2315

International Maritime
   Organization
4 Albert Embankment
London SE1 7SR
England

International Research and
   Evaluation
21098 IRE Control Center
Eagan, MN 55121
(612) 888-9635

International Research
   Communications System
St. Leonard's House
St. Leonardgate
Lancaster LA1 1PF
England

International Technical
   Information Institute
Toranomon-Tachikawa Bldg, 6-5
1 Chome
Nishi-Shimbashi, Minato-ku
Tokyo, Japan

I.P.C. Electrical-Electronic
   Press Ltd.
Quadrant House, The Quadrant
Sutton, Surrey SM2 5AS
England

IRE
    See
International Research and
    Evaluation

ISI Press
3501 Market Street
Philadelphia, PA 19104
(215) 386-0100

Johns Hopkins University Press
Baltimore, MD 21218
(301) 338-7861

J. J. Keller and Associates,
    Inc.
145 West Wisconsin Avenue
P.O. Box 368
Neenah, WI 54956-0368
(800) 558-5011

Knowledge Industry
    Publications, Inc.
701 Westchester Avenue
White Plains, NY 10604
(800) 248-KIPI

Lab Safety Supply Co.
P.O. Box 1368
Janesville, WI 53547-1368
(608) 754-2345

Labelmaster Customer Service
    Dept.
5724 N. Pulaski Avenue
Chicago, IL 60646
(800) 621-5808

Library of Congress National
    Referral Center
John Adams Building, Rm. 5228
Washington, D.C. 20540
(202) 287-5670

Mary Ann Liebert, Inc.
157 East 86th Street
New York, NY 10028
(212) 289-2300

Alan R. Liss
41 E. 11th Street
New York, NY 10011
(212) 475-7700

MacMillan Publishing Co.,
    Inc.
866 Third Avenue
New York, NY 10022
(212) 935-2000

Manufacturing Chemists
    Association
        See
Chemical Manufacturers
    Association

March of Dimes Birth Defects
    Foundation
1275 Mamaroneck Avenue
White Plains, NY 10605
(914) 428-7100

Materials Research Society
9800 McKnight Road
Suite 327
Pittsburgh, PA 15237
(412) 367-3003

McCoy and Associates
13131 West Cedar Drive
Lakewood, CO 80228

McGraw-Hill, Inc.
1221 Avenue of the Americas
New York, NY 10020
Orders: (609) 426-5254

Medcom, Inc.
Distributed by Williams &
    Wilkins Co.
428 E. Preston Street
Baltimore, MD 21202
(301) 528-4221

Medical Publications, Inc.
5002 Lakeland Circle
Waco, TX 76710

MEDLARS (Medical Literature
  Analysis & Retrieval System)
National Library of Medicine
8600 Rockville Pike
Bethesda, MD 20209
(301) 496-6217

Michigan State Department of
  Public Health
Division of Occupational
  Health
3500 North Logan
Lansing, MI 48909

Microwave News
P.O. Box 1799
Grand Central Station
New York, NY 10163
(212) 725-5254

Millbank Films
Thames House North
Millbank, London SW1P 4QG
England

MIT Press
28 Carleton Street
Cambridge, MA 02142
(617) 253-2884

National Academy Press
2101 Constitution Avenue, NW
Washington, D.C. 20418
(202) 334-3313

National AudioVisual Center
Washington, D.C. 20409
(301) 763-1872

National Center for
  Atmospheric Research
P.O. Box 3000
Boulder, CO 80302
(303) 497-1000

National Conference of State
  Legislatures
1125 Seventeenth Street
Suite 1500
Denver, CO 80202
(303) 292-6600

National Fire Protection
  Association
Batterymarch Park
Quincy, MA 02269
(617) 770-3000

National Information Center
  for Educational Media
P.O. Box 40130
Albuquerque, NM 87196
(505) 265-3591

National Institute for
  Occupational Safety and
  Health
5600 Fishers Lane
Rockville, MD 20857
(301) 443-2140

National Library of Medicine
8600 Rockville Pike
Bethesda, MD 20209
(301) 496-6095

National Safety Council
444 North Michigan Avenue
Chicago, IL 60611
(312) 527-4800

National Solid Wastes
  Management Association
1730 Rhode Island Avenue, NW
Suite 512
Washington, D.C. 20036
(202) 659-4613

National Technical Information
  Service (NTIS)
5285 Port Royal Road
Springfield, VA 22150
(703) 487-4785

National Trade Publications,
    Inc.
8 Stanley Circle
Latham, NY 12110
(518) 783-1281

New England Legal Foundation
55 Union Street
Boston, MA 02108
(617) 367-0174

NewsNet, Inc.
945 Haverford Road
Bryn Mawr, PA 19010
(800) 345-1301

NIH/EPA Chemical Information
    System (CIS)
User Support Group
Computer Sciences Corp.
P.O. Box 2227
Falls Church, VA 22042

North Atlantic Treaty
    Organization/Committee on
    the Challenges of Modern
    Society (NATO/CCMS)
c/o NATO
B-1110
Bruxelles, Belgium

W. W. Norton & Company, Inc.
500 5th Avenue
New York, NY 10110
(212) 354-5500

Noyes Data Corp.
Mill Road at Grand Avenue
Park Ridge, NJ 07656
(201) 391-8484

Nuclear Technology Publishing
P.O. Box No. 7
Ashford, Kent
England

Occupational Health Services,
    Inc.
P.O. Box 1505
400 Plaza Drive
Secaucus, NJ 07094
(201) 865-7500

Occupational Safety and
    Health Administration
Department of Labor
Washington, D.C. 20210
(202) 523-8148

Ontario Ministry of Labour
400 University Avenue,
    10th Floor
Toronto, Ontario M7A 1T7
Canada
(416) 965-1641

Organization for Economic
    Co-Operation and Development
1750 Pennsylvania Avenue, NW
Suite 1207
Washington, D.C. 20006
(202) 724-1857

Oelgeschlager, Gunn & Hain
    Publishers, Inc.
1278 Massachusetts Avenue
Cambridge, MA 02138
(617) 437-9620

Oxford University Press, Inc.
200 Madison Avenue
New York, NY 10016
(212) 564-6680

Pacific Radiation Corp.
9827 Daines Drive
Temple City, CA 91780
(818) 286-8222

Paragon House Publishers
2 Hammarskjold Plaza
New York, NY 10017
(212) 223-6433

Paramount Publishing Ltd.
17-21 Shenley Road
Borehamwood
Hertfordshire WD6 1RT
England

Pennsylvania Environmental
  Council
225 South 15th Street
Philadelphia, PA 19102
(215) 735-0888
(215) 735-0966

Peter Peregrinus (Imprint of
  Institution of Electrical
  Engineers)
P.O. Box 26
Hitchin Herts SG5 1SA
England

Pergamon Press, Inc.
Maxwell House
Fairview Park
Elmsford, NY 10523
(914) 592-7700

The Philadelphia Inquirer
P.O. Box 8263
400 North Broad Street
Philadelphia, PA 19101
(215) 854-2000

Plenum Publishing Corp.
233 Spring Street
New York, NY 10013
(212) 620-8000

Polecon Publications
2 Wyverne Road, Chorlton
Manchester M21 1XR
England

Premier Press
P.O. Box 4428
2914 Domingo Avenue
Berkeley, CA 94704
(415) 841-2091

Prentice-Hall
Route 9 W.
Englewood Cliffs, NJ 07632
(201) 592-2352

Preston Publications, Inc.
7800 Merrimac Avenue
Niles, IL 60648
(312) 965-0566

Princeton Scientific
P.O. Box 3159
Princeton, NJ 08540
(609) 924-6044

Proceedings in Print, Inc.
1026 Massachusetts Avenue
Arlington, MA 02174
(617) 646-0686

PSG Inc.
545 Great Road
Littleton, MA 01460
(617) 486-8971

Public Affairs Information
  Service
11 West 40th Street
New York, NY 10018
(212) 736-6629

Albert R. Pudvan Publishing
  Co.
1935 Shermer Road
Northbrook, IL 60062
(312) 498-9840

Questel, Inc.
1625 Eye Street, NW
Suite 818
Washington, D.C. 20006

Radwaste News
P.O. Box 7166
Alexandria, VA 22307

Random House, Inc.
201 E. 50th Street
New York, NY 10022
(212) 751-2600

Raven Press
1140 Avenue of the Americas
New York, NY 10036
(212) 575-0335

Reddy Communications, Inc.
537 Steamboat Road
Greenwich, CT 06830
(203) 661-4800

Regulatory Science Press
P.O. Box 7166
Alexandria, VA 22307
(703) 765-3546

D. Reidel Publishing Co.
c/o Kluwer Academic
190 Old Derby Street
Hingham, MA 02043
(617) 749-5262

Rimbach Publishing, Inc.
8650 Babcock Boulevard
Pittsburgh, PA 15237
(412) 364-7234

Royal Society of Chemistry
Burlington House
Piccadilly
London W1V 0BN
England

Ryan Research International
1593 Filbert Avenue
Chico, CA 95926
(916) 343-2373

SCA Services, Inc.
60 State Street
Boston, MA 02109

SDC Information Services
2500 Colorado Avenue
Santa Monica, CA 90406
(800) 421-7229

Science Reviews Ltd.
Northwood, Middlesex
England

Scranton Gillette
  Communications
2250 E. Devon Avenue
Des Plaines, IL 60016
(312) 694-2410

Seacoast Anti-Pollution League
5 Market Street
Portsmouth, NH 03801
(603) 431-5089

Sierra Club Books
2034 Fillmore Street
San Francisco, CA 94115
(415) 981-8634

The Sierra Club Radioactive
  Waste Campaign
78 Elmwood Avenue
Buffalo, NY 14201
(716) 884-1000

Slack, Inc.
6900 Grove Road
Thorofare, NJ 08086
(609) 848-1000

Society of Nuclear Medicine,
  Inc.
475 Park Avenue, South
New York, NY 10016
(212) 889-0717

Society for Occupational and
  Environmental Health
2021 K Street, NW, Suite 305
Washington, D.C. 20006
(202) 737-5045

Society of Photo-optical
   Instrumentation Engineers
405 Fieldston Road
Bellingham, WA 98225
(206) 676-3290

Society for the Prevention of
   Asbestosis and Industrial
   Diseases
38 Drapers Road
Enfield, Middlesex EN2 8LU
England

The Soil and Health Foundation
Now:
Regenerative Agriculture
   Association
222 Main Street
Emmaus, PA 18049
(215) 967-5171

Southam Business Publications
1450 Don Mills Road
North York, Ontario M3B 2X7
Canada
(416) 445-6641

Springer-Verlag New York, Inc.
175 Fifth Avenue
New York, NY 10010
(212) 460-1500

Sten-O-Press
1862 23rd Street
Box 207
San Pablo, CA 94806

Strategic Assessments, Inc.
5000 Butte Street, Suite 132
Boulder, CO 80301
(303) 444-1343

Superintendent of Documents
      See
Government Printing Office

Syracuse Research Corp.
Merrill Lane
Syracuse, NY 13210-4080
(315) 425-5100

System Development Corp. (SDC)
SCD Search Service
      See
SDC Information Services

System Safety Society
14252 Culver Drive
Suite A-261
Irvine, CA 92714
(714) 551-2463

Technomic Publishing Co.
851 New Holland Avenue
Box 3535
Lancaster, PA 17604
(717) 291-5609

The Techrite Company
P.O. Box 7928
Marietta, GA 30065
(404) 973-0190

C. C. Thomas
2600 S. First Street
Springfield, IL 62717
(212) 789-8980

Thunderbird Enterprises Ltd.
102 College Road
Harrow, Middlesex HA1 1BQ
England

Toxic Waste Assessment Group
      See
California, Toxic Waste
   Assessment Group

Transportation Research Board
National Academy of Sciences
2101 Constitution Avenue, NW
Washington, D.C. 20418
(202) 334-3220

TUC Centenary Institute of
   Occupational Health
London School of Hygiene and
   Tropical Medicine
Keppel Street (Gower Street)
London WC1 7HT
England

Union of International
   Associations
Rue Washington 40
B-1050 Brussels
Belgium

UNIPUB
205 East 42nd Street
New York, NY 10017
(212) 916-1659

U.S. Committee for Energy
   Awareness
1735 I Street, NW
Suite 500
Washington, D.C. 20006
(202) 293-0770

University Microfilms
   International
P.O. Box 1764
Ann Arbor, MI 48106
(313) 761-4700

University of Pennsylvania
   Press
Blockey Hall
418 Service Drive
Philadelphia, PA 19104
(215) 898-6261

University of Washington
Department of Environmental
   Health
461 Health Sciences Building
Seattle, WA 98195
(206) 543-4252

University of Wisconsin
Extension Services in Pharmacy
155 Pharmacy Building
Madison, WI 53706

Urban & Schwarzenberg
7 E. Redwood Street
Baltimore, MD 21202
(301) 539-2550

Van Nostrand Reinhold Co.
135 W. 50th Street
New York, NY 10020
(212) 265-8700

Van Nostrand Reinhold
   Information Services
115 Fifth Avenue
New York, NY 10003
(212) 254-3232

Vance Bibliographies
P.O. Box 229
112 N. Charter Street
Monticello, IL 61856
(217) 762-3831

Virginia Water Resources
   Research Center
Virginia Polytechnic Institute
   and State University
617 North Main Street
Blacksburg, VA 24060-3397
(703) 961-5624

Wakeman/Walworth, Inc.
P.O. Box 1939
New Haven, CT 06509

Water Information Center, Inc.
The North Shore Atrium
6800 Jericho Turnpike
Syosset, NY 11791
(516) 921-7690

Water Pollution Control
   Federation
2626 Pennsylvania Avenue, NW
Washington, D.C. 20037
(202) 337-2500

Water Well Journal
   Publishing Co.
500 W. Wilson Bridge Road
Worthington, OH 43085
(614) 846-4967

Westview Press
5500 Central Avenue
Boulder, CO 80301
(303) 444-3541

John Wiley & Sons, Inc.
605 Third Avenue
New York, NY 10158
(212) 850-6418

Williams & Wilkins
428 East Preston Street
Baltimore, MD 21202
(301) 528-4221

H. W. Wilson Company
950 University Avenue
Bronx, NY 10452
(800) 367-6770

World Health Organization
    distributed by
Q Corp.
49 Sheridan Avenue
Albany, NY 12210
(518) 436-9686

World Information Systems
P.O. Box 535
Harvard Square Station
Cambridge, MA 02238
(213) 532-6730

Yale University Press
302 Temple Street
New Haven, CT 06520
(203) 436-7582

Year Book Medical
    Publishers, Inc.
35 E. Wacker Drive
Chicago, IL 60601
(312) 726-9733

Index

ABOUT THE EDITOR

JAMES K. WEBSTER is Associate Librarian in the Science and Engineering Library, State University of New York, Buffalo. His earlier books include *What Every Engineer Should Know about Engineering Information Resources* and *The Bibliographic Utilities: A Guide for the Special Librarian*. He has contributed numerous articles to *Special Libraries, Education Libraries* and other journals.